Solar Power

A Practical Handbook

Consulting Editors **Matthew Williams** and **John Deacon**

Consulting editors
Matthew Williams and John Deacon, Orrick, Herrington & Sutcliffe LLP

Managing director
Sian O'Neill

Solar Power: A Practical Handbook
is published by

Globe Law and Business Ltd
3 Mylor Close
Horsell
Woking
Surrey GU21 4DD
United Kingdom
Tel: +44 20 3745 4770
www.globelawandbusiness.com

Printed and bound by CPI Group (UK) Ltd, Croydon CR0 4YY

Solar Power: A Practical Handbook

ISBN 9781787421325
EPUB ISBN 9781787421332
Adobe PDF ISBN 9781787421349
Mobi ISBN 9781787421356

Table of contents

Part VI – Broader perspectives

Preface

Matthew Williams
John Deacon
Orrick, Herrington & Sutcliffe LLP

The impetus for this book arose from two sources – the success of another title by the same publisher which focused on the renewable energy sector as a whole, and our personal enthusiasm for a project which turned the spotlight on solar, a market whose time has truly come. By any measure, the growth of solar power has been exponential, and the demand for sector-specific commentary and insight has increased to such an extent that addressing it specifically has become an imperative.

As co-editors the proof was always going to be in the pudding. Not only were our law firm colleagues swift and generous in lending their considerable expertise, but all of our many other contributors were unstinting in their enthusiasm and commitment to the project. Their contributions are as thought-provoking as they are well-crafted, and we are immensely grateful for every long hour that has been devoted to the task.

If this book can to any degree reflect the dynamism and industry of the solar sector on which it comments, then we have achieved our aim. It is our sincere hope that the enjoyment of contributing to and editing the book is matched by the satisfaction of those who have taken the time to read it.

Introduction

Matthew Williams
Orrick, Herrington & Sutcliffe LLP

Sunlight fell upon the wall; the wall received a borrowed splendour. Why set your heart on a piece of earth, O simple one? Seek out the source which shines forever.
(Rumi, Masnavi II, 708–709)

Challenging as it might be to provide a comprehensible and meaningful description of the sun's vast potential as a source of energy, it is surely worthy of an attempt to do so. Total solar energy absorbed by the earth is estimated to be 3.8 million exajoules (EJ) per year, which is equivalent to more energy being absorbed in one hour than the global population consumes in a year. Put another way, researchers at Stanford University calculated that the amount of solar energy reaching the earth's surface in a year is about twice as much energy as human civilisation can reasonably be expected to extract from all the world's known and estimated resources of fossil fuels and mined uranium combined. Of course, not all of that solar energy is available for harvest. The United Nations Development Programme in its 2000 World Energy Assessment estimated the annual potential for harnessing solar energy by various means being between 1,575 EJ and 49,837 EJ (taking into account insolation levels, cloud cover and usable land resources). Even this lower figure is substantially higher than global energy consumption in 2012 of 560 EJ.

While energy from the sun has been utilised in various ways by humankind for thousands of years, the first commercial adaptations such as water heating, steam generation and water treatment emerged in the 19th century. Solar energy as a means of electricity production is a more recent phenomenon. The first photovoltaic (PV) cell was constructed by Charles Fritts in the 1880s, with further technical advances leading to American engineer Russell Ohl patenting the first modern solar PV cell in 1946 and the manufacture of the first crystalline silicon solar PV cell in 1954. From humble beginnings, the oil crises of the mid-1970s led to the development of the first commercial-scale solar PV power generation plants. Concentrated solar power (CSP) systems, which employ lenses or mirrors to focus sunlight into a beam which is used as a heat source in thermal power generation, employ technology which harks back to the early days of solar energy exploitation. However, realisation of CSP generation on a commercial scale did not occur until the mid-1980s.

In recent years, solar power generation has enjoyed a period of unprecedented growth. Since 2000, global solar power installed capacity has increased from 1 GW to over 300 GW and the curve is not linear; the last five years have seen a marked acceleration as technologies have become more efficient, manufacturing costs have fallen and more governments adopt low-carbon power generation policies in which solar power plays a key part. Moreover, in contrast to some forms of renewable power generation where deployment is constrained by topographical, environmental and other physical factors which are often beyond the influence of technological progress, developments in solar power technology have led to an expansion in its use across a diverse range of geographies.

Much of this growth has been driven by China, which alone saw an increase of solar power installed capacity of 33 GW in 2016 (and where the China PV Industry Association has predicted substantial additional growth in both the short and long term). Together with China, the United States and Japan make up the top three solar power producers globally and these too have seen significant growth over the last decade. It is true that in some 'mature' solar markets, perhaps most notably in Western Europe where there has been a marked reduction in subsidies and other incentives for renewable power generation, the pace of growth has slowed somewhat since the high watermark of 2015. However, any such slowing is likely to be more than offset over the next few years by the expansion of solar power in developing markets such as Africa, where several countries enjoy some of the highest insolation rates in the world and demand for technology suitable for use in distributed generation systems is increasing.

The purpose of this book is not to dispense a history lesson on solar power so, having used the introduction to indulge in a little 'scene-setting', we must now turn to our main task – to provide a practical guide to various aspects of the solar power market. However, the subject area is vast, and we have had to focus on particular facets which we think will be of most relevance to those interested in the development of solar power on a commercial scale. For that reason there is no coverage of the residential solar market or other small-scale applications. We also readily admit to employing editorial favouritism towards discussion of PV projects at the expense of those involving CSP technology. There is no doubt that the latter will continue to play an important, and perhaps growing, part in the project landscape of the future, but the absolute dominance of PV technology in terms of global installed capacity demands a privileged position in a book of this size.

Aside from this introduction (Part I), the remainder of the book is divided into five parts. While these will bear reading chronologically, we have set out below a brief description of each remaining part in order to assist the reader in taking a more bespoke approach to navigating the book's content, should they wish to do so.

Part II seeks to analyse the fundamental drivers of the solar power market. In most respects these are shared with the other renewable power generation technologies, particularly in terms of the regulatory framework in which they operate at both an international and domestic level. They all have their genesis in the policy maker's desire to incentivise low-carbon power generation, even if many markets are now transitioning towards a low, or no, subsidy regime. This section also looks in detail at the technology deployed in solar power projects and the way in which both the regulatory and technological aspects drive the economics of solar power project development.

Having explored the market's fundamentals, Part III goes on to dissect solar power at the project level. This begins with an overview of a typical project structure, a description of the key participants and the contractual framework which binds them together, and an analysis of how common internal and external risks are allocated or otherwise mitigated. The remaining chapters of this section look in greater detail at particular elements of a project which are critical to its implementation and success, including establishing real estate rights, negotiating construction and related contracts, securing project income and understanding the fundamentals of debt financing in the project context.

Project development is only one important element of a more complex marketplace. Part IV focuses on corporate activity involving the buying and selling of development projects, operating assets and the companies which own them. In common with other forms of renewable energy assets, a high proportion of solar power projects are developed by land owners or small development companies, often with a view to selling them on to larger market participants at either a pre or post-construction stage. Likewise, some of these buyers might in turn sell on portfolios of projects to larger asset aggregators, such as infrastructure funds or similar financial investors. All of these activities play a role in generating an active mergers and acquisitions market in solar power.

While solar power generation is, in general, more straightforward than its thermal power equivalent in terms of technology, construction, operation and economics, the global market in which it functions is, nonetheless, complex. As well as advances in technology and manufacturing processes, and variations in how such technology is deployed, the regulatory and fiscal policies adopted in each country, and indeed even within individual countries, can vary widely. It is for this reason that Part V, a significant proportion of the book, is dedicated to a series of regional perspectives. We have sought to balance this section by providing the commentary one would expect on the largest national solar power markets, while also leaving space for countries of particular interest by dint of other metrics, such as the speed of their emergence as promoters of solar power or the relative growth of solar power in a market otherwise dominated by other forms of power generation. Once again, the scope of this section has been limited by the size of the book and we readily apologise if we have omitted a regional perspective deserving of a seat at the table.

The book concludes with Part VI, which considers solar power's place in an ever-changing global power market. In particular, we look at how solar power integrates with power storage, smart grids and other emerging technologies. And we also look to what the future might hold for a technology which, in many respects, can still be regarded as being in its infancy.

To draw this introduction to a close we invite the reader to consider solar power as not just another sub-sector of renewable energy but, in fact, as transcendent. The sun had been shining on earth for 4.5 billion years before our ancestors gazed heavenwards in fear, celebration and worship of the great golden orb. Our modern civilisation, blessed with a degree of scientific understanding, has become more familiar, perhaps even at ease, with our star. But if familiarity has not bred contempt, we are sometimes guilty of neglecting or underestimating the sun's immense potential as a source of energy, our attention drawn instead to finite and irreplaceable terrestrial resources. Only in the last century have we truly started to appreciate, and even more recently to realise, how the sun can energise our civilisation in the millennia to come.

Legal, regulatory and industry frameworks

Munir Hassan
Andrew Sawbridge
CMS Cameron McKenna Nabarro Olswang LLP

1. Introduction

Solar power appears to be at the cusp of fulfilling its promise of providing utility-scale clean power at an affordable price. The plunge in solar technology costs, among other factors, has catalysed significant deployment of solar projects across the globe. Investors are keen to take advantage of the efficiencies offered by these developments while looking for a sufficient measure of stability to enable the required financial decisions to be made. Part of that stability is afforded by the legal and regulatory framework within which projects progress. But the framework may also seek to achieve a number of other objectives, such as:

- to progress and implement social, policy and environmental objectives;
- to ensure proper integration of solar projects within the wider industry matrix of legislation, codes, contracts, permits and guidance documents;
- to provide routes to market for the power produced from the project;
- where subsidies exist, to ensure that the project is, and remains, eligible for such subsidies.

The dynamic nature of the solar industry and its rapid technological advances have made governments and regulators sometimes feel, in responding to demands from investors for certainty and clarity, that they are seeking to establish enduring, stable and reliable policies on ever-shifting sands. Nevertheless, left to themselves, price signals from liberalised electricity markets generally tend to favour the safest investments. This disadvantages projects with high upfront capital expenditure (compared to operating costs) such as solar projects. Therefore, despite the difficulties, supportive national policies – in particular, specific renewables targets, arrangements for stable revenue streams and (in many cases) formal procurement programmes for solar projects – have been crucial to the success and growth of the solar sector. With the advent of subsidy-free solar, even in northern climes, the key to continued future growth will be for these policies to provide a sufficiently stable landscape for supply chains to be established and bed down, and for the economic benefits of greater scale to be fully realised.

The chapter provides an overview of the main considerations for solar projects in respect of the overarching legal, regulatory and industry frameworks, with specific details drawn from the regime in the United Kingdom and selected other jurisdictions to illustrate the approaches taken. What it shows is that the solar industry is in a period of major transition. The industry has long looked to legal, regulatory and industry frameworks to help to give solar projects preferential access to the system, to provide a policy backbone of decarbonisation and, in most cases, to provide subsidies to help solar projects achieve financial returns that justify investment. The solar industry is now having to adjust to a new reality, one in which, particularly in Europe, projects need to proceed without the benefit of a highly supportive policy blanket. In this more Darwinian world, revenue streams are potentially fixed at below market prices, or left entirely for projects to agree with offtakers. While formal procurement programmes for new solar projects remain in many countries, for those governments and regulators that have invested in solar programmes over the past years, many are now looking for the sector to become less reliant on government policy. It should be noted that legal frameworks can both support and actively destroy value in the sector, as was demonstrated by the examples of retroactive changes to subsidy arrangements for renewables a few years ago. However, with solar projects becoming increasingly self-sufficient, their exposure to such intervention is diminishing.

2. International legislation indirectly promoting solar projects

Climate change legislation has been seen as an important driver for policies promoting solar projects. Such legislation can be applicable on an international, continental, federal, state or national scale.

Public international law can come in a variety of forms, and does not automatically have the same legal effect or characteristics as legislation promulgated under the sovereign authority of a nation state. The need for a global response to the challenges of energy security and climate change has led to coordinated international legislative measures that are of varying legal effect. A well-known example is the Kyoto Protocol, an international agreement linked to the United Nations Framework Convention on Climate Change. The major feature of the Kyoto Protocol is that it set binding targets for 37 industrialised countries and the members of the European Union at that time for reducing greenhouse gas (GHG) emissions. These amounted to an average reduction of 5% against 1990 levels over the five-year period from 2008 to 2012. Levels vary by country: the United Kingdom accepted a legally binding target to reduce GHG emissions by 12.5% from 2008 to 2012.

The Kyoto Protocol requires countries to meet the targets mainly through national measures, although three market-based mechanisms are also offered to meet targets: emissions trading, the clean development mechanism (CDM) and joint implementation. The policy behind these mechanisms is to encourage green investment globally at least cost to the countries. For example, Brazil has implemented measures allowing the use of credits under the CDM in connection with renewable energy projects to incentivise investment in green energy.

The Kyoto Protocol works by monitoring the emissions made in each country and keeping a record of the trades that each country carries out to ensure that all transactions are within the rules prescribed. Countries submit annual emissions inventories and national reports for this purpose. The Kyoto Protocol has also established a compliance system that can assist countries with meeting their targets if they are having difficulty in complying.

In addition, the Kyoto Protocol helps countries to adapt to the negative impacts brought about by climate change and an Adaption Fund has been established to finance projects within this area. The Kyoto Protocol drove the development of national emissions reduction schemes and legislation in many countries worldwide before being superseded by the Paris Agreement.

The Paris Agreement entered into force on 4 November 2016. The text[1] was agreed by 195 countries including a number of prominent polluters who were not signatories to the Kyoto Protocol, such as China, India and, initially, the United States.

The overall objective of the Paris Agreement was to hold global average temperatures to well below 2°C above pre-industrial levels. The message of the Paris Agreement was clear: it is the shared responsibility of the global community to mitigate the impact of climate change, and those with the broadest shoulders should take on the largest burden.

Developed countries are required to provide nationally determined contributions (NDCs) which consist of economy-wide absolute emission reduction targets. Article 6 of the Paris Agreement provides a framework for bilateral cooperation between countries to lower GHG emissions; countries may cooperate in implementing their NDCs by internationally transferred mitigation outcomes. This does not replace the CDM established under the Kyoto Protocol although the idea of a CDM transition has been floated, under which CDM projects could be incorporated into the structure of the Paris Agreement, rather than the Kyoto Protocol.

The longer-term commitment enshrined in Article 4 of the Paris Agreement is "to achieve a balance between anthropogenic emissions by sources and removals by sinks of GHGs (greenhouse gases)". The development of renewable sources of energy remains crucial to achieving these goals.

The Paris Agreement provides for a two-track approach. Developing countries have more time to transition to a low-carbon economy than developed countries.

Public international law, such as the Paris Agreement, is relevant for those projects that specifically use the mechanisms established under it. For most projects, the Kyoto Protocol and the periodic summits to discuss the climate change agenda are more important as indicators of trends, and drivers of government policies to promote the green agenda, of which renewable generation (and solar generation) is a part.

At an EU level, legislation can have direct effect on solar projects or require the implementation of measures that have such an effect. The principal piece of legislation is the 2009 Renewable Energy Directive (2009/28/EC), enacted by the European Parliament and the Council to promote the use of energy from renewable sources. It establishes an EU-wide framework for the promotion of renewable energy and sets legally binding targets on member states for the use of renewable energy. In particular, the directive requires:

- 20% of the European Union's overall energy consumption (electricity, heat and transport fuels) to come from renewable sources by 2020; and
- 10% of each member state's transport energy consumption to come from renewable sources by 2020.

1 See https://unfccc.int/files/meetings/paris_nov_2015/application/pdf/paris_agreement_english_.pdf.

Under the directive, each member state is required to submit a national renewable energy action plan (NREAP), following a template produced by the European Commission. The United Kingdom's NREAP requires the United Kingdom to reach its target for 15% of energy consumption in 2020 to be from renewable sources by setting out certain measures and action items. The directive also requires member states to implement measures for guarantees of origin, information and training and providing access to the grid for renewable energy projects.

The directive allows member states to achieve their targets by receiving a transfer of a certain amount of another country's renewable energy to contribute towards their national targets. Further, the directive provides for member states to enter into joint projects on renewable energy with other member states or third-party countries.

The directive must be implemented by each EU member state. For example, in the United Kingdom, the Promotion of the Use of Energy from Renewable Sources Regulations 2011 (SI 2011/243) implements the target set by the Renewable Energy Directive for the United Kingdom to achieve 15% of its energy consumption from renewable sources by 2020, together with interim indicative targets. Separate legislation then introduces the measures intended to achieve these targets.

In terms of broader organisations, a number of development banks and international organisations are active in the sector, to differing degrees. The World Bank has at its core, in relation to sustainable power, the Sustainable Energy for All initiative. This aims to promote sustainable energy across the world, particularly developing countries, while still allowing states to be autonomous in their energy policy decisions. The International Finance Corporation (IFC) appears more focused on renewables as a policy decision than the World Bank at large. In 2015, the IFC launched the 'Scaling Solar' project to attract investment in two large-scale solar projects in Zambia with a total capacity of 76 MW. The auction attracted 48 solar power developers, seven of whom submitted final proposals which secured the lowest solar power tariffs in Africa up to that point. In June 2016, one of the successful bidders offered to sell power at just US$0.06015 per kWh for 25 years.[2] The European Bank for Reconstruction and Development takes a more geographically focused view on the renewable energy market as relevant to its reconstruction remit.

3. National legislation

3.1 Legislation indirectly promoting solar projects

A solar project developer should be cognisant of the overall international legislative landscape for its project, even if the developer is one step removed from day-to-day implementation, because it drives the impetus towards clean energy projects in general. However, the key requirements and determinants of success will be in the national-level legislation that directly applies to the project. A solar project is a complex venture requiring compliance with a wide variety of legislative requirements. It is beyond the scope of this chapter to detail the full scope of rules

2 See https://renewablesnow.com/news/financing-agreed-for-475-mw-solar-project-in-zambia-595421/.

applicable to a solar project. This section therefore deals with two types of legislation: that dealing with climate change generally and that specifically dealing with solar or renewables projects.

Some countries have implemented national legislation to deal with decarbonisation in order to address climate change and to encourage investment in renewable energy. In the United Kingdom, the main piece of climate change legislation is the Climate Change Act 2008. The explanatory notes outline the purpose of the act as implementing a framework for the United Kingdom "to achieve its long-term goals of reducing greenhouse gas emissions and to ensure steps are taken towards adapting to the impact of climate change".

Some elements of the act include:

- establishment of emissions reduction targets and carbon budgeting in statute by the creation of an emissions reduction programme and medium and long-term targets;
- annual reporting by the government on the United Kingdom's GHG emissions, and the granting of powers so that domestic trading schemes can be established through secondary legislation; and
- the establishment of the Committee on Climate Change, an independent body set up to advise the government and devolved administrations.

3.2 Legislation directly promoting solar projects

The principal way that countries have supported solar projects is through the creation of financial support mechanisms. In Europe, such mechanisms tend to take the form of either a fixed payment or a top-up payment per megawatt hour (MWh) of electricity produced, the level of which tends to differ for different types of renewable generation (further detail on which can be found in the chapter on financial incentives and revenue streams). By way of example, policy in Morocco has tended to focus more on providing attractive tax incentives for renewable generation. There is no feed-in tariff in Morocco. Instead, there is a total exemption from corporate tax for any new industrial business and a number of Export Processing Zones have been established which are exempt from customs regulation, foreign trade and exchange control and other restrictive regulation. Likewise, in the case of India, the central government has tended to grant exemptions from customs and excise duties on certain machinery and systems needed to set up a renewable energy project. See the chapters on Morocco and India for further details of these solar markets.

Other governments have adopted legislation to limit foreign direct investment in order to protect the country's home-grown industries. In the United States, the government has approved tariffs of up to 30% on solar equipment made outside the United States,[3] while high tariffs on solar technology remain in Brazil, despite the reduced import tax for solar PV equipment introduced in 2011, provided there is no equivalent local production.

In addition to providing financial incentives, governments have taken other

3 See http://time.com/5113472/donald-trump-solar-panel-tariff/.

actions to promote solar and other renewable projects. The European Union has been active in this area – for example, by requiring member states to impose an obligation on suppliers to disclose information about their fuel mix in the hope that this will enable consumers to exercise their purchasing power in favour of renewable electricity. It has also imposed a requirement on EU member states to ensure that renewable generation has priority access to the grid.

More broadly, governments can also make a difference to renewable projects by ensuring that the wider regulatory regime does not present an obstacle to their success – for example, by ensuring that planning laws balance local and national interests appropriately, and that the interests of renewables are taken into account in the development of industry arrangements, such as the network charging regime. In general terms, the perceived stability and nature of the regulatory regime in a particular jurisdiction can have a significant effect on the success of the renewables industry and the terms on which investors are willing to finance projects there, and therefore a stable legislative regime is important. Of course, this is important to all investors, but the fact that solar projects have high upfront capital costs and have traditionally depended on subsidies makes it all the more so for them.

3.3 Sector statutory framework

Countries almost invariably have electricity sector-specific statutory frameworks and these will usually extend to apply to solar projects, which are often treated the same as more conventional technologies. Sector-specific legislation establishes an overarching framework within which both the highly regulated and more competitive aspects of the electricity sector operate. A clear, well-developed sector legislative framework is important to give investors confidence. The sector framework is important both in its direct effect on solar projects and to understand the obligations and constraints within which other key players, such as the regulated network utilities, must operate.

The overall sectoral framework comprises primary legislation passed by a parliament, secondary legislation produced by governments or other public bodies, licences issued by government or a regulator, industry codes, contracts and guidance. Different jurisdictions have different practices, but as a general rule there is an increase in flexibility in governance, rule changes and interpretation as one moves down this framework. The relative rigidity, importance and inflexibility of primary legislation mean that it is appropriate for the most important concepts to be hardwired into the primary legislation. Where the arrangements may need to be more dynamic to respond to shifting priorities or requirements, they may be dealt with in subsidiary documents. Similarly, where an element of judgement may need to be exercised on the implementation of rules, it may be more appropriate for such rules to be in a licence or other document overseen by the regulator.

The licences issued by the regulator can be characterised as a regulatory (administrative law) contract, and often comprise standard conditions applicable across all similar licensees and special conditions unique to the specific licensee.

The legislative framework also deals with a wide variety of other matters that can be relevant for solar projects. In particular, it sets out provisions by which licence holders may be able to exercise compulsory purchase powers – that is, the power to require a land holder to grant rights required to allow the solar project to carry out its licensed activities.

4. Key bodies

4.1 Introduction

The government (through an energy ministry), sectoral regulator, transmission system operator, distribution network operator, wholesale market operator and the 'route to market' (via a retailer or electricity trader) are the key functions that usually need to be understood and interacted with in promoting a solar project in most jurisdictions. This section outlines the role of the most fundamental entities. Beyond these, there can be a number of other relevant jurisdiction-specific entities, such as regional municipalities, public-private partnership bodies, environmental agencies, planning consenting bodies, health and safety regulators, network system owners, power exchanges and various stakeholders interested in or affected by solar developments. The section draws on a number of examples, with the United Kingdom (and within that, Great Britain) being the principal framework drawn on for illustration purposes given its influence on the design of frameworks across many jurisdictions.

4.2 Role of government 1 – general

In the 1980s and 1990s, power markets took an orthodox view of the primacy of competitive markets which were thought to drive optimal least-cost decisions by developers as to the amount and type of new generation capacity that they should build. Despite ardent efforts by economists to create a perfectly level playing field in the electricity markets, the industry remained closely regulated and sector rules required regular tweaking (or stronger interventions, such as restructuring utilities or forcing divestments of assets) by the regulator or government to ensure that it functioned in the interests of consumers while also promoting appropriate investments.

In particular, the design of many markets loaded the dice against solar developments. Gas-fired power stations were often the price-setting technology in the wholesale market for electricity. Power prices were likely to track increases and decreases in gas prices, and power markets therefore provided a natural hedge against volatility of gas prices. Markets also tended to favour technologies that were flexible to help the system match fluctuating demand for electricity.

Governments have maintained a central role in setting the policy in relation to solar development. As an example, in Great Britain, government policy for the electricity sector is the responsibility of the Department for Business, Energy and Industrial Strategy (BEIS). Certain responsibilities for the sector are devolved to Scottish, Welsh and Northern Irish bodies, although there remains a broad coherence of approach across the United Kingdom. Most of the legislative, trading and support arrangements in England and Wales, for example, are adopted by the devolved administrations with the appropriate adjustments to meet their slightly different policy objectives.

Many other countries worldwide have a different constitutional set-up, with federal and regional governments which share responsibility for different areas of legislation and policy. Australia has a federal (commonwealth) system of government. In the 1980s, most infrastructure was owned by the different Australian states. However, over the last few decades, the states have withdrawn from involvement and nowadays the commonwealth government has more legislative control over the electricity sector; agreements have been put in place between the states and the commonwealth government to transfer regulatory powers to institutions governed by commonwealth legislation.

The government has a much greater role in the sector than defining financial support. It will often also intervene in the electricity sector to deal with, for example, reforming the wholesale electricity market arrangements, establishing the regulatory frameworks, redefining (within limits) the role of the regulator or guiding its activities in specified areas, ensuring that adequate spare capacity is maintained to avoid blackouts or security of supply issues, promoting particular competing high-cost or innovative technologies such as carbon capture and nuclear power, and in effect prohibiting some technologies such as unabated coal-fired power. In the United Kingdom, the 2017 Autumn Budget stated that the government will continue to target a fixed carbon price of approximately £24 per tonne of CO_2 until unabated coal is no longer used.[4] However, significantly it also pronounced a moratorium on new carbon levies for eight years under current forecasts and froze the fuel duty escalator for liquefied petroleum gas, casting doubt on the UK government's political will to drive forward new renewables projects.

4.3 Role of government 2 – subsidies

Many solar sector investors previously considered the key role of government as being to define the level of financial support for solar projects and other technologies. This could be indirect, such as taxation on carbon emissions, or direct, providing grants or additional revenues for solar projects. With the advent of lower subsidy or zero subsidy projects, this role has diminished.

The variety of different frameworks and approaches makes it difficult to claim global trends in the solar industry as there are always exceptions and examples that do not fit the model. Nevertheless, it seems clear that direct financial subsidies to solar projects are either reducing dramatically, are falling away or have been fixed at negative levels (ie, below the market price of power for the usual price-setting technology, eg gas-fired generation).

This move away from subsidies is in large part attributed to the dramatic fall in solar panel costs over recent years, with costs expected to continue to fall for the next decade. Alongside the fall in solar and wind costs, other costs are also falling. For example, the costs of equity, of debt and of operation and maintenance of solar projects have all fallen considerably over the past few years. In addition, the costs of system access are also falling as new options for managing the intermittency risks are

4 See www.gov.uk/government/uploads/system/uploads/attachment_data/file/661480/autumn_budget_2017_web.pdf.

emerging, in particular energy storage. The impact of these cost reductions on the overall development of the solar market is explored further in the "Economic drivers" chapter.

Government-led subsidy regimes, in particular feed-in-tariffs (FiTs) or Green Energy Certificates granted to solar power producers, are therefore generally in retreat. There are still examples of countries implementing such regimes. Hungary, for example, has introduced an offtake regime that offers mandatory FiTs and premium subsidies to new-build renewables plant. However, generally, these subsidies are now being amended or withdrawn in favour of more market-led approaches.

The amount of support available has come under scrutiny, with formal caps on the total amount of financial support for renewables (often with sub-caps for solar and other technologies) being introduced in a number of countries. In Great Britain, this takes the form of the 'levy control framework'.

Subsidy regimes that are capped or secured through a selection process have been criticised as sometimes leading to market inefficiency and a policy of 'picking winners'. For example, the European Commission has stated that member states should adopt measures to control 'overcompensation' in respect of some of their support schemes.[5] In the United Kingdom, Oxford Professor Dieter Helm's scathing assessment of the UK energy industry argued that the UK subsidy system has led to an inefficient culture of lobbying and crystal-ball gazing.[6] Perhaps the clearest manifestation of the overcompensation caused by a subsidy regime is in Portugal. In February 2017, the government ordered that promoters of power plants who had received FiTs and public support funds should return 'excessive' amounts to the National Electric System.[7]

The merits of auction processes have also been noted. Those introduced in Slovakia, Spain and Mexico are seen as indicative of a transition to a more competitive energy industry. However, in many countries, this transition is still at an embryonic stage. Although Peru's first renewable energy auction took place in 2009, it was only in 2016 that the guaranteed prices awarded to bidders fell dramatically, and government subsidies could be curtailed. Equally, some countries, such as Slovenia, have avoided the transition to a more competitive renewable energy market altogether, instead choosing to reinforce and renew existing subsidy schemes.

Rather than adopt an auction process, some countries have developed tax incentives to encourage the development of renewable technologies. Luxembourg's government programme 2013–2018 provides for the introduction of an accelerated tax write-off of investments in assets contributing to energy efficiency in buildings.[8] Luxembourg has also implemented exemptions from income tax for the sale of electricity generated from solar PV sources.

5 See https://ec.europa.eu/energy/sites/ener/files/documents/com_2013_public_intervention_swd04_en.pdf.

6 See https://assets.publishing.service.gov.uk/government/uploads/system/uploads/attachment_data/file/654902/Cost_of_Energy_Review.pdf https://assets.publishing.service.gov.uk/government/uploads/system/uploads/attachment_data/file/654902/Cost_of_Energy_Review.pdf

7 See www.globallegalinsights.com/practice-areas/energy-laws-and-regulations/portugal.

8 See https://ec.europa.eu/energy/sites/ener/files/documents/NEEAP_LU_EN.pdf, p86.

Singapore has resisted a subsidy or FiT regime for renewable energy altogether, instead generously funding R&D and waiting until technologies are commercially viable under usual market conditions. While the Singapore model may be seen as attractive to many countries, the larger economies will need to show leadership in helping to commercialise immature technologies if momentum on these is to be maintained.

4.4 Role of government 3 – auction mechanics

As developers, investors and supply chains have bedded down in different regions, governments have sought to bring competitive tension into the development of new solar projects. A typical example would involve the government or a designated procurement entity issuing a tender for the delivery of a new project of a particular capacity within defined time frames, typically awarded to a qualifying bidder bidding the lowest-price (although not always). Auctions have become increasingly favoured because they notionally provide a level playing field (ie, all bidders must meet the qualification requirements) and allow governments to control the extent of the investment pipeline and the financial commitments made in relation to it. For discussion on economic considerations and details of the United Kingdom's CfD and RO regimes, see the "Economic drivers" chapter.

Engaging in such auction processes has been a significant part of the development activity for many project developers in both emerging and highly mature jurisdictions. The requirements of the process can, however, be significantly different. Key considerations for bidders include:

- Minimum level of development of the project: the qualification requirements can be very basic, particularly where the public sector may deliver appropriate sites to be developed (potentially with preparatory work concluded), or can require consents and connection/land rights to have been secured by the bidder, or potentially for the project to be capable of achieving financial close in a short timeframe. Stricter requirements reduce the number of bidders, but also reduce the perception of non-delivery risks from winning bidders.
- Pricing their bid: this may be complicated for auctions of early stage projects due to the uncertainties arising from the need to take the development forward into the consenting process, the need to contract for key equipment, installation, operation and maintenance and the need to secure debt financing terms in the future.
- Ancillary requirements: these can include commitments relating to use of domestic supply chains (sometimes called 'local content') and the provision of credit security (such as bid bonds) for delivery of commitments. The scale of the latter, in particular, can have a significant impact on the types of entity bidding and on their bidding behaviour.
- Performance requirements: some procurement arrangements can penalise successful projects by adjusting their revenue streams both if the actual capacity contracted is not delivered, but also on the basis of output being outside the range of anticipated output from the project.

- Technical requirements: in some auction arrangements, particularly in developing jurisdictions, minimum specifications may be set out to ensure that the technology deployed meets public policy.

Of course, real-life examples are rarely straightforward. The long-term auctions (LTAs) in Mexico are a good example. They set out a competitive tender process, with a long-term, fixed-price power purchase agreement being offered by a state-owned body, CFE. The auctions simultaneously allow bids by market participants to buy or sell energy, capacity (a commercial product which is granted to energy generators for making electricity available to the grid at peak times) and Clean Energy Certificates (CELs). The auctions are neutral between qualifying technologies, and all transactions are completed under the terms of standardised Electricity Coverage Contracts of 15 years for electricity and capacity, and 20 years for CELs. The LTAs provide a regulated and transparent mechanism for selecting the most cost-efficient projects. The wide range of possibilities to optimise bids makes the bidding process complex and highly sophisticated.

Despite the uncertainties introduced by auction mechanics when compared to open access subsidy regimes such as feed-in tariffs, auction processes have been credited with spurring the dramatic price reductions of recent years. The trend looks set to continue, with procurement in 2018 of 4 GW of solar announced through auction mechanics in Saudi Arabia alone,[9] and many further gigawatts of new solar projects set to be procured through auction mechanisms worldwide.

4.5 Electricity regulator

International best practice, which to some extent could be said to be embodied in the law of the European Union, recognises the need for an independent regulatory body (ie, independent of both government and industry) to oversee the orderly functioning of electricity markets. Initially, regulators were seen as a necessary evil, tolerated only because newly established electricity markets stacked the odds in favour of incumbent utilities and against new entrants. It was thought that they would gradually diminish in importance as markets matured, until eventually a small regulatory office might remain to regulate particular aspects of monopoly network businesses. In fact, the opposite has occurred.

In Great Britain, the regulator is the Gas and Electricity Markets Authority, exercising day-to-day functions through the Office of Gas and Electricity Markets (Ofgem). Rather than diminishing in relevance since liberalisation of the market in 1990, Ofgem has mushroomed in size, with its functions now radiating into a huge variety of areas. In a typical month, Ofgem might publish around 100 documents, many of them consultations, guidance documents or new rules, and it can be a challenge for the industry simply to process and manage this level of regulatory activity.

The broad scope of Ofgem's role is partly a reflection of the increasing amount of

9 See https://cleantechnica.com/2018/01/29/saudi-arabia-set-auction-around-4-gigawatts-solar-wind-capacity-year/.

government involvement in the sector, with Ofgem being the delivery vehicle for many policy interventions, such as managing the subsidy arrangements for solar and other renewables projects. It is partly a reflection of the maturity of the sector, which requires significant and experienced teams to oversee highly intricate and complex industry and regulatory arrangements such as the technical network codes, market trading rules and network incentive arrangements. Also, contrary to some expectations, the electricity market has not stabilised, but continues to evolve, constantly giving rise to new industry and regulatory issues that need to be resolved – among other things, the market has had to deal with the implications arising for, and from, the increasing amount of solar and other so-called 'intermittent' power on the system.

In a number of jurisdictions, the main weapon in the regulator's armoury has been the electricity licence. Major electricity companies generally need a licence to own or operate a network, operate a power station or supply electricity to customers, and this requires applying to the regulator which is responsible for issuing and subsequently modifying, enforcing and revoking licences. A solar project's size often meant that it was able to take advantage of formal exemptions to the need to hold a licence. With the increase in the size of solar projects, such exemptions may not always be available. Where a licence is required, it should be noted that they can be lengthy documents. The UK examples run to more than 100 pages. However, much of the document is standard and in practice, in our experience, the regulator is principally concerned with ensuring that the solar project complies with the wider industry framework, and that network utilities provide the solar project with appropriate access to their system. Such access usually takes the form of a connection offer, being the commercial and technical terms on which the generator can connect to and use the wider electricity network. See section 4.6 of this chapter for further detail about the role of the network operator and grid connection issues for solar projects.

The regulator also has the power to take enforcement action and impose significant fines. Ofgem has enforced payments of over £220 million on energy companies since 2010, arising from compensation and penalties.[10] Solar investors and developers must also consider the requirements of competition regulators. The competition authorities of Estonia and Spain have both imposed significant fines for breach of the EU Regulation on wholesale energy market integrity and transparency (REMIT), which governs the use and disclosure of inside information and market manipulation relating to the energy industry. The Estonian Competition Authority fined the electricity Transmission System Operator €10,000 for failure to disclose inside information to the market in "sufficient time", while the Comisión Nacional de la Competencia, the Spanish competition regulator, imposed a €25 million penalty on a Spanish energy company for manipulating the price of electricity it generated from several hydroelectric plants.

The interventionism of regulators clearly varies on a jurisdiction-by-jurisdiction basis. While some regulators adopt a light-touch approach, pro-actively engaging

---

10 See www.ofgem.gov.uk/news-blog/our-blog/compliance-and-enforcement-action-our-future-approach.

with the industry to improve compliance processes, they are often invested with considerable enforcement powers. The REMIT Enforcement Regulations[11] give regulators the power to impose unlimited fines, enter premises and seize documents (within the scope of the inquiry and in accordance with national legislation). Although many regulators have kept their powder dry in this regard, stakeholders in solar projects should take steps to ensure their compliance programmes are up to the required standards.

4.6 Network utilities – system owners and operators

The structure of the electricity market and management of the network has particular relevance when it comes to having solar generation on the system. There are both predictable and unpredictable periods when there is little to no solar power being exported to the grid. Networks are grappling with the issue, including the increased need for standby power sources (often called 'standby' or 'reserve' capacity) and reconfiguring networks to make them smarter and more able to withstand periods of low or high sunlight. The network operator must manage this issue by actively assessing and forecasting meteorological data in its forward planning.

A physical network is an essential element of every electricity industry. The division between what constitutes transmission and what constitutes distribution is generally determined by reference to voltage. Transmission grid voltages are generally high, while distribution network voltage levels are normally low; but this is not a hard-and-fast rule. One or more bodies are licensed to own and/or operate the transmission or distribution sectors, overseeing issues such as system management and connections. The management of networks is particularly important to the solar industry, due to the intermittent nature of solar power. To operate efficiently, solar projects need to be connected to distribution systems (or occasionally, transmission systems) that can balance and settle electricity effectively, so that any increase or decrease in the solar project's generation does not have an adverse effect on the overall network.

In Great Britain, the national electricity transmission system operator is National Grid, which also owns the transmission system in England and Wales. Scottish Power Transmission (SPT) and Scottish Hydro Electric Transmission Limited (SHETL) each own part of the transmission system in Scotland. In the United Kingdom, there are 14 licensed distribution network operators (DNOs), each responsible for the ownership and operation of the distribution system in a specified distribution service area. The 14 DNOs are owned by six different groups. In contrast, in many countries outside the European Union the transmission and distribution systems are still owned and operated by a single entity, often state-owned.

There are a multitude of reasons why network utilities are important for a solar project. First and foremost, interaction with the network utilities is required to access the market (except in the case of generation solely for use on the same site as the solar project). The terms on which connection to and use of the system is provided are crucial for a commercially viable solar project. Second, network utilities vertically

---

11 The Electricity and Gas (Market Integrity and Transparency) (Enforcement etc) Regulations 2013.

integrated with retail businesses may, in some countries outside the European Union, be designated as the required or likely offtaker for the power output from the project – if this is the case, the terms of the power purchase agreement or FiTs offered will be key for the project's financial viability. Further, the network utilities will assess compliance with system technical requirements during and after the commissioning of the project (including, crucially, the connection point itself and related system reinforcements) and manage the flows onto their system from the solar project in its operational phase. These requirements are commonly set out in a grid code (which, depending on the country, may be a distribution system code, a transmission system code, a requirement to adhere to both or an overall aggregated network code).

The system operator which operates the relevant network may also have rights to require the project to cease to export in specified circumstances. It may be able to require, or contract for, services from the solar project required for the network to operate safely or to electrically balance the overall system. If the solar project is connected at distribution voltages, there may be a need to contract with both the distribution network operator/owner and the transmission network operator/owner, not only to access both networks but also to ensure that the requirements of each network are respected. Except in some countries where this is dealt with as part of the broader power purchase arrangements, network owners and operators would also charge for the cost of connection and ongoing use of the system by the solar project and these charges can have a material influence on the siting or cost-effectiveness of individual solar projects.

The broadly monopolistic nature of electricity networks has led a number of jurisdictions to legislate for a formal separation of electricity transmission businesses from generation and/or supply businesses. This can take the form of business separation through to a prohibition on ownership of both. The latter situation requires careful consideration where the owner of a generation business is looking to invest in transmission assets or vice versa.

4.7 Wholesale market 1 – general

The wholesale market describes the main forum in which electricity suppliers purchase the electricity they sell to consumers. There can be mandatory participation for certain generators in the wholesale market. Markets can take various forms, including a 'pool' market where the output of most of the larger generators is sold into a pool out of which electricity suppliers buy their power, or a bilateral contracts market, where market participants contract directly with each other and notify the volumes and time periods of their agreed contracts to the market operator.

4.8 Wholesale market 2 – market operator

The network utility may operate a market for the wholesale buying and selling of electricity. In some instances, for example where a market is overlaid on a single buyer model, the utility may purchase power placed on the market for its retail arm, or sell it on to other market participants. Alternatively, as is more common in fully liberalised sectors, the market may be a platform provided to market participants for the trading of power, and a separate entity may be designated as a market operator.

In setting up such markets, there is often an aspiration for the market to provide price signals for investment, and a basis for investors to determine forward curves for power prices that could support investment decisions. This is effectively the case where, for example, solar projects enter into a power purchase agreement where the price paid is linked to the wholesale electricity price in the market. Where there is insufficient confidence in the forward power price, investors in solar projects are likely to seek a power purchase agreement with more pricing certainty for all or most of the power produced.

The wholesale market typically does not interfere with the day-to-day physical operation of the system, but overlays that with arrangements to facilitate trades between buyers, sellers and traders and, among other things, requires payments to be made in relation to surplus power produced or consumed. The markets in many jurisdictions are being reviewed as they were not designed with the requirements of solar and other renewables projects in mind.

Solar projects typically participate in a passive manner from a market perspective, producing power when the weather permits with their output depending on the available solar resource. Under the 'balancing market' where charges are levied by the system for imbalances (where actual power produced from the project does not match up to that anticipated), solar projects can be at a disadvantage. A variety of mechanisms have been developed to help to mitigate these impacts on solar developers, including passing the risk of imbalances onto the offtaker who can better manage that through a portfolio of generation assets on its books.

The challenge for markets and for network utilities has continued to increase as the volume of solar projects connected to the system has increased significantly over the past decade. Capacity markets have been introduced to procure and pay for spare capacity on the system to deal with situations where the system may be stressed by too much or too little power generation on the system to meet the demand at that time.

5. Licences

The main principles set out in primary legislation can in some cases be very detailed and there has been a distinct trend towards having significantly longer statutes promulgated by Parliament. Similarly, government departments, under powers delegated in primary legislation, have been producing significant volumes of secondary legislation. While practice differs from jurisdiction to jurisdiction, the day-to-day regulation of the sector is often conducted through licences issued (or regulations overseen) by an electricity regulator, which also then oversees the various codes by which the industry operates.

In the United Kingdom, licences are required for many solar projects, particularly large-scale projects. Licences are usually required for electricity sector activities such as generation (which covers generation from solar), transmission, interconnection, distribution and supply.

A common legal formulation is for the carrying out of a regulated activity without a licence or exemption to be prohibited, often as a criminal offence. For example, in the United Kingdom, a generation licence or exemption is required under the Electricity Act 1989 "to generate electricity for the purpose of giving a supply to any premises or enabling a supply to be given". A solar project will therefore need to consider whether it needs a generation licence or whether it is exempted from the need to hold a licence. Similarly, exemptions may apply for certain electricity transmission, distribution or supply activities.

Although a licence is often not strictly required by a solar project, it confers certain advantages, such as the right to apply for powers relating to compulsory purchase of land required for the purposes of the licensable activities. Accordingly, some solar developers may consider applying for a licence.

In the European Union, from the perspective of a solar project, a key point to note is that any transmission assets (ie, high-voltage connection lines) associated with the project cannot be owned or operated by the generator. The requirement arises from EU law, and the requirement to keep the activities separate may extend further across the generator's corporate group. Different EU jurisdictions have dealt with this issue in different ways.

A retail licence (sometimes called a 'supply licence') allows the licensee to supply electricity to premises – that is, retail electricity to customers. In the United Kingdom, a retail licence cannot be held in conjunction with an electricity distribution licence. There are a number of exemptions to the requirement to hold a retail licence in the United Kingdom. For example, generators who supply their own energy benefit from a class exemption, provided that the energy provided is less than 5 MW, of which no more than 2.5 MW is supplied to domestic consumers. Under this exemption, households that harness energy from solar panels on their roofs do not need to apply for a retail licence.

While there is no specific prohibition on a generator holding a retail licence, few electricity generators do so. Even vertically integrated utilities carrying out both activities would normally have separate corporate entities carrying out the generation and retail activities.

Solar project generators will typically sell their electricity to a licensed electricity supplier, unless they wish to trade directly in the wholesale electricity market and take responsibility for balancing their contractual position with their output. As a result of the difficulty in obtaining long-term fixed-priced power purchase agreements with licensed suppliers that would allow project financing, there has been a trend for some generators to enter into corporate power purchase agreements directly with creditworthy end users with large electricity needs – although as such end users do not have a retail licence, they need to have suitable arrangements with a licensed supplier to ensure compliance with the industry framework. Solar project generators can also sell some of their power to a customer located on the same site, thus avoiding the need to use electricity networks or interact with the market. See the "Securing project revenues" chapter for further detail on corporate power purchase agreements and private wire commercial arrangements.

6. Conclusion

Many of the legal, regulatory and industry frameworks discussed in this chapter were developed at a time when fossil fuels were king. Despite the increasing significance of renewables, it should be recognised that the consumption of energy from fossil fuels still dwarfs the amount of energy consumed from renewable sources. Nevertheless, a material change is occurring. Regulatory frameworks which were built on the predictable generation and low fixed cost/high marginal cost of fossil fuel generation now have to cater for unpredictable, intermittent generation and the high fixed cost/low marginal cost of renewable energy.

Some jurisdictions have struggled to adapt to this transition. The Capacity Market in the United Kingdom, for example, was introduced to procure and pay for spare capacity on the system to deal with intermittent capacity problems. However, the results of the most recent T-4 Capacity Auction, particularly the small proportion of new-build generation-securing capacity agreements, raises questions about whether the Capacity Market is meeting its original policy intentions, including encouraging investment in new-build generation. A more obvious failure of the legal, regulatory and industry frameworks in some jurisdictions has been the over-subsidisation of renewable energy through FiT or Contract for Difference support mechanisms. Bulgaria is just one example which shows the dangers and difficulties of such policies. A FiT scheme introduced in May 2011 was so attractive that it resulted in a boom in renewable energy production and significant revenues for renewable energy generators. The Bulgarian government has tried to claw back these revenues in recent years. In September 2012, the Regulator adopted a decision which introduced a retroactive temporary grid access fee for all renewable energy sources. Furthermore, from 1 January 2014, Bulgaria introduced a fee on the revenues of PV and wind farms at 20% of their FiT (without VAT). Despite this fee being pronounced unconstitutional by the Constitutional Court in July 2014, some developers were unable to recover their revenues, affecting investor confidence for renewables in the country. Bulgaria is an extreme example of the difficulties that have bedevilled many jurisdictions – striking the right balance between incentivising renewable energy production and securing value for money for the taxpayer.

Instead of grafting low-carbon technologies onto the existing system, some commentators have called for the development of entirely new legal, regulatory and industry frameworks which embrace new technologies. One such technology is Artificial Intelligence. Some network control systems are already experimenting with using Artificial Intelligence to improve the performance of networks. In 2017, Google's DeepMind was in discussion with the United Kingdom's National Grid to deploy Artificial Intelligence to help balance energy supply and demand in Great Britain.[12] However, new technologies have the potential to not only improve, but also disrupt existing frameworks. The commercialisation of electric vehicles will require additional generation and network capacity to be made available at certain times to cater for peaks and troughs in demand. Combined with lower fixed costs for solar panels, smart meters and blockchain (a digital platform that allows an individual party to conduct a transaction with another party without the need for an intermediary), have the potential to revolutionise the electricity industry. More and more people are capable of generating and selling electricity directly, altering the role

12 See http://uk.businessinsider.com/google-deepmind-wants-to-cut-ten-percent-off-entire-uk-energy-bill-using-artificial-intelligence-2017-3

of intermediaries. Many traditional gatekeepers are increasingly being bypassed and a new type of consumer – the 'prosumer' who both consumes and produces energy is emerging. Existing legal frameworks are racing to catch up with these industry developments.

A further challenge faced by regulators and governments is the increasing size of solar projects. For example, the Ouarzazate CSP power plant in Morocco is expected to have a capacity of 500 MW when constructed.[13] As a comparison, the Vanguard I satellite, launched in 1958, used a 1 W solar panel to generate electricity for its radios. The sheer scale of new renewables projects, including solar projects, necessitate additional obligations being placed on developers, whose projects acquire national significance. This adds to the ever-more complex regulatory web that developers and investors face when promoting solar projects.

In many respects, the frameworks designed to promote the production of renewable energy can be seen as an attempt to paper over the cracks of an anachronistic system. Layer upon layer of rules have been laid on top of each other to create increasingly Byzantine systems. It remains to be seen whether governments will persevere with such policies, or wipe the slate clean to create legal, regulatory and industry frameworks which put renewables at their heart.

---

13 See www.afdb.org/en/news-and-events/morocco-works-on-worlds-largest-solar-plant-financed-by-afdb-go-underway-11775/.

Solar power technologies and how they work

Alberto Cuellar
James Larmuth
Jose Muñoz
Oscar Velasco
Mott MacDonald

1. Introduction

In recent years renewable energies (RE) have been gaining increasing prominence in the global energy mix and, within RE, solar technologies are progressively emerging as the highest growth component. The two main strands of solar technology for power generation are photovoltaics (PV) and concentrating solar power (CSP). PV technology has dominated the solar power landscape, the main reasons being:

- Modularity – PV systems range from small kW domestic sizes to hundreds of MW PV plants, while CSP plants are large-scale developments (usually 25 MW and above);
- Solar resource utilisation – due to the technology's intrinsic characteristics PV can be installed almost anywhere, while CSP requires high direct irradiation locations to operate and to be cost efficient, which reduces its geographical area potential;
- Technology characteristics – CSP plants require a large footprint for the solar field and utilise a traditional steam plant power block which includes several heavy engineering components (eg, turbines, pumps and exchangers). On the other hand, PV plants are quite simple to install, operate and maintain and typically have short construction times, making the technology appropriate for most territories globally.

All the above in conjunction with the steep reduction in manufacturing costs of the main PV components over the past decade has led to PV technology being chosen over CSP in terms of new project development.

CSP's main advantage over PV is its dispatchability potential, due to its storage integration capability through the use of molten salts, a significant attribute which PV systems are only now beginning to approach through the use of batteries, albeit at higher costs. A recent trend in large-scale projects is hybridisation of CSP and PV under a single generating facility project.

2. Solar PV generation systems

Solar PV generation systems are comprised of the PV array, the inverters, the balance of system components and the utility or loads. The application or type of system (ie, grid-connected or off-grid) dictates the elements of the balance of system, namely mounting structures, batteries, cables, combiner boxes, transformers, energy meters, power controller, etc as depicted in Figure 1 below.

Figure 1: Overview of a typical PV generation system – key components

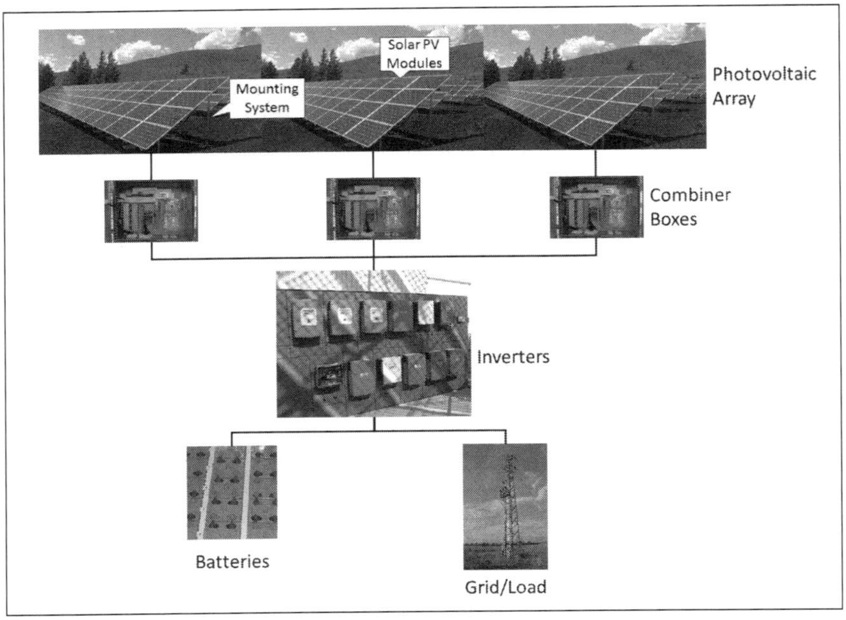

Figure 2: PV system foundation – from the PV cell to the PV system

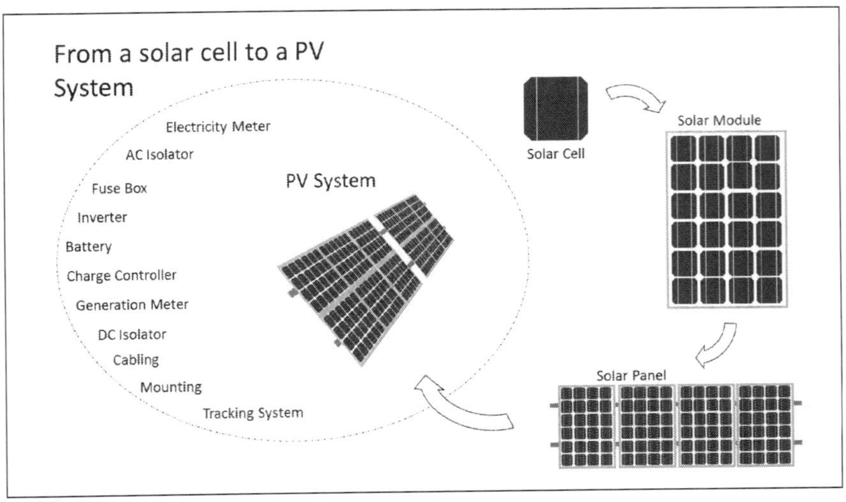

A PV array is a complete unit of PV modules connected together, while a PV module is comprised of several solar cells electrically connected and assembled together. The solar cells, which convert the solar irradiance into direct current (DC) electric power, are electrically connected in either series or parallel formation to increase the voltage and/or current output.

2.1 PV Systems main components

(a) PV modules technology

There are two main types of PV module technologies commercially available on the market:

- crystalline silicon (c-Si), which includes polycrystalline silicon (p-Si) and monocrystalline silicon (m -Si); and
- thin film, which includes amorphous silicon (a-Si), cadmium telluride (CdTe) and copper indium gallium selenide (CIGS).

Other, less commercially available, solar PV technologies include concentrating photovoltaics (CPV), and organic and holographic technologies that are at different stages of maturity and development.

Crystalline silicon, thin film and CPV are referred to as first, second and third generation modules, respectively. Figure 3 below shows examples of these three PV technologies.

Figure 3: Crystalline silicon, thin film and concentrating PV modules

Silicon-based semiconductors (monocrystalline) were first developed in the 1950s, and have seen substantial development over the last six decades. Polycrystalline silicon entered the market in 1981. Both crystalline silicon technologies are quite similar, with the former associated with higher efficiency and manufacturing costs.

Technological and manufacturing advancement has caused module efficiency to increase to over 20%, while reducing the semiconductor material requirements to less than 6g for every watt of peak power output of the PV module (ie, a PV module with a nameplate capacity of 260W would require less than 1,560g of semiconductor material) has resulted in a shorter energy payback period (time required to generate as much energy as is consumed during the production process of the PV module) of up to three years[1] depending on the region and climatic conditions.

Thin film technology entered the market after crystalline silicon and has gained momentum in recent years due to its lower manufacturing costs and its ability to perform better in higher temperatures (low temperature loss). However, it has a lower efficiency than crystalline silicon thereby requiring a larger footprint per Wp.

CPVs are referred to as high efficiency technologies as the technology concentrates light onto a reduced solar cell area using lenses. These systems show very high cell conversion efficiencies of over 38%.

The table on the next page summarises the main advantages and disadvantages of the three PV technologies.

(b) *PV inverters*

PV inverters, also known as converters, are used to convert the DC power produced by the solar PV modules (or accumulated inside a battery system) into alternating current (AC) to be injected into the grid (or consumed by a local off-grid load).

PV Inverters are classified as either grid-tied or off-grid depending on their application. Grid-tied inverters are used in PV systems connected to the utility grid and have the capability of automatically turning off if the grid fails (anti-islanding capability) thereby averting damage to the solar PV system equipment. Other technical features for grid-tied inverters include maximum power point tracking (MPPT), optimising efficiency and minimising losses by increasing thermal performance. Moreover, grid-connected inverters integrate grid support capabilities such as frequency control and reactive power support.

There are three types of grid-tied inverters: string, central and micro inverters. The most commonly used inverters for utility-scale application are the central and string inverters.

Figure 4: Central and string inverters

Source: Pixabay.com

1 Solar Photovoltaics Technology Brief, IEA-ETSAP and IRENA© Technology Brief E11 – January 2013.

PV module technology	Module efficiency range	Advantages	Disadvantages	Percentage of global annual production (commercial use)
Crystalline silicon (c -Si)	< 24%	• Higher efficiencies compared to thin film • Mature and proven technology • Elemental abundance • High durability	• Polycrystalline has lower efficiency compared to monocrystalline • Higher energy consumption during the manufacturing processes compared to thin film • Relatively thick material, not suitable for flexible applications • Higher temperature losses compared to thin film	94%
Thin film (CdTe, CIGS)	< 21%	• Lower temperature losses than c-Si • Generally lower prices (€/Wp) and manufacturing costs than c-Si • Suitable for different applications, BIPV, flexible, etc	• Lower efficiencies than c-Si, but improving • Less field experience and track record • More land area needed for the same installed capacity	6%
Concentrating photovoltaics (HCPV, LCPV)	<38%	• High efficiency • Higher energy production in regions with high direct normal irradiation (DNI) values (>6 kWh/m2 day)	• Requires use of accurate solar trackers • Only uses direct normal irradiation (DNI) • Long-term behaviour uncertainty • Higher costs	—

Off-grid inverters, also known as battery inverters, are used in off-grid systems and connected to a storage system, typically DC batteries, in order to convert the DC power either from the batteries or from the PV modules to AC power, to then be usable by most loads and appliances. The batteries in the solar PV system are charged directly by the PV system. Off-grid inverters are typically used in small-scale off-grid solar PV systems like households and mini-grid systems.

(c) *PV mounting structures*

The mounting structure selected for a PV system depends on application, environment and required energy production output. A solar PV system can be mounted on an existing structure (eg, rooftops) or alternatively on the ground using a stand-alone support.

With respect to environmental drivers, mounting structure choice is influenced by its capability to withstand the site-specific weather conditions over the duration of the PV project. The environmental conditions usually determine the material used for the mounting structures, with steel being the most commonly used material and aluminium used in certain situations due to its light weight.

The commercially available structures are either fixed structures or solar trackers. While fixed structures offer a low-cost solution, tracking systems enhance the energy generation capacity of the PV system. Commercially available tracking systems are characterised by their motion capability and are distinguished as either single-axis or two-axis tracking. Single-axis, by its name, implies that its motion is along one axis, usually tracking the sun from east to west. Two-axis tracking allows for bi-directional movement – vertically and horizontally – which increases the solar irradiation received by the PV modules' surface during the course of the day, thus increasing energy production.

Figure 5: Ground-mounted fixed tilt structures

Figure 6: Single axis and two axis tracking systems

The project's energy production requirements versus land availability and system cost influence the decision between the use of either fixed or tracking systems, a key decision for developers of utility-scale PV systems where maximising return on investment is critical.

(d) ***Batteries and battery management system (BMS)***

There are different battery technology options to store energy, however the two main technologies used in PV applications are lead acid and lithium ion technologies.

Lead acid batteries are used typically in small and medium-scale mini-grid applications. The standard lead acid battery operates with moderate cycle efficiencies higher than 75%, which means that the battery will only be able to deliver (discharge) 75% of the energy that was required to charge the battery, a loss of 25% during each charge–discharge cycle. Further, depending on the operating and ambient conditions, their life can vary between 1,500 to 2,500 cycles at 60% of the depth of discharge (DoD), where DoD is the charge that the battery delivers in each charge–discharge cycle and hence influences the number of charge–discharge cycles. The DoD of the batteries depend on the project specific requirements such as demand profile and the configuration of the generation equipment. Battery manufacturers provide information on this aspect in the battery manuals.

Li-ion batteries are suitable where the application requires fast response times, small dimensions and reduced weight. The systems can achieve cycling efficiencies higher than 90%, maintain low standby losses and deliver good tolerance towards cycling and their life is around 3,000 cycles at 85% DoD depending on the operating and ambient conditions that can be achieved.

For medium and large-scale plants, the battery system will require a first level of supervisory control known as the battery management system (BMS). The main function of the BMS is to monitor the battery cells to maximise performance and ensure the safe operation of the system by monitoring the state of charge (SoC) of the batteries, the DoD and in most cases the state of health (SoH).

The Li-ion technology is costlier than lead acid technology; however, the higher cycle life and better technical characteristics such as higher DoD, lower weight, ability to maintain constant voltage level are stimulating the use of Li-ion, particularly for large-scale plants.

(e) *Energy management system (EMS)*

The energy management system (EMS) is the third monitoring level in PV systems such as hybrid systems with diesel generator and/or battery storage and grid-connected PV with batteries system for off-grid and on-grid projects.

The EMS controls the energy output of the system so as to synchronise with the load requirements and/or comply with the grid code. It gathers electrical information (current, voltage, power) from the main generation equipment (PV inverters, diesel generators), the battery inverters, and the loads to apply the best energy strategy that meets the demand by prioritising the cheapest generation resource (photovoltaics) and maximising the battery lifetime.

The main controls of the EMS depend on the application and while standard software allows quick configuration of the system in the field, customised software may be required to address specific requirements of the plant.

2.2 PV generation application

(a) *Grid connected PV systems*

A grid-connected solar PV power system is an electricity generating plant directly connected to the utility grid, whether at medium or high voltage level. Grid-tied, on-grid, utility-interactive and grid back-feeding are different terms used to describe the same concept.

A grid-connected solar PV system comprises the following main components:

- PV array;
- inverter;
- mounting structure;
- combiner boxes and other safety disconnects;
- SCADA and monitoring system; and
- cables and other BoS components.

The main application of grid-connected PV systems is to supply bulk power into the grid. Centralised systems feed electricity directly into the grid and are usually connected at medium or high voltage. A bi-directional interface needs to be established between the centralised PV AC output circuits and the electric utility network. The inverter plays a major role in ensuring that the output power of the PV system meets the power quality requirements of the grid in terms of frequency limits, harmonics level and voltage thresholds.

Several research and development activities on the different components of grid-connected systems are ongoing with one recent technological advancement being the use of 1,500 Vdc systems. This advancement has a direct impact on the system topology as it enables series connection of more PV modules and hence fewer combiner boxes, cabling and associated civil works as compared with traditional

1,000 Vdc systems. This optimisation has enabled grid-connected systems to achieve significant installation cost reductions.

Grid-connected PV systems have relatively simple deployment and installation methods and their modularity allows for power generation plants to be built in phases, while their operation and maintenance (O&M) activity is much simpler compared with other power plants. The lifetime period of commercial grid-connected PV plants can be for at least 25 years.

In recent years, due to grid operator compliance requirements, PV plants have developed new grid integration features such as:

- Active power control – if necessary, PV plants power output can be remotely curtailed or reduced to zero;
- Voltage and frequency regulation – reactive power is the 'non-working' power in a network that exists due to the current and voltage wave forms being out of phase, and the power factor is a measure of how out-of-phase the current and voltage are. PV inverters have the capability to change the power factor, and through dynamic control systems they are able to generate lagging or leading reactive power (current flowing before or after voltage) according to the grid's needs;
- Night-time reactive power generation – load management on the grid to ensure stability can be enforced through reactive power generation during night-time periods; and
- Low voltage ride-through – inverter capability allowing them to remain connected in case of short grid voltage dips, supporting the grid by feeding in reactive power.

A second type of grid-connected PV system is the grid-connected distributed PV system, which provides electricity directly to a specific grid-connected load (eg, a domestic consumer). Such systems utilise net metering or net billing, which enables the consumer to evacuate the excess solar power produced by the consumer's individual solar PV power system into the grid with the effect of rotating the consumer's electric meter backwards. At any billing period the consumer is billed the 'net amount' of electricity – being the difference between the amount of energy generated and the amount of energy consumed. Normally the consumer is incentivised through credits by the offtaker (eg, the local power company) for the amounts of electricity produced.

(b) *PV grid-connected and batteries (capacity firming)*

Some grid-connected large-scale PV systems integrate the use of batteries in order to be able to achieve the requirements of the power purchase agreements (PPAs), such as energy dispatchability or grid stability support.

One of the options in terms of dispatchability is capacity firming, which means that the generation source (PV system plus battery or also generically named as energy storage system (ESS)), is able to dispatch a committed power output during a defined period of time. This committed power output level will vary depending on the PPA specifications. As shown in Figure 7 below, part of the energy provided by the PV system in the central hours of the day will be stored in the batteries and exported to the grid by the batteries during low irradiance periods (ie, early mornings and late afternoons).

Figure 7: Capacity firming generation curve

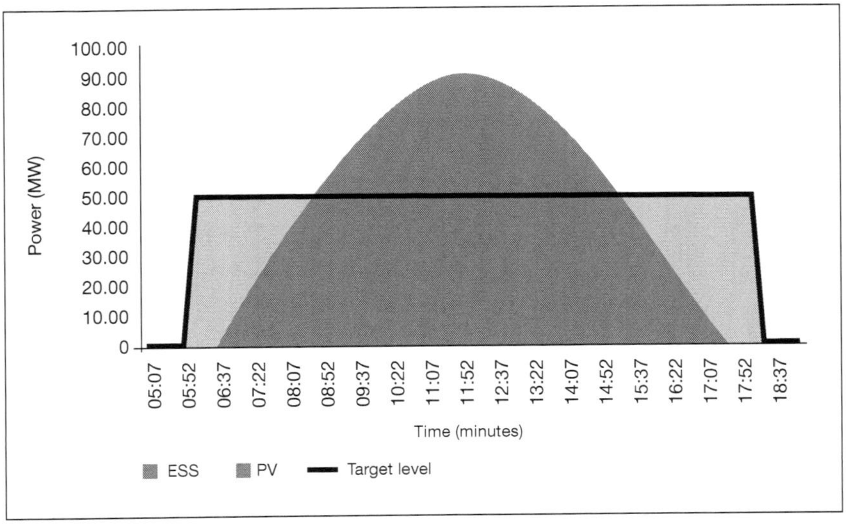

Capacity firming applications may be required by grid operators through the PPA where grid instability is likely to occur. In these cases the energy tariff offered will need to be sufficiently incentivised for the investment to be appealing.

The topology for this application is shown in Figure 8 below.

(c) ***Off-grid PV systems***

Off-grid PV plants are designed to meet the demand requirements of specific loads they are connected to, such as isolated homes, health centres, villages or industries.

There are two common topologies for off-grid PV systems, depending on how the generators and consumers are interconnected (which are shown in Figure 9 below):

- AC coupling – generators and consumers are connected using an AC bus. This approach provides less flexibility on the PV peak power and some limitations in the storage capacity. It requires a more complex control logic and generally more BoS equipment. It is commonly used when the proportion of the energy provided from the PV modules to the loads is high.
- DC coupling – the solar modules and battery system are interconnected using a DC bus by a solar charge controller. An inverter is then used to convert the DC power from the PV modules or the batteries to AC power to meet the load demand. This topology requires a less complex control system and is commonly used when the energy supply to the loads is required during periods of no solar irradiance (ie, during the night).

Figure 8: Capacity firming topology

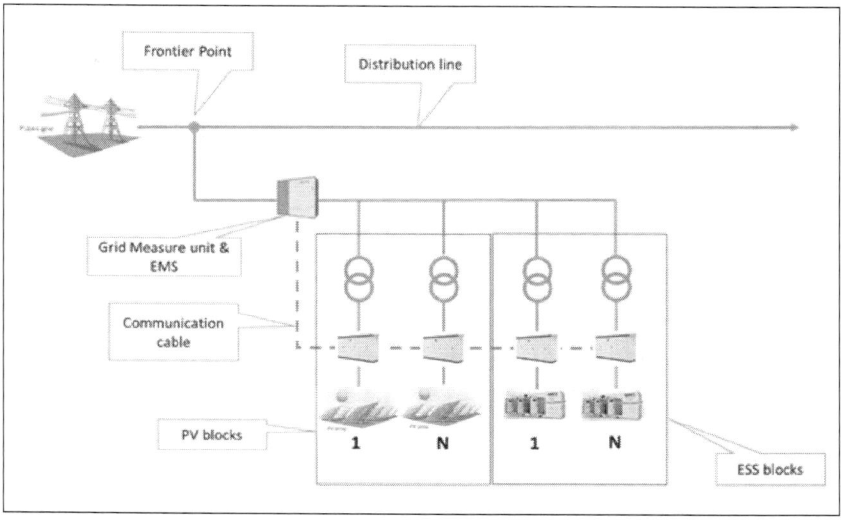

Figure 9: AC and DC coupled off-grid PV systems topology

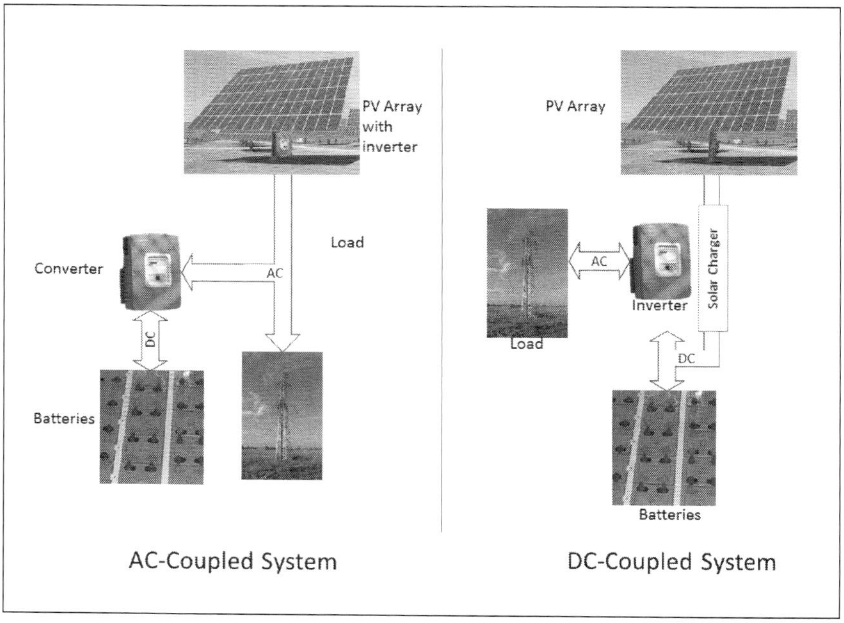

(d) ***Hybrid PV systems***

Off-grid PV systems including only PV modules, batteries and inverters, would be commonly used in small-scale applications and as the solar irradiance would be the only energy source they are weather-sensitive. Another possible option to meet medium and large isolated loads includes the hybridisation with other energy sources. These applications are described in the following sections.

Hybrid PV systems with diesel generators: This fuel-saving application, which does not typically require battery integration, is suitable for isolated heavy industries or villages that use diesel generators to meet the load demand. With typical requirements from hundreds of kW to tens of MW, shifting from diesel-only generation to hybrid solutions helps achieve economic and environmental benefits, increasing the diesel generators' service life and decrease the periodicity of required maintenance.

Figure 10 below shows how the PV generation helps reduce diesel consumption for a daily load profile. Typical PV penetration in this type of application is around 10–15% of the energy generated.

Figure 10: Hybrid PV system with diesel generation curve

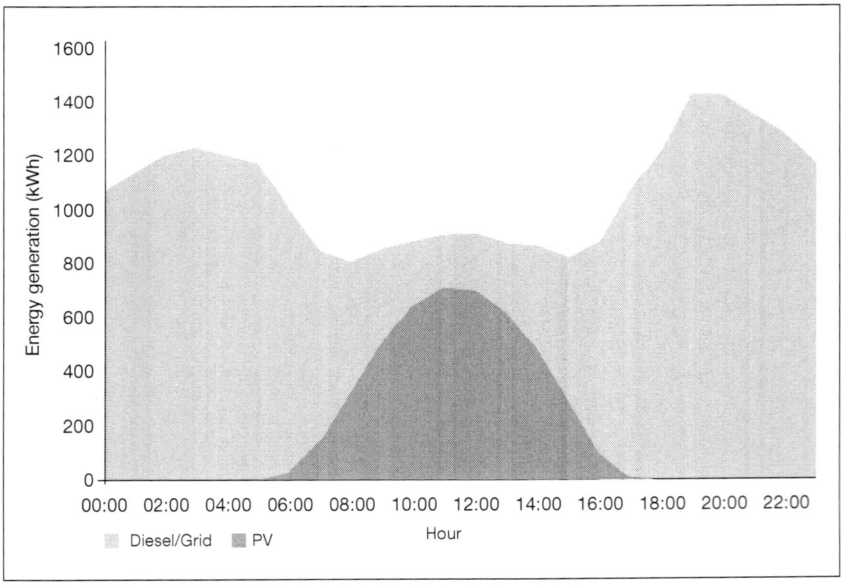

Hybrid PV Systems with battery storage: Hybrid PV systems with batteries typically have two applications – to overcome grid challenges (eg, grid stability) and/or to increase PV penetration.

Where battery storage is integrated to control grid stability, batteries are dimensioned to cater for severe drops in the PV generation normally caused by cloud transients, allowing a smoother energy generation curve as shown in Figure 11 below. Typical PV penetration in this type of application is around 20–25% of the energy generated.

Figure 11: Hybrid PV system with battery storage generation curve

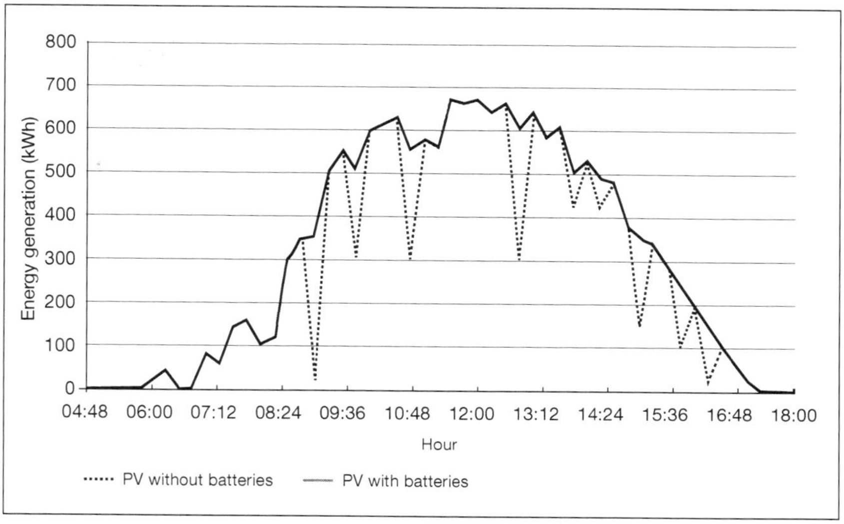

To achieve higher PV penetration rates, batteries are used to shift the surplus PV energy generated to meet the load requirements during periods of low solar irradiance. As shown in Figure 12 below, the energy generated by the PV system is used to support the diesel generator and also to charge the batteries during daytime in order to power the loads during periods of low irradiance. Consumption from diesel generators is limited to periods when neither the PV system nor the batteries are able to meet the demand. Typical PV penetration in this type of application is around 30–70% of the energy generated.

3. Concentrating solar power (CSP)

3.1 Technology options

Concentrating solar power (CSP) offers an alternative means to harness the solar energy resource, as it has the potential to achieve higher capacity factors and provide less fluctuating generation profiles when compared to other RE technologies. A key benefit of CSP is its ability to incorporate highly efficient, high capacity thermal storage systems. These systems allow CSP plants to produce dispatchable energy and generate power for peak load, or even provide base load power. The use of thermal storage systems also reduces the impact of transient solar resource fluctuations on the generation profile. CSP technologies can also be combined with fuel-firing systems to increase the capability of generating dispatchable energy. The degree of hybridisation can range from small back-up firing-up systems to large conventional power plants with an augmented solar field.

CSP comprises a suite of technologies that use mirrors to concentrate the incident solar radiation onto a smaller surface called a receiver, where high-temperature heat is generated and transmitted to a working fluid. The working fluid is then used to drive a heat engine and generate electric power or can alternatively be used as a heat source for industrial processes. There are four primary formats of CSP technology which are defined by the method used to concentrate the sun's rays onto the receiver – line-focus systems (parabolic trough and linear fresnel) where the mirrors track the sun along a single axis, and point-focus configurations (central receiver and parabolic dish) where the mirrors track the sun along two axes. The receiver may be fixed, as in linear fresnel and central receiver, or mobile as in parabolic trough and parabolic dish.

Figure 12: Primary commercial CSP plant formats

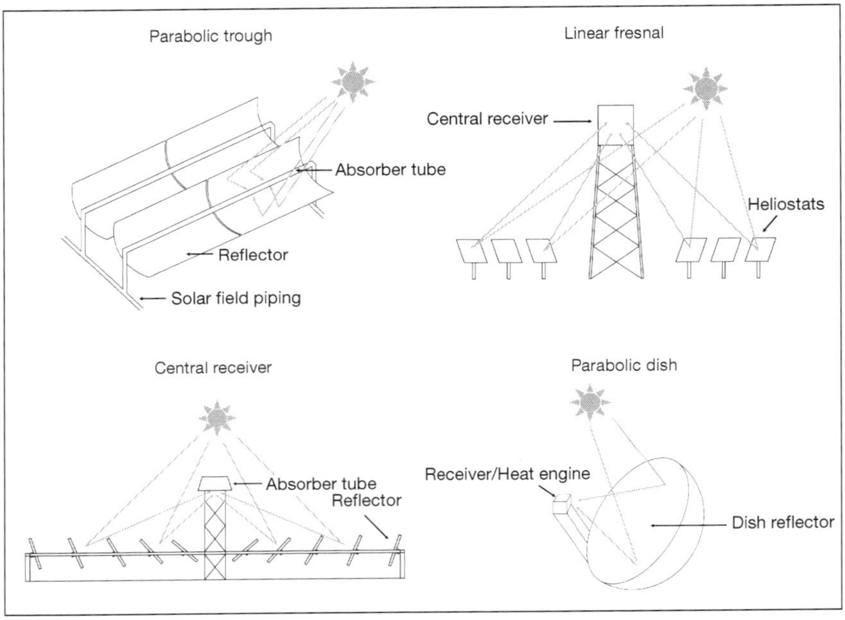

A CSP plant consists of several subsystems, namely: the solar field, the heat transfer fluid (HTF) system, the steam generation system, the power generation system, and an optional thermal storage and/or fossil-fired backup system.

The most widely implemented technology is parabolic trough, however in recent years central receivers have become more recognised as the benchmark CSP technology due to their ability to achieve higher temperatures and increased efficiencies, while offering the most cost-effective implementation of thermal storage. As a result, most of the new CSP plants under construction and development are based on central receiver technology.

There are various thermal energy storage (TES) technologies which can be augmented into the above CSP plant formats. Depending on size and configuration, TES systems are able to buffer short load fluctuations, or shift or extend the plant's energy supply into peak demand times when energy is most needed and typically has a higher tariff. TES systems can store heat by either:

- sensible heat storage;
- latent heat storage; or
- thermochemical storage.

Technology	Concentration method (concentration factor in commercial applications)	Maximum working fluid temperature (in commercial applications)	Storage capabilities	Commercial readiness level
Linear fresnel	Linear concentrators (<50)	400°C	Poor	Poor
Parabolic trough	Linear concentrators (<100)	400°C	Good	Very good
Central receiver	Point concentrators (<3,000)	560°C	Very good	Good
Parabolic dish	Point concentrators (<4,000)	750°C	Very poor	Very poor

The most implemented technology to date is parabolic trough technology with sensible heat storage in the form of molten salt storage systems.

(a) ***Parabolic trough***

A parabolic trough CSP configuration consists of a modular solar field, composed of multiple parallel rows of single-axis-tracking parabolic trough collectors. These collectors are typically oriented in a north–south direction and concentrate direct solar irradiation onto a receiver tube running along the focal point of the collector. Heat transfer fluid (HTF) is circulated through the receiver tube where it is heated prior to reaching the steam generation system. After reaching the steam generator it generates steam that is used to drive a steam turbine and generate electrical power.

Parabolic trough technology is considered the most mature among the CSP plant formats. The technology has a track-record in excess of 30 years. Its first implementation was in the SEGS (solar energy generating systems) plants built in California in the early 1980s. The SEGS experience and knowledge was of paramount importance for the resurgence of CSP development in 2005, which resulted in many more parabolic trough plants being built internationally. Currently there are approximately 70 operational commercial parabolic trough plants with a cumulative installed capacity of ~4,200 MW or ~84% of the total worldwide CSP installed capacity.

Parabolic trough plants have seen significant technical evolutions, mostly in terms of plant size and use of storage. New projects are typically in the 150–200 MW range, allowing economies of scale and associated cost reductions. The use of storage systems has become standard in new projects, and ~40% of the parabolic trough installed capacity has storage capabilities in the range of 5–9 hours.

The key disadvantage of parabolic trough technology is the limitation on the maximum temperature achievable, which is limited to 400°C for the synthetic oil typically used as HTF, due to chemical degradation issues. To increase the HTF temperature, and therefore achieve higher power cycle efficiencies, alternative HTF mediums are being explored, such as molten salts and water/steam. The use of these fluids would also provide other advantages in terms of safety and environmental benefit, since they are not flammable and will not lead to soil contamination in case of spills. Furthermore, the direct use of salt and water/steam can potentially allow for a simpler configuration with fewer heat exchangers, thus reducing the capital investment cost of the system. However, there are important technical issues mostly related with freeze prevention and piping dimensions, which still need to be addressed and these configurations are not yet commercially proven.

Currently research and development into parabolic trough technology is focused on new HTFs and on improvements in the solar field, aiming to achieve lower cost and higher efficiencies. Increased collector dimensions are being deployed, thereby reducing the number of collector parts per square footage to reduce solar field costs. Improvements are also seen in the mirrors, for example thin glass mirrors and reflective polymer mirrors, however, these have not yet been proven.

(b) Central receiver systems

Central receiver plants operate by using a solar field consisting of large numbers of heliostats to concentrate the solar rays onto a receiver located at the top of a central tower. These heliostats track the sun by means of a two-axis mechanism, and are often distributed around the tower in a circular array. HTF is circulated through the receiver where it absorbs the concentrated radiation reflected by the heliostats and converts it into thermal energy which is used to generate steam, drive a steam turbine, and ultimately generate electrical power.

Early central receiver technology was developed in the United States in the 1980s and 1990s with the installation of two pilot projects called Solar One and Solar Two. However, the first commercial central receiver plant only came online several years later in 2007 in Spain. Since then central receiver technology has undergone substantial improvement and innovation. Currently there are eight operational commercial central receiver plants with a cumulative installed capacity of ~620 MW which represents ~13% of the total worldwide CSP installed capacity. The CSP market has seen a clear shift towards central receiver technology in recent years.

Central receiver plants have seen significant technical evolution, mostly in terms of optimising component and plant size and the use of storage. New projects are typically in the 100–150 MW range, allowing for economies of scale. Heliostat sizes have notably increased up to 170 m^2. However, there is not currently a clear consensus in the industry regarding the optimal heliostats size and smaller heliostats in the range of 20–40 m^2 are being explored by different industry players. Molten salt is currently the preferred HTF option since it allows for a relatively high operating temperature of ~560°C and offers a cost-effective integration of large storage capacities. Contrastingly, molten salt has a high freezing point of ~230°C which adds certain operation complexities.

The degree of system and component design standardisation is still limited when compared with that of parabolic trough technology, as there are several different variants of receivers, towers, TES systems, heliostats and solar field layouts.

The primary benefit of central receiver systems is the higher operating temperatures gained from its point-focus configuration, which allow it to achieve higher thermal to electrical efficiency. The higher temperatures further allow for a reduction in the size of the TES resulting in significant cost reduction.

The key disadvantages of the technology are the larger plant footprints required and the limitation of the maximum capacity per tower, which is associated with attenuation losses caused by the long distances between the heliostats and the receiver. Attenuation losses are particularly significant in locations with high prevalence of aerosol, water vapour and dust.

Currently research and development in central receiver technology is focused on improvements in HTF and the solar field in order to achieve lower cost and higher efficiencies using linear drive systems, structural optimisation techniques and autonomous tracking technology, among others.

3.2 Thermal storage systems

The most widely used TES system is sensible heat storage systems. These systems operate by simply raising or lowering the heat of a storage medium. The amount of the storage medium typically determines the quantity of energy stored. The current state of the art in storage medium technology is molten salt. Molten salt thermal storage was first implemented in the mid-1990s and is now featured in the majority of commercial CSP projects. More than 30% of projects in operation and more than 80% of CSP projects in construction incorporate a sensible heat TES.

Molten salt is a relatively inexpensive compound, which has high maximum operating temperatures, in excess of 565°C, but freezes at 228°C and therefore requires potentially expensive anti-freeze systems such as heat tracing. Molten salt can also cause significant corrosion on certain plant equipment. The two-tank molten salt sensible heat storage configuration has been widely employed in commercial CSP systems to date, and two variations of this system currently exist – direct storage systems and indirect storage systems.

Figure 13: CSP TES showing (a) direct heat storage (b) indirect heat storage

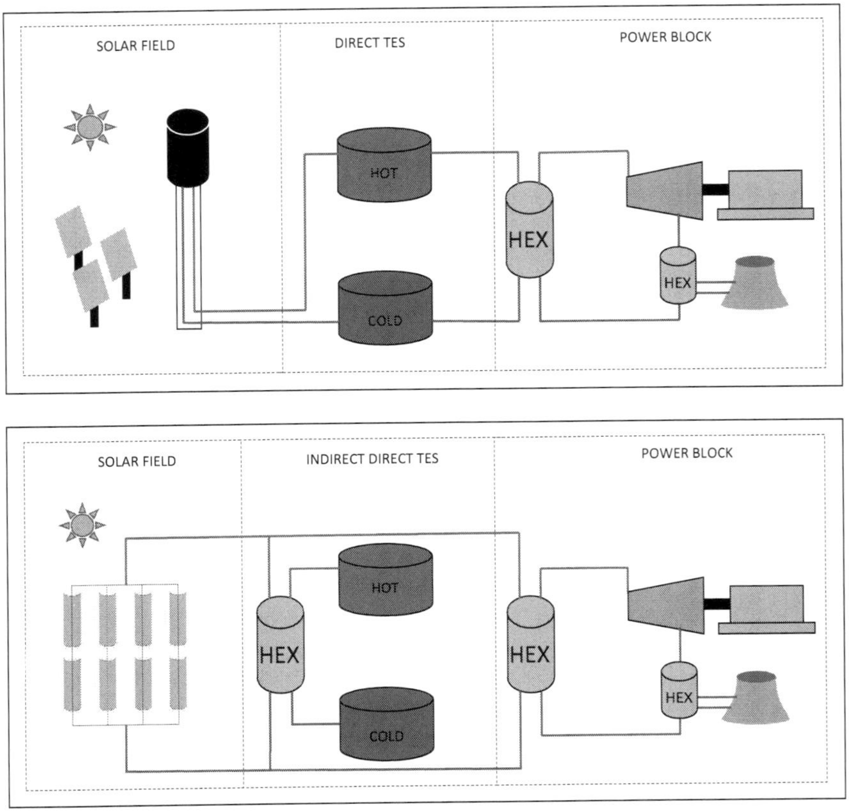

Direct thermal storage systems use the HTF directly as the storage medium. These systems are cheaper and more efficient than indirect systems and fit better with central receiver than parabolic trough configurations.

The indirect storage system uses a different storage media to the HTF. As a result, a heat exchanger system is needed to exchange heat between these two different fluids. This system is common in parabolic trough configurations and has been incorporated in ~40% of the commercial CSP TES systems to date. The system typically uses synthetic oil as the HTF in the solar field loop. Compared to direct systems, the indirect systems are more expensive due to the need for an additional process loop and an additional heat exchanger. These systems have the added limitation of low temperature HTF being used in the solar field as well as the additional losses associated with the extra loop.

Other forms of sensible heat storage systems are also available – steam accumulators, packed bed storage, concrete storage, thermocline systems and solid particle thermal storage. Of these technologies, steam accumulators are the most commercially viable, but are practically limited to short storage durations due to the physical size of the systems required.

An alternative to sensible heat storage is latent heat storage. A large amount of latent heat can be stored or released by changing a material from one phase to another. Latent heat thermal storage systems use phase change materials (PCMs) to store and release heat when needed by the CSP plant. Latent storage systems can use embedded structures like the concrete storage systems, as well as packed bed systems or cascading systems. These systems are currently in the research and development stage with challenges faced being poor heat transfer performance, manufacturing difficulties, safety concerns and durability issues.

In thermochemical energy storage, the energy is stored by means of a reversible adsorption process or chemical reaction. Thermochemical storage possesses high energy density with negligible heat loss. This technology potentially offers long-term energy storage using a small compact storage volume. Various thermochemical concepts have been proposed to date but when compared with sensible and latent heat storage systems, these systems are still at a very early stage of development.

The storage density varies depending on the type of heat storage used. Sensible heat storage offers the most mature and readily available technology solution but with lower available energy storage densities.

Figure 14: Thermal storage maturity and development

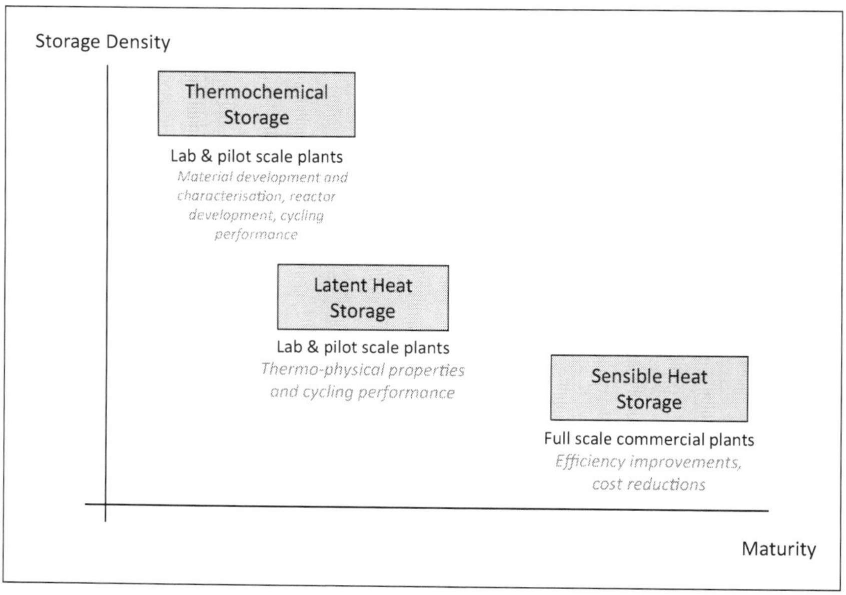

TES technologies available for implementation in commercial grid connected projects are all sensible heat systems as below:

- indirect molten salt storage with two-tank molten salt technology;
- direct molten salt storage with two-tank storage technology (central receiver applications); and
- steam accumulator systems for short-term 'buffer' storage.

3.3 Other CSP applications

Although grid-connected power generation is the main commercial application, there are other important applications of the technology, including direct use of the generated thermal energy in industrial applications.

(a) Off-grid CSP

CSP for off-grid power generation is of particular interest for remote mining sites that have limited access to the electricity grid or gas networks, and rely heavily on fossil fuels that need to be transported over long distances. As the mining industry requires a secure continuous energy supply, the use of CSP with large storage systems can be a good fit, particularly:

- parabolic trough with synthetic oil as HTF with a molten salt storage system; and
- central receiver with a molten salt as HTF with molten salt storage system.

(b) Hybrid CSP with PV

Combining CSP with PV allows for the integration of the separate advantages of the two technologies. PV offers low levelised cost of electricity while CSP with thermal storage offers dispatchability. This combination potentially enables a low-cost power system to achieve high capacity factors of up to 80–90%. The operation of a hybrid plant configuration in this manner relies mostly on PV generation during the day and CSP storing energy for the night hours and dispatching power to smooth and fill the gaps in the PV output. This configuration allows for an optimum dispatchability and reduces effects from transient weather.

These systems can be tailored to suit different plant configuration needs, depending on the demand pattern, interconnection limitations or land availability, as these factors affect the different operation modes of the plant. The optimum configuration and PV to CSP ratio also depends on the relative costs of the systems and the tariff structure, and requires a detailed techno-economic optimisation analysis.

This hybrid technology has not yet been deployed in a commercial solution to date, however several projects are in advanced stages of development and the concept does not seem to present major technical risks which cannot be managed through good engineering and design. The concept is essentially a combination of two proven technologies.

(c) Process heat

CSP technologies have the capacity to generate thermal energy and steam at temperature levels that are suitable for several applications in the food, textile and chemical industries. CSP technologies present fossil fuel alternatives for the supply of energy in the range of 150–350°C.

Due to the temperature levels and small power capacities needed, typically in the sub MW range, there are two CSP technology options that are suitable for process heat applications:

- parabolic trough with synthetic oil as HTF with or without oil storage system; and
- parabolic trough with water/steam as HTF.

(d) ***Solar cooling***

Solar cooling using CSP technologies is an attractive application for the offset of summer peaking electricity demand. In many countries, high loads are experienced from noon to late afternoon because of daytime air conditioning usage. These high loads match the window of high solar irradiation in which CSP technologies are able to generate thermal energy, which can be used as the energy source for absorption heat pumps.

(e) ***Desalination***

Desalination of seawater to meet fresh water demand in some countries is an energy-intensive process that typically relies on energy supplied from fossil fuel sources.

There are several different desalination technologies which are based on either thermal energy or electricity as the primary energy source. CSP technologies represent a solution for augmentation with desalination, especially if combined with thermal storage. This can be achieved using mainly three different design solutions which combine electricity and thermal energy generation via either thermal or membrane separation processes:

- CSP to generate electricity for the desalination process. This is suitable for a reverse osmosis desalination process;
- CSP to generate thermal energy (steam) for the desalination process. This is suitable for a multi-effect distillation (MED) and a multi-stage flash (MSF) desalination process; and
- CSP to generate both electricity and thermal energy for the desalination process. This is suitable for a MED/MSF process.

Due to the energy needs, parabolic trough is seen as the most suitable option for the thermally driven desalination processes (MED and MSF) and both parabolic trough and central receiver systems are suited for the electricity-driven desalination processes (RO).

(f) ***Enhanced oil recovery (EOR)***

As an oil field ages, the decline in pressure and oil production requires the injection of steam, chemicals or gas into the well to facilitate and improve the extraction of the remaining oil. Such a technique is known as enhanced oil recovery (EOR) and is typically used in mature fields. Most of the EOR techniques applied worldwide use natural gas to produce steam, which could be replaced using CSP technologies.

Due to the temperature levels and the need for a continuous energy supply there are two CSP technology options that are suitable for EOR applications, namely:

- parabolic trough with synthetic oil as HTF with a molten salt storage system; and
- central receiver with a molten salt as HTF with molten salt storage system.

Economic drivers: support mechanisms, technology pricing, and other market dynamics

Ricardo Pineiro
Foresight Group

Solar photovoltaic (PV) installations have experienced a decade of exponential growth globally, driven by climate change commitments and a significant reduction in technology costs.

According to the International Energy Agency, solar PV reached 302 gigawatts (GW) of global installed capacity at the end of 2016.[1] This figure compares to approximately 7 GW of global installed capacity at the end of 2006,[2] an increase of more than 4,000%.

A similar pattern was displayed by the UK market, a nation that had virtually no solar PV installations in 2009, yet reached 11.9 GW of total installed generation capacity by the end of 2016.[3]

During the same period, the broader market has experienced a rise in regulatory measures to tackle climate change and the impact that human activity is having on the environment. Historically, the adoption of support mechanisms was necessary to allow solar PV investment to be economically viable and compete with traditional energy sources and fossil fuels. According to the International Energy Agency, approximately US$750 billion in economic incentives have been provided to renewables over the past decade. However, this trend is slowly changing as solar PV becomes economically viable without support mechanisms, especially in European markets that benefit from higher solar irradiance profiles.

In this chapter, we are going to explore the economic instigators for the growth of investment in solar PV. There are several key areas to which we can attribute a significant role in the build-out of new solar PV installations, decreases in technology costs, the increasingly competitive generation costs of PV per unit of electricity and, finally, the climate change commitments and subsequent renewable generation support mechanisms offered by governments. In respect of the latter, we have used extensive examples of subsidy mechanisms from the UK market, illustrating their impact on the scale and rate of deployment of new-build assets, although parallels can be drawn with other markets operating similar mechanisms.

1 Source: "2016 Snapshot of Global Photovoltaic Markets", IEA 19 April 2017.
2 Source: IEA.
3 Source: BEIS.

1. Climate change commitments and support mechanisms

An instrumental contributor to the rapid growth in installed capacity of solar PV is the popularity and prominence of 'green' environmental agendas over the years. More recently, the international participation in the Paris climate agreement, at an overwhelming majority, demonstrates that being a 'clean energy' nation has come very far from being a footnote at the bottom of the political agenda, to becoming an issue that governments globally are taking seriously.

The Paris climate agreement, signed on 4 November 2016, is the most recent attempt to create a unified, global response to climate change, by creating a network of government bodies with the objective of lowering carbon emissions. The agreement, within the United Nations Framework Convention on Climate Change, has the primary objective of limiting the rise in global temperature to 2 degrees Celsius and to pursue efforts to limit the temperature increase to 1.5 degrees Celsius.

These objectives are expected to be achieved by implementing regulatory and commercial frameworks between the participating nations and delivering an increased level of transparency on the degree of success of adopted measures in each nation in terms of the emission levels achieved. Despite the United States initiating a withdrawal process in June 2017, an economy which represents approximately 15% of the global CO_2 emissions in 2015[4] (whether this is an actual statement of intent or an attempt to renegotiate terms), the ongoing support of the remaining nations is expected to continue to act as a catalyst for the renewable energy sector growth.

Governmental support for renewable energy was operating for decades before the Paris climate agreement, with varying degrees of success. In Europe, the adoption of the Climate and Energy Package in 2009, as part of the ongoing reform of the European Energy Market, focusing on emissions cuts, renewables and energy efficiency, is the most relevant governmental support policy of the last decade.

As part of the Climate and Energy Package, the EU introduced the Renewable Energy Directive, requiring member states to derive a pre-agreed proportion of energy generation from renewable sources, such that the EU collectively will source a minimum of 20% of its total energy consumption from renewables by 2020. This represented the first-time binding renewable energy targets agreed between EU member states.

The commitment in the Renewable Energy Directive to source at least 20% of total energy consumption from renewables by 2020 saw the introduction of widespread incentive schemes offering various types of support mechanisms in an attempt to encourage investment into renewable energy generation. Based on the growth of the installed capacity of renewable energy generation in Europe since the implementation of the incentive schemes, it is safe to conclude that the adopted policy instruments had a positive impact on the penetration of renewable energy generation. Countries such as Spain, Italy and the United Kingdom took full advantage of the adoption of subsidy mechanisms to increase the renewable energy installed capacity over a limited number of years, at the same time as the European solar industry expanded.

The primary support mechanism used to promote the objectives of the

---

4 Source IEA.

Renewable Energy Directive in Europe is the Feed-in Tariff (FiT) mechanism. The FiT mechanism offered generators of renewable energy a predictable revenue stream by guaranteeing a per unit tariff for electricity generated over a specific period (eg, 25 years in the UK market). The tariff is obtained in addition to the wholesale power price resulting from exporting electricity to the grid. The tariff is set by regulatory bodies and, in most cases, linked to inflation.

A second mechanism is the introduction of quota obligations via tradable renewable certificates, such as the Renewable Certificate Scheme in the United Kingdom, where certificates would be allocated to renewable generators per unit of electricity generated. Energy suppliers in countries where this scheme applied were mandated to source a certain level of their electricity from renewable sources, which was regulated by providing renewable certificates gained from generators. The scheme created a supply-and-demand market for certificates, where they could be traded in open markets or purchased from regulatory bodies. The revenue obtained by generators from the sale of renewable certificates complements the revenue resulting from the sale of electricity to the grid.

The two mechanisms described above were responsible for most of the development of the renewable energy sector in Europe. However, changes to the EU guidelines on state aid caused the introduction of new mechanisms based on competitive auctions. The mechanism has been adopted for renewable technologies in countries such as Germany, the Netherlands, Denmark, Spain and the United Kingdom.

This third mechanism is known as the Contracts for Difference (CfD) auctions in the United Kingdom. The mechanism sets levels of capacity, with fixed-price offtake contracts entered for a specific period of time (eg, 15 years), effectively removing revenue volatility resulting from wholesale power price fluctuations. To the extent that the wholesale power price is below the CfD strike price, generators will receive a 'top-up' payment equal to the level of the shortfall, and to the extent the wholesale power price is above the CfD strike price, generators will be obliged to make a payment equal to the level of the excess. The CfD mechanism also provides an inflation indexation, however this indexation follows CPI rather than RPI as per the previous UK support mechanisms.

Other support mechanism examples include investment grants and tax incentives, among others.

The adoption of a support mechanism, either to promote the transition to a low-carbon economy or to eliminate vulnerabilities in terms of security of energy supply, is seen by governments as a temporary financial commitment that is expected to deliver lower technology deployment costs over time due to economies of scale and technological and supply-chain improvements.

For this reason, support mechanisms are adjusted over time as a reflection of the perceived cost-effectiveness associated with each mechanism and changing economic conditions. This is an important metric to measure the success of a support mechanism considering the financial cost of implementing such mechanisms, either directly funded by governmental bodies or by companies operating in the electricity distribution markets (and passed through to electricity consumers).

Establishing a support mechanism involves identifying the right balance between achieving the regulatory stability and cash-flow predictability required by investors to support the development of renewable technologies and the introduction of sufficient flexibility in the regulatory framework to ensure the support mechanism remains economically efficient as installation costs decrease.

Support mechanisms that cease to be efficient, especially from an economic perspective, tend to trigger a revision of the mechanism by the relevant governmental body, revision that can cause varied levels of market disruption, which can have tangible impacts on the long-term penetration of renewable generation.

Changes to support mechanisms have historically been focused on FiT, as this was the mechanism generally adopted in Europe when the greatest reduction in installation costs took place, between 2010 and 2012. The retroactive changes to the Spanish FiT implemented in 2010, resulting in a tariff reduction of up to 45% for large, ground-based photovoltaic plants, is the most notorious example of how a change to a subsidy regime can create significant disruption to the solar PV market and a drastic slowdown in investment levels. This decrease in tariff was a result of the unprecedented level of new installations of nearly 2.8 GW[5] in the 2008 calendar year, and associated financial cost for a country in a financial crisis. Although the retroactive change was only approved by Royal Decree in 2010, the government's intention to change the tariff level had been known since 2009, impacting the rate of new PV plant deployment from that year onwards. At the end of 2017, Spain presented a total installed solar capacity of approximately 4.7 GW, which compares with a total capacity of 4.2 GW in 2011.[6]

In the example of the United Kingdom, it resulted in three different incentive schemes for solar PV being implemented consecutively over a period of four years (FiT in 2010, Renewable Obligation scheme (RO) in 2012 and CfD in 2014), in an attempt to identify an effective incentive scheme with sufficient flexibility to adapt to changes in the market.

At the time the first solar incentive schemes were announced, the UK electricity generation mix was dominated by coal, natural gas and nuclear powered plants. The UK energy market was facing the dual challenge of replacing life-expiring coal and nuclear generation capacity together with complying with binding EU policy requirements to significantly cut carbon emissions and increase the proportion of energy supply generated from renewable sources.

This led the UK government to place a strong emphasis on incentivising investment in renewable generation, including solar PV, in order to comply with EU and domestic targets under two main support mechanisms:

- the RO, a 20-year incentive scheme which mainly applied to large installations (above 50 KW); and
- FIT, a 25-year incentive scheme which mainly applied to installations with capacity of less than 50 KW (initial maximum installed capacity per system was set at 5 MW, resulting in a limited number of large-scale projects to be installed up to August 2011, the first date the FiT reduction was announced).

5 Source: REE.
6 Source: REE.

The high tariff level per unit of electricity generated together with the rapid decrease of installation costs during that period created regulatory challenges as the rigid structure of the mechanism caused over-subsidisation and consequent expense for electricity consumers to support. This led to the transition to the RO mechanism between 2011 and 2012, creating a hiatus in large-scale solar PV deployment during that period.

The RO scheme, in operation since 2002, was the primary mechanism to support electricity generation from renewable sources. Support for solar PV was originally announced in December 2012 under an annual degressive incentive mechanism up to 2017. The quota obligations mechanism proved successful for the industry as it allowed the sector to boom in terms of installed capacity, from 1.7 GW in 2012 to 11.9 GW by the end of 2016, peaking at 2.5 GW of new installed capacity in 2014/15 (April to March).[7] This was in part possible as the mechanism – as its predecessor FiT – allowed for a rapid deployment of new eligible installations as long as certain criteria were met. These included technological criteria (eg, solar PV technology), planning consent being granted and confirmation the facility had reached the start of commercial operation.

Figure 1: Solar PV Annual New RO Accredited Capacity (MW)

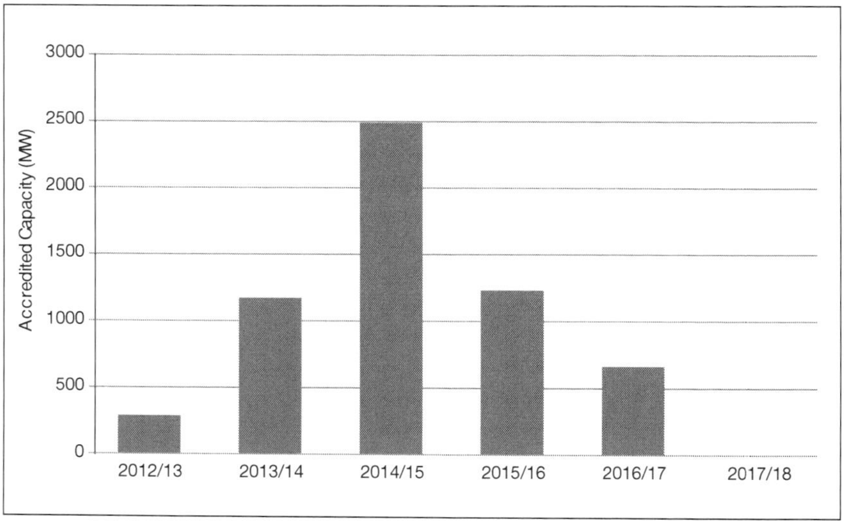

Source: BEIS

The RO scheme was closed to new installations at the end of March 2017, giving way to the CfD auction mechanism, a mechanism better aligned with the EU state aid guidelines and governmental budgetary requirements.

The auction mechanism represents the most cost-effective scheme as it introduces competition between electricity generators to secure a CfD contract. The mechanism is expected to be implemented under annual auctions for the allocation

7 Source: BEIS.

of 15-year contracts, where different technologies compete for a limited CfD annual budget, set annually under the wider Levy Control Framework, which governs the UK government's energy policy expenditure. Bids under the auction mechanism are expected to be structured to provide sufficient return to equity investors. This should result in those offering the most competitive cost of capital securing contracts, effectively maximising the cost-effectiveness of the incentive scheme and delivering economic value to taxpayers and electricity consumers.

By allowing different technologies to compete in the CfD auction, the UK government anticipates an overall reduction in the incentive scheme costs, as the level of competition should support a reduction in project installation costs over time. The concept is based on a levelised cost of energy, a concept we will explore in more detail in the next section.

The CfD mechanism also presents some challenges, as the focus on the most efficient technology from a levelised cost of energy perspective can impact the development of a diversified electricity generation mix, as the ongoing reduction of technology pricing can be limited if only a single type of technology benefits from incentive schemes.

In the United Kingdom the results of the first CfD auction were announced in February 2015, with 27 contracts worth up to £315 million being offered to projects expected to deliver over 2 GW of renewable energy.[8] Solar PV only represented 72 MW across five contracts, in an auction dominated by onshore wind technology. Of the five solar CfD awarded, only two projects were commissioned due to the relatively low strike price of the CfD compared with the expected installation costs.

The second auction was due to be held in December 2015 but was postponed until March 2017. Offshore wind technology dominated the auction as onshore wind and solar PV were excluded from the allocation policy. The results were published in September 2017 with 11 contracts worth £176 million awarded, representing 3.3 GW of installed capacity once operational.

The exclusion of onshore wind and solar PV was justified by the UK government's recognition of the mature status of these technologies and the perception that solar PV and onshore wind could become economically viable without the support of incentive mechanisms. For this situation to materialise, the industry will require installation costs to continue to decrease over time.

2. Technology pricing

As with most types of technology, the installation costs of solar PV have been decreasing over the years as global installed capacity increases. This is a result of developments in solar technology and optimisation of the supply chain. What is unique about these circumstances, however, is how rapidly the cost has decreased during the last decade. As an example, total solar PV installation costs in the United Kingdom have decreased from approximately £2.5 million per MW peak installed in 2011 to approximately £800,000 at the end of 2016.[9]

---

8 Source: BEIS
9 Source: Foresight Group.

This reduction in installation cost was primarily caused by the introduction of support mechanisms on a global scale, as this allowed for a higher and faster penetration of solar PV. As the size of the solar industry increased, the manufacturing industry for the main components of a solar park benefited from economies of scale and enhanced efficiencies. Over time, it allowed for solar technology to become competitive against base load energy sources with lower electricity generation costs.

The main cost components of a solar park are solar modules, inverters, balance of system and third-party services (including, among others, legal and financial services). Solar modules represent 35% to 40% of the total installation cost of a solar park and have been the main driving force behind the reduction of overall solar PV installation cost. Solar module cost presented a sharp reduction from US$0.61 per watt in 2012[10] to US$0.37 per watt in early 2018,[11] mainly driven, again, by economies of scale and increased production efficiency.

The solar manufacturing industry has experienced transformational changes during the last decade with China emerging as the dominant manufacturer of PV cells and modules. The success of the Chinese industry was a result of significant investment in the sector over more than a decade, with China now representing around 60% of total annual solar cell manufacturing capacity globally.[12]

Despite the decrease in solar modules prices in recent years, the industry is facing some head winds in the form of anti-dumping and anti-subsidy policies that have been adopted in Europe and the United States in respect of components produced by Chinese manufacturers. Such policies have been introduced to offer some protection to local manufacturers of PV modules, however it creates a challenging balance between the interests of the local manufacturing industry and the overall response to climate change.

In Europe, Chinese solar panel imports have been subject to counter-dumping measures and subsidies since 2013. This was the result of two anti-competition cases against Chinese manufacturers of solar modules and associated products such as wafers and cells in 2012. In the anti-dumping case, Chinese manufacturers were accused of selling products below manufacturing costs. As part of the anti-subsidy case the European Commission also investigated whether Chinese companies received subsidies from the Chinese government which distorted competition.

Since 2013, importers of China solar modules are required to pay anti-dumping duty if the price is above the Minimum Import Price (MIP). An 18-month extension was agreed by EU countries in 2017, however the MIP for Chinese solar panels is expected to be reduced to reflect the reduction of local production costs.

In 2016, anti-dumping duties on solar modules and components were extended to similar imports from Malaysia and Taiwan. A similar case took place in the United States when the panel manufacturers Suniva and SolarWorld filed a complaint with the US International Trade Commission claiming that imported cells and modules had caused significant financial damage to the US solar manufacturing industry. The

10 Source: BNEF.
11 Source: BNEF.
12 Source: IEA.

proposed action is to introduce import tariffs on imported solar cells and panels. Although the tariffs are meant to apply to any foreign company, the objective is to focus on solar modules and cells imported from Chinese companies.

The expected impact of such measures will be to limit the potential decrease in installation costs and affect installed capacity growth in the short term, as governmental actions of this nature tend to create uncertainty in the market.

In relation to other main solar power installation costs, inverter technology experienced a 55% decrease in price from 2010 to 2017.[13] However, inverters represent only approximately 5% of the overall project installation cost, meaning that its reduction in price is important for sustaining the overall trend, but not a key driver in cost reduction.

The Balance of System (BoS) elements are the remaining components used in solar projects such as mounting systems, cabling and other physical equipment. BoS components were subject to a decrease similar to inverter technology during the same period of analysis.

As component costs in the market continue to fall, and the industry continues to build market share, it is likely that cost competitiveness against other electricity generation technologies will no longer be the main concern for PV.

The cost of developing installed capacity has decreased steadily over the past 20 years, but in the past seven years the cost reduction and competitiveness as against more traditional technologies has been dramatic, having a significant impact on the build-out rate of projects.

3. Levelised cost of energy (LCOE)

As solar installation costs decrease, the competitiveness of solar against other energy generation technologies, from thermal to renewable generation, became a wider point of discussion in global markets. The assessment of energy generation costs is a critical part of energy market analysis for industry participants and policy makers, assessment that is usually performed under a LCOE analysis.

As defined in "Electricity Generation Costs" published by BEIS in November 2016, "the levelised cost of a particular generation technology is the ratio of the total costs of a generic plant (including both capital and operating costs), to the total amount of electricity expected to be generated over the plant's lifetime. Both are expressed in net present value terms". This results in a comparison of the relative cost of energy produced by different technologies, regardless of the project's scale or operating time frame. Key inputs to calculating LCOE include capital costs, fuel costs, fixed and variable operation and maintenance costs and financing costs.

According to Bloomberg New Energy Finance, the LCOE for solar PV has reduced from $225/MWh in Q1 2010 to below $100/MWh at the end of 2016, with a further 15% reduction in the first half of 2017 to $86/MWh. Some projects in locations such as the UAE, Israel, France, United States and Chile are boasting LCOE below $50/MWh due to a combination of optimum solar irradiance, low financing costs and decreasing installation costs.

---

13 Solar Market Inverter Outlook – BNEF, 6 April 2017.

The decrease in the solar PV LCOE in recent years coincided with the rapid penetration of solar globally, with an inverse correlation between the decrease in solar PV LCOE and the solar PV cumulative global installed capacity.

Figure 2: Global PV LCOE ($/MWh) – Installed Capacity (GWp)

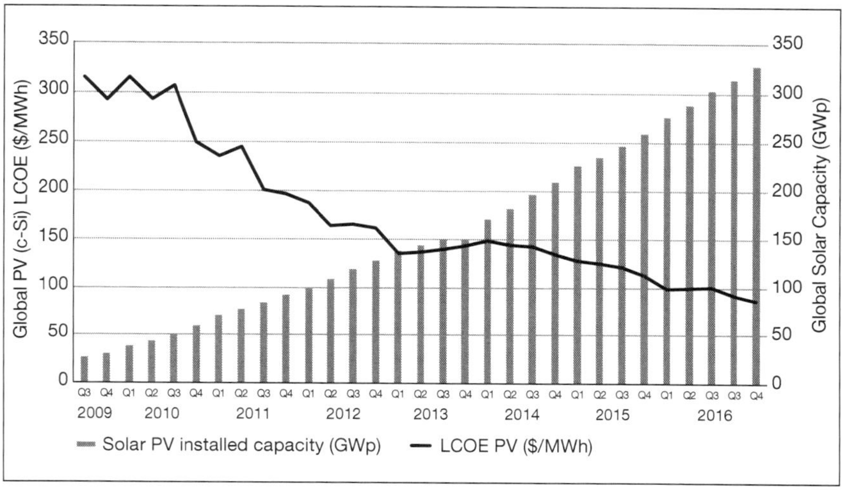

Source: IEA, BNEF

While LCOE is a useful tool for comparing the cost of energy from different technologies, its results are highly sensitive to the underlying input assumptions and how these can change in the future. However, based on most of the analysis available to date in the market, LCOE suggests that solar PV remains a cost-effective energy generation technology to tackle climate change.

4. Other market dynamics

The value of solar must be considered in the context of the generation system in its entirety, and increasing the possibility for using flexibility options on the grid is a fundamental part of this analysis.

Flexibility options are becoming more and more relevant because of the intermittent nature of solar PV generation, resulting in an intermittent energy source that cannot provide the same regular output of traditional fossil fuels. As more renewable energy is commissioned into the grid, the more relevant grid balancing requirements become. Advances over the last few years in battery technology have demonstrated that energy storage has the potential to address the variable nature of the generation profile of renewable assets, however, at this point in its development battery storage technology (predominantly lithium-ion) is still not sufficiently developed to optimise the generation profile of most renewable generation technologies and regulate energy demand.

Although renewable energy is increasingly cost-competitive and storage technology holds great potential, renewable energy technology on its own will not be able to satisfy the base-load generation needs of most nations for the foreseeable future. Therefore, the optimal solution for many countries is a combination of conventional and renewable energy technology in the energy generation mix.

Another relationship to consider in the solar landscape is that between utility-scale versus residential (rooftop) installations. Prices of residential installations are not as competitive as large-scale projects, which can take advantage of higher economies of scale. While the costs of installation may not be as competitive as large-scale solar PV, residential installations can still be revenue generating. Residential installations can avoid the transmission costs that utility-scale solar plants face as they can deliver power directly to the end user without a connection to the wider network. In addition, residential installation has the benefit of not requiring additional land – a scarce resource – for the solar installation to operate.

Finally, the expected growth in renewable energy deployment can also be positively affected by the recent trend of some of the world's largest oil companies making investments in the renewable energy market. Recent examples include Statoil, BP and Royal Dutch Shell, among others, announcing initiatives in the sector:

- Royal Dutch Shell created a new business unit in 2015 that aims to invest over $1 billion in energy transition technologies by 2020, with investments expected to include solar and wind technologies;
- Statoil (soon to become Equinor) is becoming one of the largest offshore wind developers with 1.1 GW of operational capacity and 4.8 GW of new projects consented. The company also piloted the world's first floating wind farm. Statoil's long-term objective is to move towards low carbon, having created a new business unit focused on new technologies and energy efficiency which is expected to represent 25% of the total research funds invested by the company by 2020, and 15–20% of group CAPEX by 2030. In addition, Statoil has created a $200 million fund to invest in charging, storage and solar companies;
- BP announced the objective to have full participation in the energy transition process from 2030 onwards. The company currently owns 2.3 GW of wind installed capacity and is investing in EV charging stations in New Zealand, with the intention of developing expertise to leverage across the network. More recently, the company announced a $200 million investment in the UK solar developer Lightsource with the objective of increasing its presence in the solar sector.

A potential transition from the oil and gas industry into renewable energy markets, either partially or in its entirety, could act as a catalyst for a faster deployment of renewable generation capacity on a global basis.

5. Future

Despite the decrease in incentive schemes available in more mature solar markets, the development of solar globally is showing no signs of slowing down. In 2016, 75 GW of solar PV was installed globally with a further 83 GW to 91 GW expected for completion in 2017. It is expected that 2018 will be the first year that PV new-build will hit 100 GW globally, setting a new record for the solar industry.[14]

The growth to date in the solar PV sector in developed markets reflects only a small proportion of the potential global solar PV market, with a significant number of regions with high solar irradiance levels such as Africa, the Middle East, Southeast Asia and Latin America currently representing modest figures of solar PV installed capacity.

Today, solar makes up just 2% of the world's electricity generation mix,[15] but with the ongoing reduction of installation costs and the increasingly competitive LCOE against other generation technologies, this number is expected to increase over time.

In addition, solar PV is now at the early stage of being commercially viable without incentive schemes in markets such as Spain, Italy and Portugal, but undoubtedly a reduction in regulatory support will continue to affect the growth rates across many markets. However, support mechanisms are expected to become less relevant as technology costs decrease and unsubsidised solar PV installations become the norm, eventually reaching a grid parity status. The increasing ability of solar investments to generate sustainable, unsubsidised returns is a generally positive indicator for the future.

14 3Q 2017 Global PV Market Outlook, BNEF, 18 August 2017.
15 Source: EIA.

The anatomy of a solar power project

Matthew Williams
Orrick, Herrington & Sutcliffe LLP

1. Introduction

The purpose of this chapter is to provide an overview of the various components which make up a solar power project: the key project participants; the contractual framework which provides legal structure and binds those participants together; how the project development vehicle is constituted and organised; and finally, how risk is allocated between project participants or is otherwise mitigated. The details of the main project contracts themselves will be addressed in later chapters in Part III, as will the way in which projects are financed with equity and debt. However, in order to better appreciate how each individual element of a project operates, it is important to understand how each works in concert, hence our interest in the overall anatomy of a project.

Solar power projects are relative newcomers to the energy market; very few projects have reached the decommissioning stage and most are less than 20 years old. However, in many respects they closely resemble power generation projects utilising much older technology, both in terms of participants and structure. A thermal independent power producer (IPP) project developer from the mid-1990s would immediately recognise the contractual structure and allocation of risk adopted by the vast majority of modern solar power project developments. That said, in several respects the anatomy of a solar project is simpler than its IPP forebears, with no requirement for fuel supply arrangements and with many of the construction and operational elements greatly simplified compared to, for example, a combined cycle gas-fired plant.

Figure 1 (overleaf) provides a simplified diagrammatical overview of a typical solar project development structure, illustrating the role of key participants within the project, the primary project contracts and how they relate to each other. The diagram contemplates a project being developed by an individual project developer in a market such as the United Kingdom where private land ownership rights generally apply, and also that the project is owned and operated through a corporate special purpose vehicle (SPV), itself owned by a holding company (Holdco) with multiple shareholders. In many jurisdictions an additional element not shown in the diagram would involve the granting of a concession and ancillary contracts with a government or governmental body. The solid lines in the diagram link the key participants and the most common contractual arrangements, while the dashed lines link participants and contractual arrangements that may be required on a project-specific basis, such as when third-party financing is involved, or where an equipment supplier is used in addition to a construction contractor.

Figure 1: Typical solar project development structure

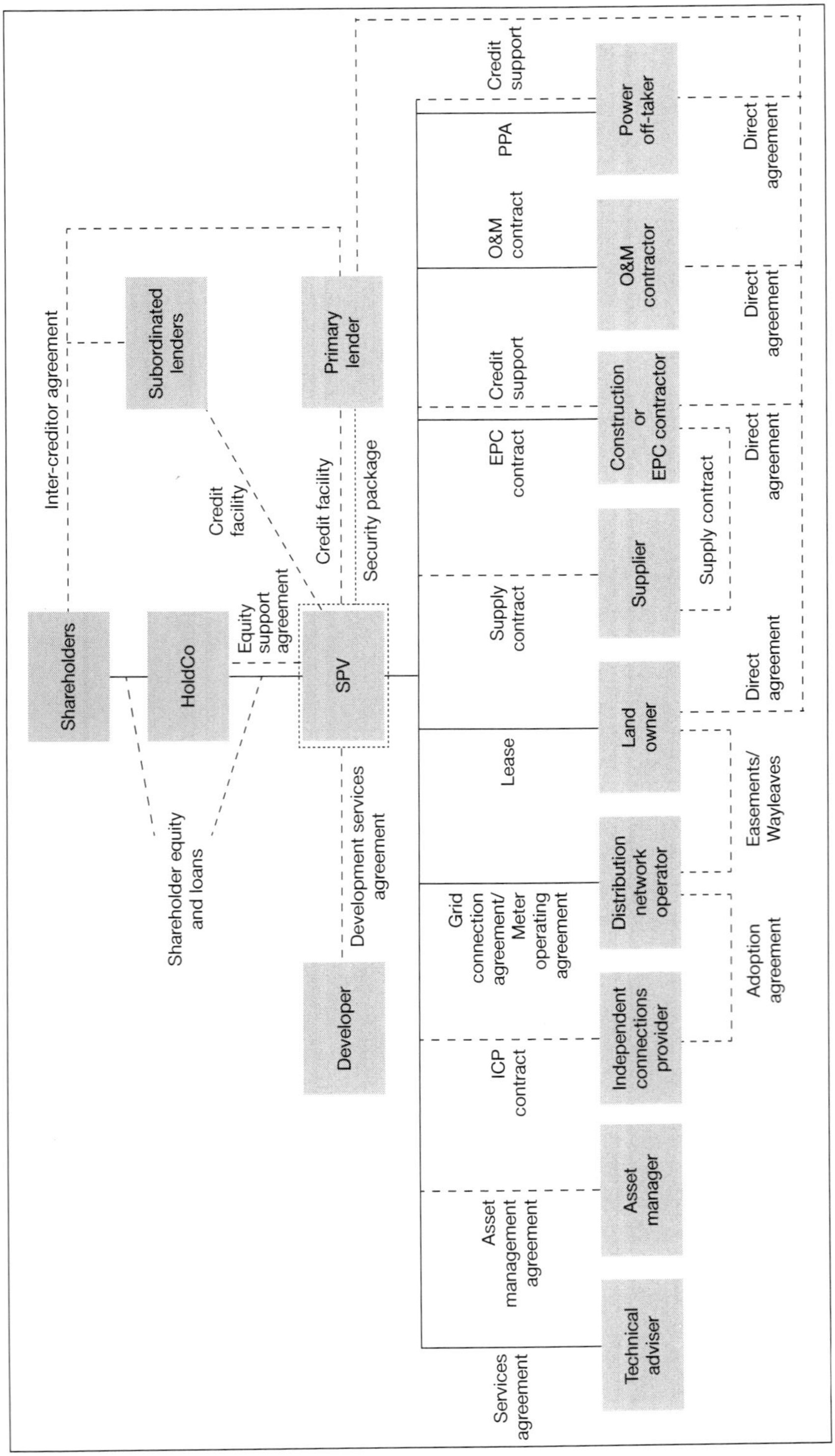

As well as the key participants and contracts involved, this chapter considers the constitution and organisation of the SPV and certain issues relating to corporate structuring in general. This is an important consideration as it impacts the basis on which the participants own and operate the project, how contractual arrangements are entered into, how liability is apportioned between participants (ie, how they operate in a manner limiting their liability to one another, and to third parties) as well as considerations such as tax treatment and how project revenues are utilised and profit distributed.

Finally, the chapter gives an overview of some of the common risks which arise in solar project developments and how either the project agreements are used to allocate those risks between the key participants or the risks are otherwise mitigated. Risks can arise at different stages of a project – early stage development (usually risks relating to securing land rights, obtaining authorisations and grid connection), risks associated with construction, and performance risks throughout the operational life of a project. External factors may also constitute project risks, such as foreign currency exchange and interest rates, power price volatility, counterparty creditworthiness, and political/regulatory risk (which can vary greatly by jurisdiction).

2. Key project participants

Every project will have a common set of key participants whose identity and role are generally well understood. The part played by these key participants, be they developer, lender, power offtaker, grid operator or construction, operation and maintenance contractors, tends to vary little between jurisdictions as these parties are fundamental to the way in which projects are developed around the world. In certain circumstances, however, a broader range of participants can become involved in a project, often dependent on the jurisdiction. Examples include electricity supply companies in markets with licensed supply regimes, corporate offtakers and captive offtakers in 'private wire' projects, governments and governmental bodies in concession regimes, and asset managers working with project portfolios.

Below we have set out further details of the key participants in solar power projects as identified in Figure 1, together with a description of other entities which are likely to have an impact on a solar power project development.

2.1 Developer

The developer is an individual or corporate entity that undertakes early-stage development activities and may also see the project through the construction phase to commercial operation. Very often, and usually for tax-efficiency reasons, it is not the party which owns the development assets themselves, which are held in an SPV, and it may provide its services to the SPV through a development services agreement. On a solar project it is usual for the developer to obtain the necessary land rights, planning consent, grid connection offer and other regulatory consents necessary to build the project. If third-party debt is to be used for construction then it is important that these early-stage development activities are carried out in accordance with the 'bankability' criteria that prevail in the relevant market. Following early-stage development the developer can either manage the construction of the project itself, or seek to sell the project on a 'shovel ready' basis to a third party. The developer will usually charge a development fee for its services to the SPV, payment of which is often funded from the first draw-down of the construction loan in debt-financed projects.

2.2 Lenders and funders

Lenders and funders provide debt or equity financing to a project. There are a broad range of debt and equity investors active in the solar power market. In general terms, the earlier that funds are invested in a project, the more risk the investor carries and the greater the required return on the investment. Commercial third-party debt is usually only available for the construction and/or operational phase of a solar power project, although there are specialist investors who will invest both equity and debt in early-stage developments where attractive investment conditions exist. The role of debt and equity investors in solar power projects is examined in greater detail in the "Financing perspectives" chapter.

2.3 Distribution Network Operator

The Distribution Network Operator (DNO) (or any similarly characterised body in a given jurisdiction) is an entity whose role is to provide an electrical connection for the solar power plant to allow it to export power onto 'the wires'. Sometimes state-owned, a DNO will usually operate the local distribution system on a monopoly basis, in which case its operations will normally be highly regulated to ensure there is no abuse of its status. In the United Kingdom and other similarly structured markets, the DNO is the only body authorised to determine the point of connection of a plant. Further details about the role of the DNO and similar bodies can be found in the "Securing project revenues" chapter, and are commented on further in a market-specific context in Part V.

2.4 Power offtaker

The power offtaker purchases electricity from the SPV and may also be the same entity purchasing renewable certificates or similar carbon 'credits' from the generator, providing further revenue for the project. In many non-European jurisdictions, the offtaker is a DNO, national grid operator or other entity whose role is defined by a specific regulatory purpose in that jurisdiction. In Europe, unbundling rules mean that grid operators cannot act as power offtakers – this role is, however, undertaken by licensed suppliers. Other markets have seen the development of so-called 'corporate PPAs', whereby the offtaker is a commercial entity with no link to the market other than its status as a customer for the power produced. Or the offtaker might be what is often described as a 'captive customer', as is the case with 'inside the fence' or 'private wire' projects. Because of its crucial role in providing revenue to a project, the creditworthiness of the power offtaker and the terms of supply are often crucial factors in the ability of a project to obtain long-term third-party debt (particularly in low or no subsidy environments). Further details about the role of the power offtaker can be found in the "Securing project revenues" chapter, and are commented on further in a market-specific context in Part V.

2.5 Construction contractor

The construction contractor is charged with the timely construction of the plant in accordance with an agreed specification. Where the contractor also provides design and engineering services and equipment supply on a 'turnkey' basis they are generally referred to as the 'EPC contractor' and this model is usually adopted on more substantial projects where developers prefer to put the risk of designing a viable project in the hands of an experienced contractor. In some projects there may be a separate equipment procurement contract between either the SPV and the equipment supplier, or the EPC contractor and the supplier. The supplier will supply the equipment necessary to construct a project, including transformers, inverters and panels. Both the creditworthiness and operational track record of the construction contractor are important factors in the ability for a project to secure third-party construction debt. Further details about the role of the construction contractor can be found in the "Project agreements" chapter.

2.6 Operation and maintenance contractor

The operation and maintenance (O&M) contractor is contracted to ensure the solar plant is operated and maintained in accordance with agreed metrics, such metrics being driven by the performance standards required to secure the projected revenues of the plant. O&M services are sometimes performed by in-house teams, particularly with larger owner/operator companies, but where the owner does not have sufficient resources to undertake the work itself, it will be contracted to a third party provider with the requisite technical expertise. In the solar power context it is commonplace for the same party to be contracted to provide EPC and O&M services, this being both commercially and operationally efficient. Contracting with a third party means that an owner has contractual recourse to the provider if services are not performed to the required standard, usually in the form of a claim for liquidated damages under the O&M agreement. Further details about the role of the O&M contractor can be found in the "Project agreements" chapter.

2.7 Technical adviser

The technical adviser provides advice to clients, usually the owners or lenders, on the technical risks associated with a project and generally reports on the viability of the project from a technical perspective. As well as advising on the technical aspects of the equipment to be installed at the procurement stage, the technical adviser will monitor construction work on site and the commissioning process once construction is complete. The results of the testing and analysis will usually be set out in the form of a technical report which will be relied on by the owner for the purpose of EPC/O&M contract management and as part of the due diligence process for lenders to a project.

2.8 Land owner

The land owner provides the SPV with the necessary legal rights to build the solar project on a given site in accordance with the legal regime applying in that jurisdiction. As this would imply, land owners can take multiple forms, including corporate entities, private individuals and, in many jurisdictions, national or local government bodies. The nature of the rights granted will also vary significantly according to jurisdiction, but will generally fall into the broad categories of full ownership transfer, leasehold rights, licences or other forms of temporary rights and concessions. The rights granted also need to encompass access rights to the site during construction and operation and the necessary rights for any connection to a neighbouring grid. It is common for developers to seek an option over real estate during early-stage development which can then be exercised once all the other elements of a viable project have been secured. Further details about the real estate aspects of solar power projects can be found in the "Real estate" chapter, and are commented on further in a market-specific context in Part V.

2.9 Planning and environmental authorities

Various national and local bodies are empowered to ensure that project developments comply with their particular planning and environmental regimes. Planning and environmental consents will be required for the construction of all solar projects, irrespective of jurisdiction. Where consents have been granted, permission will also be required for amendments to the agreed scope. The authorities will be responsible for reviewing applications requesting permission to build projects at specific sites, which permissions will usually be granted subject to conditions which must be satisfied at different stages of project development – pre-construction, pre-operation, and in the operational phase. While common to all jurisdictions, their scope and application varies. See Part V for a more detailed analysis on a jurisdiction-by-jurisdiction basis.

2.10 National regulator

All mature energy jurisdictions actively regulate the behaviour of participants to ensure the correct functioning of the market. The responsibilities and powers of the regulator will derive from applicable local, and sometimes international, law (such as within the European Union). As most subsidy regimes designed to support the development of solar power projects are sponsored by central governments, the national regulator will usually be tasked with the role of reviewing and granting applications for applicable subsidies, as well as ensuring compliance and administering sanctions where there is a failure to comply. Active and constructive engagement with a national regulator is often a hallmark of successful project developers and owners. The role and power of national regulators varies widely by jurisdiction.

3. The contractual framework

The main contractual agreements for a solar power project relate to:

- real estate (establishing the land rights upon which the solar project will be built, and additional rights or consents that may be required for effective operation);
- electrical connection (agreements setting out the terms on which the solar project will be connected to the local grid system);

- construction and maintenance of the plant (services contracts setting out the terms and specification of construction works to be carried out on the project (EPC contract), and the ongoing operation and maintenance of the plant throughout its operational life (O&M contract));
- power offtake arrangements (a key contract for any power generation project as the power purchase agreement will define the basis on which the project receives revenues for the electricity generated);
- financing contracts (where a solar project requires debt financing to be developed, agreements that set out the terms upon which credit will be provided to fund a project); and
- necessary security arrangements related to the provision of credit.

Below is a summary of the key terms which one would expect to find in the main project contracts in a solar power project context. A more detailed analysis of the terms of EPC and O&M contracts can be found elsewhere in this book.

3.1 Real estate

In the context of private land ownership, an SPV will either secure a full legal transfer of title to the project site or, more usually, a leasehold interest. A lease will set out the SPV's rights to the land and its obligations to the landowner including payment of rent and the obligation to decommission the plant and remediate the site at the end of project's life. Irrespective of the nature of the legal interest in the main project site, solar projects will usually require rights of access (such as easements) for cabling and grid connection purposes across land owned by third parties, which will be in the form of agreements between the SPV and/or the DNO and the relevant land owner. In a situation where the landlord may have concerns about the ability of the SPV to properly decommission the plant and remediate the site at the end of the lease, a decommissioning bond may be required to set aside funds at certain stages throughout the duration of the lease, usually in an escrow account, to contribute towards decommissioning and remediation costs. Further details about the terms of real estate agreements in the solar power context can be found in the "Real estate" chapter.

In certain jurisdictions, particularly in the developing world, it is common for the right to construct a plant on a particular site to be part of the package of rights granted by way of a concession agreement entered into with the SPV by a national government or governmental body. While certain common terms exist in concession agreements, such as concession payments, decommissioning and remediation obligations and term, many provisions are market-specific and these are explored further in Part V.

3.2 Electrical connection

The SPV will enter into an agreement with a DNO (or any similarly characterised body in a given jurisdiction) which contractually defines the connection point between the plant and the grid and establishes the rights and obligations of the SPV and DNO in respect of the connection through general operating terms and conditions. Any engineering or construction works necessary to establish the connection, and which party is responsible for conducting and/or paying for such works, will either be set out in the connection agreement itself or in ancillary engineering and construction contracts. In general, it is the responsibility of the plant operator to install and maintain the meter which measures the export of electricity from the project and which is used for calculating payments under the PPA and, where appropriate, receipt of subsidy payments.

3.3 Construction

The construction contract is the principal agreement between the SPV and the construction contractor under which the contractor agrees to carry out certain construction services for the project in accordance with a detailed scope of work and technical specification and in return for payments made by the SPV. The majority of construction contracts in the solar sector are entered into on a 'turnkey' basis, whereby the contractor is obliged to deliver an operating project complying with a technical specification, on an agreed timescale and at a fixed price. These are generally referred to as EPC contracts. Contract payments are usually structured to be made once specific procurement and construction milestones have been met, concluding with a final payment once commissioning of the plant has been achieved in accordance with pre-defined performance tests.

The construction contract will provide for the payment of liquidated damages by the contractor to the SPV for delay where the contractor fails to perform the works in accordance with the schedule defined in the contract, where there is a delay in commissioning the plant or if the plant does not perform to the operational specification. In the majority of cases, the SPV will seek some form of financial guarantee to back up the contractor's obligations, which can vary from a bank guarantee to retentions from contract payments and parent company guarantees.

In comparison to fossil fuelled plants the construction of solar plants is straightforward from an engineering perspective and claims for failure by the contractor to perform are relatively rare. Problems sometimes arise with technical equipment faults or issues to do with the interface between different elements in the technology chain such as panels, inverters, controllers and storage (particularly if different manufacturers are involved). The extent to which the contractor will be liable for such claims will largely depend on whether it has procured the equipment itself or the SPV has done so, and whether it has complied fully with manufacturers' instructions during installation.

3.4 Operation and maintenance

The SPV will enter into an O&M agreement with a suitably qualified contractor who will operate and maintain the solar plant according to an agreed scope of work and performance parameters and in return for a services fee. Some larger developers/owners have an in-house O&M function but, even where this is the case, services are likely to be subject to an 'arm's-length' commercial contract, particularly if third-party debt financing is in place. In the solar power context the operational scope of work is relatively narrow, consisting primarily of performance monitoring and ensuring optimal parameters are met. The maintenance obligation extends to both scheduled works and resolving unplanned outages. The contract will usually provide for liquidated damages to be paid by the contractor to the SPV where a failure to properly operate or maintain the plant in accordance with the contract leads to a loss of revenue by the SPV under the terms of its offtake arrangements and subsidy payments.

Usually provisions will link the O&M contract to the EPC contract, setting out when the O&M services will commence depending on the stage of works under the EPC contract, and how services between the two contracts will be coordinated between the contractors. This interface is often the source of legal conflict where operational performance issues arise on a project, particularly in the early stages of operations.

3.5 Power Purchase Agreement (PPA)

Aside from any subsidy related revenues, the Power Purchase Agreement (PPA) is the primary contractual mechanism by which a project secures income, hence its terms are usually keenly negotiated. The term of the contract can vary considerably. Short-term contracts may be preferable for an offtaker if they wish to 'shop around' in a market where power prices are prone to volatility; longer-term contracts may be preferred by sellers in instances where third-party financing is needed to fund the project as lenders will require long-term revenue certainty to service the debt. Likewise, pricing may be fixed (usually at a discount to prevailing wholesale prices), subject to indexation or pegged directly to market indices.

The fundamental performance obligations under a PPA are for the SPV to supply power and the offtaker to accept delivery and pay for it. Because the output of solar power plants is variable (albeit relatively predictable), the nomination and dispatching regimes in solar PPAs are usually very simple, with the SPV obliged to make available all power generated during a day and the offtaker obliged to accept delivery of it. Performance metrics are therefore usually tested on a long-term basis, with the SPV subject to liquidated damage payments to the offtaker for failure to supply based on monthly, or even yearly, output figures.

The SPV's reliance on a single source of income through the PPA (particularly in a low or no subsidy environment) will generally lead to it requiring some form of financial guarantee from the offtaker to secure its payment obligations. This can take the form of letters of credit, bank guarantees and, where a suitably creditworthy entity is available, a parent company guarantee.

3.6 Financing

Depending on the market involved, third-party debt will usually be made available to a project on a limited recourse basis – the so-called 'project finance' model. The fundamental terms of the credit agreement will comprise the amount and tenor of the loan, the interest rate applied, conditions precedent to draw-down, the obligation to repay, various financial performance tests that the SPV must meet and covenants made by the SPV concerning how it will conduct the project and its business generally for the duration of the loan. Because the debt is non-recourse, the lender will require security over the shares in, and assets and revenues of, the SPV. It will usually also require certain key project participants such as the land owner, power offtaker and EPC and O&M contractors, to enter into direct agreements with the lender which will allow the lender to 'step-in' to the project if the SPV is in default under a fundamental project contract.

Equity financing can be provided to a project in a number of different ways. Where joint-venture shareholders are involved, this may be done through the terms of a shareholders' agreement which governs the parties' investment in shares and shareholder debt in the SPV. Where third-party debt is also involved, the lender will usually require equity investors to enter into an equity support agreement (or similar) to ensure they subscribe for the shares and provide the shareholder debt which the SPV's financial model assumes for the purpose of securing the third-party debt.

In all but the simplest projects involving both debt and equity financing, it will be usual to have an inter-creditor agreement between all the investors, the primary effect of which is to subordinate the equity investors to the third-party debt providers, to ensure that third-party debt is repaid in preference to shareholder debt and equity.

4. The project vehicle

The choice of an appropriate legal entity, how it operates and whether it forms part of a larger corporate structure for the purpose of developing and/or owning a project, can have a significant impact on a project's long-term financial success. Many of the factors determining the best project vehicle for a given development will be driven by local company and tax law considerations and are beyond the scope of this summary, however some examples are explored further in Part V.

A common feature of many projects, irrespective of the jurisdiction in which they are based, is the involvement of two or more joint venture parties who develop and/or own the asset on a shared investment basis. The remainder of this section will detail various considerations which should be taken into account when establishing and structuring the corporate environment to develop and own a project, with a particular focus on how stakeholders in a shared investment can structure their commercial relationship during the life of the investment and when it comes to an end.

The most common corporate vehicle used in a privately owned project development is a limited liability company (or its jurisdiction-specific equivalent). This is attractive to shareholders in the entity as, in most instances, their liability is limited to their capital contribution, and all assets and liabilities will be held in a single corporate vehicle allowing for ease of sale and transfer of rights and obligations. This can be contrasted with an unincorporated structure, such as a contractual joint venture, where the transfer of ownership, rights and obligations is more complex.

The constitutional documents of the company will delineate the rules by which it will operate and, in some cases, simply comprise the minimum rights and protections afforded to shareholders under company law in the relevant jurisdiction.

In many instances, particularly where the level of investment is higher and with multiple shareholders, either the constitutional documents or a shareholders' agreement (SHA) entered into between them will set out the specific rights and obligations of the shareholders. The more fundamental shareholder rights and obligations found either within an SHA or a company's articles of association will include the following.

4.1 Differentiation of share rights

This may be appropriate where there are multiple equity participants in a project with different levels of investment in a company. Rights attaching to shares can differ in terms of voting rights, rights to dividends, information rights, and the right to appoint directors. The rights that will be agreed between shareholders will, of course, depend on the negotiating position of the particular investor, and how much control they will be exercising in the operation or financing of a project.

4.2 Voting thresholds required on specified matters

Although it is likely company law applicable in each jurisdiction will dictate the minimum threshold of approval required to make certain decisions (eg, in the United Kingdom, the Companies Act 2006 states that a company's articles of association may not be amended unless a 'special resolution' is passed which requires the approval of at least 75% of the shareholders entitled to vote), shareholders may wish to implement higher approval thresholds on important decisions.

4.3 Transfer of shares

It is typical to see provisions setting out restrictions on transfers, permitted transfers, and the process of how shares may be transferred by shareholders (eg, consent provisions may be inserted where transfers require approval of other shareholders). Examples include:

- majority shareholders may wish to include 'drag-along' rights, to the effect that if it wishes to sell its shares that exceed a certain proportion of ownership in the SPV, it will have the right to 'drag' the other shareholders along such that they will also have to sell their shares to the same buyer;
- minority shareholders may wish to be protected by requesting 'tag-along' provisions, meaning that if a buyer has offered to buy shares from another shareholder exceeding a certain proportion of ownership in the SPV, the minority shareholder will also have the right to sell their interests on the same terms, avoiding the situation where they may be left behind on a sale by a majority shareholder without a controlling interest in the SPV; and
- rights of other shareholders in respect of shares being disposed, such as pre-emption rights, may be included so that the existing shareholders have a right of first refusal to buy the shares of a shareholder wishing to exit, before they are offered to a third-party purchaser.

As discussed above, with more complex or higher value projects it is common for multiple equity investors to have a stake in the SPV or its Holdco, or to invest through a contractual joint venture (in either case, they are generally referred to as the 'project sponsors'). Project sponsors will either govern their relationship through an SHA, or in a contractual joint venture through the terms of a joint venture/development agreement (JVA). Terms dealing specifically with rights attaching to shares will, of course, only be found in an SHA. However, in many instances an SHA and JVA will contain broadly similar provisions, examples of which include:

- the level of funding obligation;
- general decision making;
- right to appoint management and committees;
- how conflicts of interests are handled;
- how disputes are resolved; and
- approval of, and compliance with, a project business plan.

As mentioned at the beginning of this section, one of the advantages of developing a project through an SPV is the ease in which the project can be sold to a third party. In the most straightforward form of disposal transaction, where a Holdco wholly owns the shares in the SPV (which in turn owns and operates the project), a simple share sale and purchase can be effected whereby the Holdco transfers its shares in the SPV to the third-party buyer. If the third-party buyer acquires the Holdco rather than the SPV, this reduces the risk to the buyer in terms of potential liabilities, as the Holdco will not be liable for the SPV's liabilities unless, for example, corporate guarantees are in place.

The use of a Holdco may also be an important consideration from a tax perspective. In the normal course of events a limited liability company will be taxed on its profits and capital gains, and its shareholders will be liable to tax on dividends received. However, as well as providing a useful means of shielding investors from project-level liabilities, the use of a Holdco sitting above the SPV may trigger beneficial tax treatment making it a more attractive structure to investors. For example, in the United Kingdom there is an exemption from withholding tax on payments of interest by a Holdco to shareholders if they are a UK corporate taxpayer. More generally, bringing an SPV into a larger corporate structure can often result in significant tax efficiencies such as group relief which allow companies within the same group to surrender tax losses to each other.

5. Project risk allocation and mitigation

While something of a truism, it is reasonable to say that all the risks associated with developing, owning and operating a solar power project, or indeed any form of infrastructure project, rest with the owner of the project at any given time unless, and to the extent that, it is able to contractually transfer that risk to another party or otherwise mitigate the risk through other means. Another fundamental principle of risk allocation is that providers of third-party debt will also want to transfer risk away to parties other than the borrower and, of course, themselves. The process of risk allocation and mitigation is a complex one, often subject to lengthy negotiation, and will vary depending on the particular dynamics of a given project. That said, over time a recognised 'protocol' has evolved, particularly in projects which have project finance debt as part of their funding structure.

In the following chapters we explore in more detail the process of allocating risk as between the SPV, its EPC/O&M contractors, power offtaker and lenders in the particular context of a solar power project. However, more generic principles apply to the allocation of risk in the majority of projects to a greater or lesser extent and can be categorised as follows.

5.1 Financial

In general terms the interests of owners and lenders are fully aligned in seeking to minimise project costs and reducing a project's exposure to any form of financial volatility. Examples of the way such risks can be transferred or mitigated include:

- Achieving, wherever possible, cost certainty under the main project contracts through the use of fixed prices or prices with limited variable components, for example in the EPC and O&M contracts;
- Mitigating foreign currency exchange and interest rate volatility risk through the use of contractual hedging arrangements;
- Mitigating power price volatility risk through either a long-term power purchase agreement or, in a merchant power environment, the use of contractual hedging arrangements. Long-term power purchase agreements often include 'take or pay' provisions where the offtaker contracts to pay for a minimum generation capacity over a number of years and pay for it irrespective of whether it accepts delivery of electricity from the plant;
- Reducing the credit risk posed by key contractual counterparties by requiring credit enhancement measures such as parent company guarantees, bank guarantees and letters of credit.

5.2 Technology

The technology risk posed by solar plants is, in relative terms, low compared to other power generation technologies. Such risk that does exist can be mitigated through the use of thorough due diligence of the technology itself and the specific manufacturer, including an analysis of the historical performance of plants employing the same technology.

5.3 Construction

Again, construction risk on solar projects is relatively low when compared to other power generation technologies. As well as using a fixed price, 'turnkey' contracts, the owner and lender will seek to transfer delay and initial performance risk to the contractor wherever possible, so that if there is a delay in commissioning the plant an appropriate level of liquidated damages will be payable by the contractor. At a more fundamental level, the owner will often require the contractor to be responsible for the overall design of the plant, so that any risk associated with the integration of the various technology packages also rests with the contractor.

5.4 Performance

As a general rule, the owner and lender will seek to transfer to the O&M contractor any risk associated with failure to operate, or perform maintenance on, a solar plant in accordance with agreed parameters. Solar plants are usually not prone to operational or maintenance issues, indeed it is one of the key advantages of the technology, but where problems arise the owner and lender will expect the O&M contractor to be liable if it has not performed its contractual obligations in all material respects. One exception to this rule, as is the case with the EPC contract and power offtake contract, concerns *force majeure*. The main project documents will usually stipulate that events of *force majeure* will absolve the affected party, either temporarily or permanently, from performing the obligations it can no longer perform as a result of the *force majeure* event. While market practice has identified typical categories of *force majeure* event, the owner and lender will always seek to minimise the scope of such events to prevent the relevant risk transferring back to the SPV.

5.5 Repayment

Lenders will employ a series of contractual measures and legal protections to mitigate, as far as possible, the risk that the borrower will not be able to repay its debts. Apart from the various contractual terms contained in the loan agreement concerning the control and monitoring of the financial performance of the borrower, the lender will usually require an extensive security package over the shares in, and assets and revenues of, the SPV.

5.6 Political and legal

Changes in regulation and policy which affect legal rights and obligations pose a risk throughout a project's life. In some scenarios the risks are very substantial indeed, such as retroactive changes to the subsidy regime for solar power, and can result in project owners, investors and lenders seeking legal redress through the courts. The extent of these risks, and the degree to which owners and lenders can insulate themselves from them, is largely dependent on the jurisdiction in which they occur. In some instances political risk insurance may be available, although this is often expensive to obtain and limited in scope. More commonly, project contracts will contain change of law provisions dealing with legal, regulatory and policy changes that have a material adverse effect on the performance of the contract or the project as a whole. The aim of negotiations is to identify the potential risks that can be foreseen and where losses can, to some extent, be quantified, so that provisions which set out the necessary protection and compensation can be included in the agreement.

Real estate

Brian Hegarty
Ashfords LLP

1. Introduction

This chapter considers land acquisition for the delivery and operation of a solar project and the real estate documents that will need to be put in place, particularly when either the developer does not own the land on which the array is to be constructed, or the project is being structured to enable a funder to take security over the developer's land interests. It focuses on the rights required over private land in the context of a common law jurisdiction, (drawing specifically on examples from English law practice and procedure), but many of the principles described apply equally in civil law jurisdictions and where rights over land are granted by state bodies (eg, under concession arrangements).

This chapter is not designed to explain or examine each and every provision of the key project property documents but to consider those that are, if not critical, then common to solar projects as distinct from other commercial landlord and tenant relationships.

Funding will be required or desirable in the vast majority of projects unless the developer is financing the project itself, therefore the developer will need to offer the funder security (sufficient to secure the funding) over either the project assets or other assets which are unconnected with the solar project.

Typically solar projects are appraised and funded for a defined project/investment period. A project life of between 25 and 40 years is usual.

The main property document that will need to be entered into by the developer will be the lease of the site on which the solar array is to be built and operated. The lease will be granted to the developer by the landowner (the 'landlord'). The key features of the lease are discussed later in this chapter.

In most cases the developer will want to ensure that it is entitled to a lease of the site before it speculates by investing what, in most jurisdictions, will be significant costs of site investigations, robust financial appraisals, applying for and prosecuting an application for planning permission, construction permit or other 'permission' as required in a particular jurisdiction to undertake a solar development scheme.

The developer will also usually want to secure funding before it is committed to the lease for the term of the lease.

Accordingly, the lease of the site is usually preceded by an option, although conditional contracts are sometimes used.

A conditional contract is as it sounds – a contract by which the landlord agrees to grant the lease, in an agreed form, if certain defined conditions are satisfied.

Typically the conditions will be – or include – the grant of satisfactory planning permission; the developer securing funding and/or a viability test; grid connection being secured; and, if necessary, rights over land owned by third parties being obtained. There might be others.

An option is also as it sounds – the landlord grants the developer the right to take the lease in an agreed form, usually after planning permission has been granted. The developer then has a free choice as to whether it exercises the option or not.

Both conditional contracts and options allow the developer security to invest time, people and money undertaking detailed investigations and appraisals, applying for planning permission and securing funding knowing that it can call on the lease if it is successful but without there being any obligation to take the lease if planning permission is not granted, it is granted subject to unacceptable conditions, funding is not secured, or any other conditions or requirements are not satisfied.

Both conditional contracts and options are time limited.

The essential difference between the two forms of contract is that with a conditional contract the 'conditions' are specific and, if satisfied, the contract becomes unconditional and the developer is obliged to proceed. The option gives the developer freedom to decide whether or not to proceed. Many would consider this optionality attractive in the fast-developing and uncertain world of renewable energy.

Conditional contracts are not so common in practice, not least because to satisfy the developer's need for flexibility the conditions are drafted with a great deal of freedom for the developer to decide whether the condition is satisfied. As a consequence, in practice a conditional contract will often amount to an option.

The real life operation of a conditional contract rarely meets the expectation of the landlord that the conditional contract has greater certainty. As a consequence, conditional contracts can be more contentious and costly to negotiate. Conditional contracts are also problematic for developers as they require the developer to identify all the areas of uncertainty (and therefore required conditionality) at the outset and that is not always possible.

An option is preferred from the developer's perspective and so long as the developer offers a reasonable commitment through binding obligations within the option, for example to apply for and prosecute the application for planning permission, landlords will generally take comfort from knowing that a developer is unlikely to speculate on the cost of fulfilling those obligations and the wider costs of working up the scheme, in terms of design, site investigations and reports and other costs unless it is serious about undertaking the development. Since options are the more common contractual arrangement, this chapter considers options (but not conditional contracts) in some detail. Many of the matters considered in relation to options are, of course, equally applicable to conditional contract negotiations.

If the site already has planning permission, it is likely the lease will be put in place at financial close without an option being required.

There are other property documents that may need to be entered into.

Easements – or their equivalent legal rights – might be required over land which is outside the ownership of the site owner.

A landlord can only grant rights over land that it owns. If access to the site is required over land outside the landlord's ownership, an easement will be required. Similarly, if the nearest grid connection requires cables routed across land outside the landlord's ownership, an easement will be necessary. Easements may also be required to fulfil obligations imposed in the planning permission that cannot be delivered from land within the landlord's ownership.

Even if third-party funds are not required for the project initially, perhaps because the developer is financing the project off its own balance sheet, it is important that the project is 'bankable' to facilitate subsequent third-party funding, refinancing of the project or sale of the project (which will often happen once the project has completed the higher risk construction phase and entered the lower-risk operational phase).

Consideration needs to be given to the structuring of the project ownership, the property arrangements and the planning permission at the outset to ensure a 'bankable' scheme.

2. Developer

Most solar project developers will establish a new company – a so-called 'Special Purpose Vehicle' (SPV) to enter into the option and, if the option is exercised, the lease.

Use of an SPV is important where the developer intends to sell the project when it is 'shovel ready' or where it will be looking to obtain equity investment or debt funding for development. There may also be specific financial advantages to the use of an SPV.

By way of example in the context of the United Kingdom, a sale of the SPV will attract a lower rate of Stamp Duty Land Tax (SDLT) than an assignment of the lease. As at November 2017 a sale of the shares in the SPV will attract SDLT at the rate of 0.5% of the price. An assignment of the lease will attract rates rising to 5% (above the first £250,000 of the price).

As important, if not more important, than the considerable tax saving is the fact that the SPV will be a new company with an interest in a single project making a sale considerably simpler. The SPV's trading history will be short and project-specific making buyer concerns about historic liabilities, the provision of warranties and the process of disclosure a much more straightforward process.

Also, selling a company that has a long trading history, even if the project is its only current asset, is an anxious process for buyers concerned about historic liabilities. The sale of a project-specific SPV is almost invariably the most saleable vehicle.

3. Site identification

The suitability of a particular site as a location for a solar development will be dictated by various constraints that impact the site. The main constraints relate to such things as orientation, gradient of the land (south-facing and flat or gently sloping are ideal), good access, availability of grid connection and the likelihood of obtaining planning permission. Relatively unproductive farmland is usually an attractive option as it reduces potential conflict with agricultural usage, and fewer immediate neighbours will reduce the risk of objections to the planning applications. Good road network connections are also beneficial in terms of facilitating construction traffic and deliveries.

4. Legal process

Prior to entering into any acquisition documents, the developer's lawyers should undertake all necessary due diligence. In England and Wales the expression *caveat emptor* (or 'let the buyer beware') applies. The developer takes the site as it is and the onus is on the developer to check that it is suitable in all respects for the project. The developer's lawyer will carry out searches of various public records, investigate the landlord's ownership and raise enquiries of the landlord's lawyer. This process is undertaken to find out as much as possible about the site and matters that affect it, not least to ensure that the site and the land over which rights are to be exercised are suitable for the intended use. This will include such basic steps as:

- ensuring that the land is owned by the landlord;
- checking whether it is subject to a mortgage or charge;
- checking if there any leases or other rights of occupation; and
- ensuring that there are no restrictions on the use of the site (eg, restrictive covenants) which might prohibit the solar development or the rights being exercised.

Other practical constraints will be revealed through the searches of public records. For example, it will be of little practical value to require the landlord to remove certain trees, or to keep all trees within [x] metres of the boundary of the project site lopped to below a maximum height, if there are tree preservation orders affecting the trees in question which will make performance of that obligation difficult and, in the worst case, impossible.

It is important to identify potential title issues early on, particularly those that may prevent projects proceeding or increase the project costs, as the viability of the scheme may be compromised. In particular, if areas required for grid connection or to satisfy planning conditions are located on third-party land, negotiations with third parties can be time consuming and costly, especially if these parties are completely unconnected to the project.

Typically in England and Wales solar arrays are constructed on farmland. Farming families often establish companies or partnerships which occupy and operate the farm. This often happens to facilitate the succession as Mum and Dad wind down and hand over to their sons and daughters. It is common for the freehold interest in the farm to remain with Mum and Dad (or other family members), and also common that the farming company or partnership does not have a formal lease. These are family arrangements after all.

Despite the compulsory registration of land in the United Kingdom, unwritten tenancies will override the grant of a lease and the author has lost count of the occasions when he has been told that there are no leases or tenancies that would override the grant of the solar lease, only on further probing for it to be confirmed that the 'owners' of the land are not actually farming the land.

Agricultural tenancies can be complicated things to get around – even with cooperation – and so it is important to discover them and to deal with them at the earliest possible stage.

It is also important to identify if any third-party consents are needed (eg, bank consent where the land is subject to a mortgage or local authority consent where the land is subject to existing planning agreement restrictions), as these can take time to obtain and impact on project timescales.

In addition to the property documents for the developer's land interests, documentation will be needed for the grid connection (including wayleaves and easements in the statutory undertaker's prescribed form) and there may be a need for other agreements required to obtain planning permission or to create new access to the public highway.

It is as well for the developer's lawyer to prepare early, anticipating that the funders who will almost certainly appoint their own lawyers will nonetheless want to rely on a certificate of title prepared by the developer's lawyer. Certificates of title can be time consuming to prepare and so an organised developer's lawyer will commence preparation and maintain an evolving working draft from the outset of its title investigation and other due diligence work. When funding for a project has been agreed and planning permission obtained, all parties will want to bring it to a financial close as quickly as possible and it is a simpler and more efficient task to update an existing draft certificate dealing with changes to the title (if any) and anything new revealed by an updated or refreshed search/enquiry than to start the preparation of the certificate from scratch at that stage.

It is also likely that at the time the option and lease are being negotiated the funding for the project will not have been agreed and so the funder, even if identified (and that is not certain), will not yet have instructed lawyers. That being the case, the developer will need to bear in mind generic funder requirements when negotiating the option and lease.

A developer is used to assessing risk, and making risk-based decisions. A developer will often accept a risk that a funder will reject. There can be tension between a developer's willingness to compromise and accept a risk and the adverse effect that may have on the 'bankability' of the project. In short, a developer will always need to think about whether the funders will accept the compromises that in a funding-free world the developer might accept. Returning cap in hand to a landowner after documents are agreed and signed to negotiate a variation to keep the funder happy is never a comfortable position for a developer. As well as uncomfortable, it is often an expensive exercise.

5. Option agreement

An option for lease binds the landlord so that when the developer is ready to proceed and enter into the lease, it simply has to serve a notice and the landlord will grant the lease for the site. If an option for lease is required for a solar project then the key areas to consider are set out below.

5.1 Option land

Careful consideration needs to be given to the extent of the land to be subject to the option for lease as the developer's proposals for the site may change through the planning process and this may impact on the areas required for the development. The developer does not want to be in a position where it needs to renegotiate with the landlord at the point it has planning permission and wishes to call for the lease as it will be vulnerable to the landlord looking to revisit the previously agreed lease terms.

The option will typically provide the developer with a discretion to designate the land over which the lease will be granted by identifying a larger area of land within which it has freedom to designate the land to be included in the lease.

5.2 Option period and extension of the option period

The initial option period needs to be sufficient to allow time for the developer's planning application to be submitted and decided. Certain circumstances, such as delay in obtaining planning, will necessitate an extension to the initial option period, and the developer will require the option to allow this. The length of option period including any extensions required by the developer, need to be balanced with the landlord's interests, as the landlord will not want its land to be tied up indefinitely.

5.3 Option fee

Whether a fee is paid by the developer to the landlord for the grant of an option is a purely commercial discussion influenced by the relative bargaining powers of the parties. From the developer's perspective any such fee is an unwelcome addition to its speculative spend – especially as the option fee is potentially subject to SDLT in England and Wales. An option fee is always attractive to a landlord, not least because it gives the landlord some financial incentive to enter into a speculative option thereby tying up the land for the option period with no guarantee that the lease will in due course be granted. If a fee is agreed it should, from the developer's viewpoint, at least comprise part payment of any lease premium, or treated as an advance payment of rent.

5.4 Lease premium

Similar to the payment of an option fee, whether there is a price payable for the grant of the lease is a commercial discussion. However, where a commercial rent is payable it would be unusual for a premium to be paid for the grant of the lease, unless there was a commensurate reduction in the amount of the rent payable.

5.5 Exercise and completion process

To facilitate the interface between the lease completion and other project documents, the operation of the option in the context of a solar project will generally be different from the traditional method of exercising an option and completing the lease. The corporate and real estate worlds have different practices – in England and Wales at least. In the real estate world, typically the contract for the sale or purchase – or the notice exercising the option – is served. Completion occurs some time later, either on an agreed date or an agreed number of days after notice is served. This is anathema to the corporate world where at 'financial close' the whole deal is concluded at the same time.

An option may be effective for a number of years and during that time the landlord may want to carry on using the land for its current use and therefore there will generally need to be a period of prior notice to enable it to cease operations, and vacate the land.

A farmer landlord may also want time to harvest the crop. This might be dealt with by restricting the grant of the lease so that it can only be granted outside the period between sowing and harvesting. Since that will usually make life difficult for the developer – and the corporate finance lawyers – loss of crop will usually be dealt with by compensation so that the farmer does not lose out financially and the developer is not restricted as to when it can take the lease.

In addition the extent of the land required needs to be defined.

To put control of the completion process into the hands of the developer an option will typically provide for a two-stage process to apply.

Stage 1: Once the developer has planning permission – and other necessary permissions/consents required for the project to proceed – the developer will serve a notice (often called a 'release notice') at which point the developer will designate the land it requires to be included in the lease, provide appropriate plans and the landlord and developer will execute the lease in escrow. It will be released to and held by the developer's lawyer pending Stage 2. The release will typically be time limited (release period) to ensure that the landlord has a latest date by which he will have certainty.

At this point the developer will not be obliged to complete the lease but is free to do so during the release period.

Stage 2: The lease can be completed at the discretion of the developer during the release period, which helps meet the need for simultaneous financial close on all project contracts/documents.

5.6 Assignment

The developer and funder will require the flexibility to move the option agreement within the developer's group of companies or to a third-party developer. There may be any number of reasons why this is necessary, including a change of the planned structure for the project, a particular funder's requirement or a restructuring of the developer's group, among others. The funder may also require the landlord to enter into a direct agreement which allows the funder to step into the shoes of the developer should the developer have financial difficulties or otherwise fail to perform on its obligations to the funder. This arrangement permits the funder to keep the project alive and bring in another developer, thus protecting its financial investment.

5.7 Planning obligations

The option will invariably include obligations on the developer to try to obtain planning permission and to keep the landlord informed of progress and key developments, and to consult with the landlord when appropriate. It will also require the landlord to cooperate with the developer's application for permission and sign necessary documents to obtain planning permission (with an indemnity from the developer and subject to reasonable limitations).

5.8 **Restrictions on use of land**

During the option period the developer will require the landlord to covenant not to use the land for certain purposes or, at least, not to grant other rights over the land which would either prevent the development or infringe its operation or profitability.

5.9 **Occupation of land**

The developer may require rights to enter onto the land during the option period to carry out surveys, investigations and/or to commence preparatory works. The terms of such rights need to be agreed with the landlord and set out in the option.

5.10 **Landlord's obligations**

The developer will require the landlord to enter into certain commitments during the option period including:

- an obligation to obtain any required mortgagee consent which should in an ideal world be granted before the option is entered into, and if not obtained before the option is entered into, should be a first requirement and the developer should not be obliged to do anything until it is obtained;
- to cooperate with the developer to help secure planning permission and any other necessary consents, but in doing so not to act independently of the developer;
- not to encumber the land during the option period;
- not to sell the land without novating the option – or obtaining a direct covenant from the buyer to perform the landlord's obligations;
- not to do anything or act in any way which might prejudice the developer's planning application;
- to ensure sensitive price/value information remains private and between the parties and to enable the developer to manage the publicity for the project;
- not to do anything which changes the physical state and condition of the property (normal agricultural activities are usually exempted from such a restriction); and
- not to use any other part of its land for solar-related energy generation (to the extent allowed by applicable competition law).

5.11 **Dispute resolution**

It is usual to include a dispute resolution procedure (expert determination or arbitration) to try to keep disputes out of the courts and facilitate a fast(er) track resolution of disputes.

5.12 **Form of lease**

The form of lease to be entered into by the developer and landlord in the event the developer exercises the option will be attached to the option. If the developer needs leases and easements from other land owners, option agreements will also need to be entered into with those land owners to ensure that when planning has been obtained all the options can be triggered at the same time. This point will be particularly important where a developer is looking to secure funding at the option stage as it will need to show that all necessary property rights for the project have been secured before planning permission has been obtained.

5.13 Contracting out

In England and Wales it is common to exclude solar leases from the provisions of Sections 24–28 (inclusive) of Part II of the Landlord and Tenant Act 1954. These sections, if not excluded, give the tenant a right to a new lease at the end of the contractual term. Procedurally, if such an exclusion has been agreed then the option will need to go through the same service of notice and declaration procedure as the lease (see below).

6. Lease

The lease will govern the relationship between the developer (in this section referred to as the 'tenant') and the landlord for the life of the solar project. It is a critical document. It is outside the scope of this chapter to analyse every provision that might be included within a lease of a solar project site but there follows a summary of the key provisions.

6.1 Rent

The lease will set out the annual rent payable by the tenant to the landlord as agreed between the parties. The annual rent may be subject to review (usually upwards only). A typical commercial market rent review provision does not work for a solar project lease and so another mechanism is required – often an index-linked review. The developer needs to ensure that its financial model takes account of the annual rent payable under the lease for its entire term. A landlord may request that a generation rent be payable, in addition to the annual rent, which can be linked to the turnover the developer receives from the generation of electricity, although this is often resisted by developers and funders. The developer may wish to negotiate a rent-free period at the outset of the lease to take account of the construction/installation period during which the array will not be generating electricity or, therefore, income.

6.2 Term

The lease will record the duration of the tenancy – usually between 25 and 40 years for a solar project. Whereas with many other types of development the term of the lease is not necessarily linked to any particular issue and can be freely negotiated, with a solar development the term is usually linked to the period during which the array is able to receive benefits/subsidies for generation of electricity. The lease term also needs to take account of the time it will take to build the array at the outset and decommission it at the end of the project.

6.3 Assignment

The developer will want flexibility to assign the lease without the landlord looking to request guarantees (including authorised guarantee agreements). Funders will also require the lease to be assignable in the event that the funder needs to step in should the developer run into financial difficulties and fail to meet its loan repayment obligations, or its other obligations to the funder.

6.4 Funder requirements

The funder may require a direct agreement to be entered into by the landlord in relation to the lease. A direct agreement will typically prohibit the landlord from exercising its rights arising from a breach by the tenant of its obligations without first giving the funder notice, triggering a suspension period during which the funder may:

- procure that the breach is remedied; and/or
- appoint an agent (or administrator) to step in and assume the obligations of the tenant and to remedy the breach; and/or
- assign the lease to another developer/operator.

The direct agreement will typically allow the funder to 'step out' at any time after it has 'stepped in'. The lease may include the appointment of the tenant as attorney for the landlord in the execution of the direct agreement, if the landlord refuses (when the time comes), to execute the direct agreement. This is likely to be resisted by the landlord but the tenant will be concerned to ensure that at this critical time (moving towards financial close) it can self-help and execute the direct agreement in the name of the landlord in the event rather than face a lengthy and potentially expensive action for specific performance. Indeed, the very inclusion of the power of attorney is likely to help ensure that the landlord does not drag its heels.

The lease should oblige the landlord to enter into a direct agreement, if and when required, in the form required by a funder (funders' discretion is important as they are often strict in their requirements), even if no direct agreement is required at financial close, since it may be required by future funders or investors.

The lease should also contain independent restrictions on the landlord terminating the lease in the event of tenant default where there is funding in place to allow the funder a period of time in which to remedy the breach. This provides protection for funders even where there is no direct agreement in place.

6.5 Right to renew

Notwithstanding the above, the developer may require a right to renew the lease on similar terms. The lease will set out the procedure for doing this. It will be the developer's responsibility to obtain any further planning consent permitting the continuation of the development, if required.

In England and Wales this contractual right to renew will apply even if – as is common – the tenant's statutory rights (under Sections 24–28 (inclusive) of Part II of the Landlord and Tenant Act 1954) are excluded. Such exclusion will need to be pursuant to the statutory procedure which will be set out and recorded in the lease.

6.6 **Right to break**

The developer may request the ability to break the lease, either via a rolling break right which can be exercised at any time following a certain notice period, on fixed periods (eg, every five years), or on the occurrence of specified events such as a change in the law rendering the development uneconomical to continue.

6.7 **Environmental matters**

Any developer and funder will require agreements that neither the developer nor funder is responsible for the consequences arising from any historic contamination at the site. As previously mentioned, in England and Wales the maxim 'let the buyer beware' applies, and it is likely, and certainly advisable, that the developer will have carried out environmental due diligence through desktop searches, and also preferably through intrusive site investigation, prior to entering into the lease. Nonetheless, these statements are beneficial to the developer (and therefore advisable), as it is unlikely to be able to fully 'beware' itself, because even the most thorough site investigation will not provide absolute certainty that no environmental contamination exists.

In certain cases, if the development is on industrial land, for example, environmental due diligence will need to take a higher priority and a detailed environmental deed may be required to properly allocate and record the relative responsibilities of landlord and tenant.

6.8 **Benefits**

To the extent necessary, the lease should include an agreement between landlord and tenant that the tenant has the sole right to any and all benefits or tariffs which arise from the project and to pay any that it actually receives to the tenant. The landlord should also covenant not to register itself in relation to the solar array to obtain any of the benefits/subsidies that are paid for the generation of electricity.

6.9 **Indemnities**

It is common for landlords to require indemnities from tenants for: any liability that the landlord suffers as a consequence of the tenant's use of the land; the exercise of rights under the lease; or the consequences of any breach of the tenant's obligations. It is prudent for a developer to try to avoid giving indemnities (which is unlikely to be easy to agree with the landlord), or, alternatively, at least agree a financial cap or limit on such indemnities. Funders are wary of indemnities but generally will accept them if they are capped at an appropriate amount.

Provisions that require the landlord to notify the tenant of any claim or potential claim and to enable the tenant to manage any legal proceedings, if necessary in the landlord's name, should also be included. The landlord should at the very least be prohibited from settling a claim (in respect of which it has an indemnity from the tenant) without the tenant's agreement or (in the event of a disagreement as to the merits of the claim) independent advice from counsel.

The key from the tenant's point of view is to avoid a vexatious or inflated claim being settled by the landlord without reference to the tenant and without any effective defence being raised which is subsequently passed on to the tenant under the indemnity.

6.10 Landlord obligations

The developer will require certain obligations and restrictions on the site and landlord, including:

- The lease should require the landlord to enter into any statutory undertaker's agreements or other documents or any other statutory agreements which may be required during the life of the project. These will include leases (of substations, for example), rights and wayleaves for cables, piles and stays etc, to allow the statutory undertaker to take electricity from the project site.
- Similar obligations should apply for the benefit of other third parties to enable the tenant to supply electricity through private wire.
- Protection of the solar irradiation in the form of height restrictions on growth of existing trees or shrubs, and prohibition on planting of trees, shrubs, hedges and the like, or erection of buildings within a specified distance of project site boundaries, or some of them. These restrictions should be coupled with an obligation to lop, cut or fell anything that grows beyond the height specified. The precise requirement will depend on a number of factors including the orientation of the project site, the gradient of the project site and the adjoining land. Bear in mind that the landlord can only covenant in relation to such of the adjoining or nearby land that it owns and so, if height restrictions are required on land which the landlord does not own, then the appropriate covenant should be obtained, if not before the option is entered into, certainly before the option is exercised.
- A restriction on shooting within an agreed distance of specified boundaries should also be included, where appropriate.
- Although not usual in most commercial lease situations (where the maxim 'let the buyer beware' applies), it is both common and advisable (not least because funders often require it), to require a covenant from the landlord that there are no tenancies or other rights of occupation affecting the site and the landlord has the right, power and authority to enter into the lease and to grant the rights contained within it. Also, a covenant that the landlord has obtained all necessary consents from any relevant mortgagee or chargee and from any other person who, in the absence of such consent, could prevent or impede the exercise by the tenant of the tenant's rights under the lease.
- If access to the project site is over land retained by the landlord, then a covenant to maintain it so that it is suitable for the tenant's use and to keep it unobstructed should be included.
- A tenant should endeavour to obtain a covenant from the landlord that it will not do anything on its land that has a detrimental effect on the efficiency of the solar array and ancillary equipment.

6.11 Reinstatement bond

A landlord may require a developer (or the developer may offer if its financial model allows for this) to place a sum of money into an escrow account which the landlord can call upon in the event that the developer fails to comply with its reinstatement obligations at the end of the lease. This is either put in place at the outset or, more frequently, over a number of years towards the end of the term. There are a number of ways of deciding on the amount which should be sufficient to give the landlord reasonable security without requiring the tenant to tie up an unnecessarily large sum of money. Since there is likely to be value in the equipment, it is not uncommon to agree the amount of the bond by a simple sum which starts by estimating the likely costs of removal of the equipment and reinstatement of the ground and then deducts from that the estimated value of the equipment for resale or scrappage, as the case may be. The resulting amount – perhaps index-linked – represents the target fund to be built up over a number of years.

6.12 Tenant rights

Consideration needs to be given to the practical requirements of the tenant in respect of the site. The lease needs to include all necessary rights over land owned by the landlord in order for the tenant to build and operate the array. These include:

- access to any temporary offsite areas required for site storage or compound facilities that cannot be provided on site;
- access to and from the project site;
- access onto the landlord's adjoining land to carry out works (where such access is required), and for inspection;
- to provide necessary services and facilities;
- to connect to the grid and export electricity;
- to enable the provision of private wire connection, if that is a possibility;
- light;
- signage;
- fulfilling and complying with planning conditions (eg, landscaping obligations or road improvement works); and
- anything else that the tenant requires for the project to be effectively carried out.

Since solar technology is likely continue to develop at a fast pace, it is not certain that all rights needed for the life of the project will be known when the option is entered into/lease granted and so a general grant of "... all other rights reasonably necessary or desirable for the operation of the [Project] and the Permitted Use ..." or words of similar effect should be included from the tenant's perspective. Landlords will be anxious about such a clause because of the inherent uncertainty it creates.

If rights are required over land owned by third parties then separate easements need to be entered into by the tenant with the relevant third party. It goes without saying that the landlord cannot grant rights over any land that it does not own.

6.13 Landlord rights

Due to the nature of a solar project site, the rights for the landlord over the project site will be limited. Usually the landlord will reserve a general right of support for the adjoining land, otherwise the tenant is likely to resist any other rights other than a right of access for the landlord to inspect periodically – subject to prior notice – and to enter where it has to perform its obligations. Rights for grazing (eg, sheep as anything larger is unlikely to be practical), are also sometimes agreed. There may be one or two other site or landlord specific requirements in a particular case, but these will be raised by the landlord and considered by the tenant on a case-by-case basis.

6.14 Forfeiture of lease

Most, if not all, leases in England and Wales granted at a rent contain a right for the landlord to terminate the lease it the tenant fails to perform its obligations. Notwithstanding any requirement for a direct agreement (as to which we comment earlier in this chapter), to ensure that the lease is acceptable to a funder the tenant should try to negotiate a forfeiture provision which requires the landlord to serve notice of intention to forfeit on a funder and to give the funder a period of time to remedy the breach. It will usually state that in the event that a direct agreement exists and there is any conflict between the lease forfeiture provision and the direct agreement, the provisions of the direct agreement prevail. This is necessary since the lease is likely to be agreed before the direct agreement is drafted by the funder and so there is every possibility of some inconsistency between the two and invariably the direct agreement will go further than the modification to the usual forfeiture provision.

7. Plans for property documents

It is important to ensure that accurate scaled plans are prepared at an early stage for the property documents showing the extent of the land required for the lease and the areas over which rights are to be exercised. The plans for the property documents need to be prepared in conjunction with the planning drawings as the developer will not want there to be discrepancies between its planning obligations and the land over which it has rights under the property documents.

8. Planning permission

Planning permission is usually required for the development of solar facilities. In the vast majority of projects, the process is for the developer to apply for planning permission from the local planning authority during the option period.

Once planning permission has been granted it will form a key project document, and the conditions on which it is granted will be a key project constraint.

Care needs to be taken during the application process because the statements made for, or on behalf of, the applicant and documents submitted as part of the application process often become incorporated into conditions of the planning permission.

Where there are conditions which are problematic, it is possible to make an application to vary those conditions, but this process takes time and may cause significant delay to the implementation of a project. Submitting applications to vary planning conditions can also result in the local planning authority imposing additional conditions on the development that were not imposed as part of the original planning permission. As with the initial planning application, success is by no means guaranteed and funders and investors will be alive to the risks associated with carrying on development or operations in breach of planning conditions.

It must be remembered that carrying on development or operations in breach of planning conditions can result in enforcement action being taken by the local planning authority. Enforcement action is not only costly and time consuming (and a tremendous distraction from the business at hand), but can lead to a decline in project financial viability. If the local planning authority serves 'stop notices', the developer will be faced with loss of processing time, loss of energy generation and consequently loss of income. Loss of income can cause problems elsewhere in the project as cash flow and working capital become tighter, with possible default under financing agreements and other contracts.

Where the planning permission is granted with conditions attached to it, and those conditions are either accepted by the developer or an application to vary the conditions is unsuccessful, it is critical that the project is formulated in a manner which seeks to discharge all the conditions if the project is to be fundable or 'bankable'.

9. Rooftop schemes

Although rooftop schemes are generally outside the scope of this book, which is primarily aimed at larger-scale land-based solar schemes, most of the considerations discussed in this chapter – or at least the thought processes behind them – are equally applicable to rooftop schemes.

There are of course differences. Planning permission is a little more straightforward in many cases since, in England and Wales at least, permitted development rights mean a planning application is not usually required, although there are conditions that need to be complied with.

It is likely that the building is leased by the freeholder to an occupier, and the cooperation and consent of both is likely to be required.

Where the array is funded by a third party – other than the owner or occupier – the operator will usually be granted a lease of the airspace immediately above the roof of the building, with rights to connect the array to the roof. This can only be done if the airspace has been excluded from the lease of the building, which is usual but not invariable. Rights to connect to existing meters, install new meters and areas for locating inverters and running cables are all required.

The tenant's mortgagee's consent may be harder to obtain for a rooftop scheme and in the case of residential property the Council of Mortgage Lenders in the United Kingdom has issued a set of minimum requirements for the lease of roof or airspace for the installation of a solar array on a residential rooftop.

The landlord and tenant relationship has its complications. Such complications are magnified in a multi-let property, where getting all parties aligned (the more there are, the harder it will be). This is perhaps one of the reasons why commercial rooftop schemes are not as commonplace as they might be. Small roof spaces are also a factor, as is efficiency based on the alignment of the building, roof pitch and orientation.

There is some evidence in California that the development of solar tracker systems – in which the solar array moves during the day to 'track' the sun – making an array up to 30% more efficient than a static system, may make a difference to the viability of roof-top solar, but the jury is still out. Such technology is still developing and remains expensive which results in diminished financial returns. They are also heavier than static systems and more vulnerable to high winds. That is likely to change as the technology develops.

The project agreements

John Deacon
Orrick, Herrington & Sutcliffe LLP

This chapter will address the various contracts which govern the procurement, design, construction and operation of a solar power facility – referred to as 'the project agreements'. The project agreements are contracts like any other and will, as necessary, contain all the legal requirements that any similar contracts will contain in their respective legal systems. This chapter does not address all of those legal requirements, but rather concentrates upon what makes solar power project agreements different from other contractual arrangements for the construction and operation of energy projects.

This chapter intends to take matters sequentially in the life of a solar power project. This is a rather grand way of saying that the project has to be designed before the key equipment can be purchased, before the project can be built, and built before it can be operated.

Finally, this chapter is predicated upon the assumption that some form of independent financing will be provided for the project on the grounds that independent financing is generally required before a project can be built.

1. The procurement strategy

There are a number of key features to a solar power project procurement process. More particularly:

- A large proportion of the capital cost will be consumed by three key items of equipment – panels, inverters and transformers.
- It is inevitable that the party manufacturing the panels, inverters and transformers will be different from the party actually installing them.
- The installers of equipment are unlikely to accept any risks relating to the equipment that are not completely 'back-to-back' with the equipment supplier terms.
- The operator of any solar power facility is likely to be a different legal entity to the constructor.

Bearing in mind the above bullet points, how the procurement, design and construction phase of a solar power project is progressed can take a number of different forms.

Any contract is about the allocation of risk, and what inevitably substantially increases this risk is where it is unclear who is responsible if anything goes wrong. Clearly, the skills required to procure, design and construct a solar power project are different and likely to be undertaken by different professionals. As an example, the procurement structure could look like this:

Figure 1

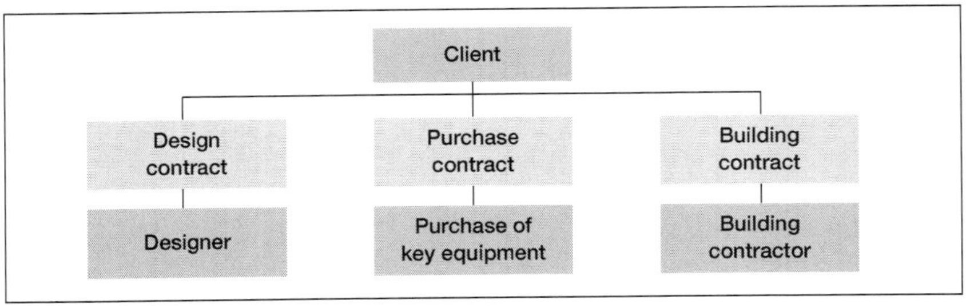

What happens if something goes wrong and how do you identify who is at fault? If an array of panels falls over is it because the supplier of the panel frames is at fault or did the construction contractor install them incorrectly?

If a solar power project is undertaken with the model set out at Figure 1, there will inevitably be an interface between each of the parties and an interface means a blurring of the lines of responsibility (again, this is a long-winded way of saying "who do you sue if things go wrong?").

Banks in particular are wary of interfaces as they prefer clear lines of responsibility. For this reason, many solar power projects are undertaken using the Engineer Procure and Construct (EPC) model which looks something akin to figure 2:

Figure 2

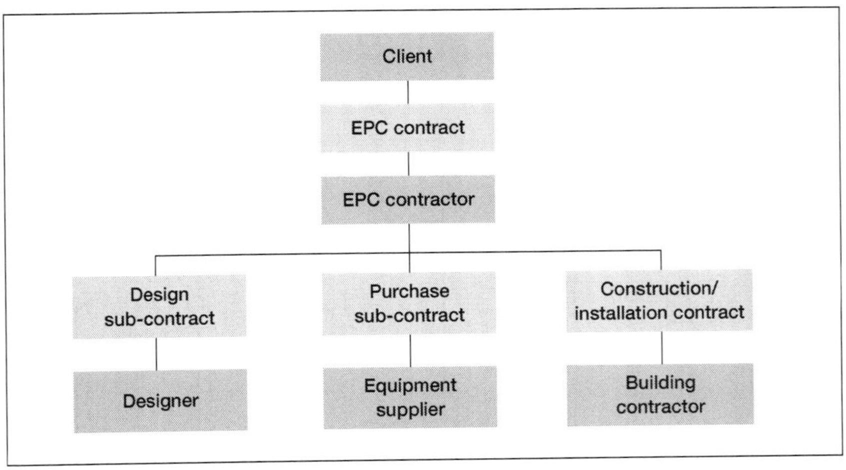

However, a nuance to solar power projects is that most are developed by companies which specifically develop solar power projects – that is, most owners of solar power projects build more than one.

Basic rules of economics apply to renewable energy projects – if 70–80% of the cost of a project is in the purchase cost of three items of equipment then you are likely to get a better price for that equipment if you purchase substantial quantities of it. This means that a large-scale developer of solar power projects can probably get cheaper panels, inverters and transformers if they are building hundreds of megawatts around the world than a small contractor. Accordingly, a common feature of solar power projects is the procurement and supply of key equipment by the owner. I will consider how this model fits in with the EPC structure below.

2. Key features of a solar power EPC contract

2.1 Site risk

There are two key elements to any site for a solar power project:

- it should be sunny; and
- the physical site should be capable of being used for solar power generation.

Instinctively one would expect the owner to take into account the sunshine and physical risk of any site, after all it is probably the party which has identified it as being suitable for solar power generation. However, the site risk is one that funders would like to see passed on to the contractor and, to a greater or lesser extent, this usually happens.

Sunshine risk – it is a rare EPC contractor which will guarantee the sunshine. However, what one should look for from an EPC contractor is a warranty in the EPC contract that the proposed site is "suitable for solar power generation". This means the contractor should have surveyed the site and confirmed that any shadow (be it man-made or natural) would not prevent the successful operation of the panels. Clearly, such a warranty would be a 'moment in time' guarantee as no contractor could stop the building of large structures on adjacent land and/or the growth of trees.

Physical site risk – some areas of land are prone to physical risks, which could be permanent or temporary and would render the site unsuitable for solar power generation. Such risks could be things like geologically weak ground, the presence of old mine markings or the risk of flooding. Inevitably the discussions with EPC contractors about them taking physical site risks can sometimes be fraught. Quite reasonably a contractor will say, "you have chosen to build your solar power facility here, why should we take the risk that the site is suitable?"

The answer to this question comes in one word – 'experience'. The contractor has (hopefully) built similar projects before and should be able to tell from a site visit that it is suitable.

An area of difficulty remains, however, in relation to things that cannot be seen. A recent example is a situation where the contractor discovered 150 fifteenth-century mine shafts on site, all of which were not visible from a simple inspection of the area and all of which had to be filled in (at great cost) before anything could be built. How can this type of risk be managed?

If time permits, the owner could commission a detailed investigation of the site which could be provided to the contractor. The contractor would then have the opportunity to factor the identified risks into the price. However, what happens if the survey misses something or there is not enough time for a survey?

If there has been a survey then the owner could transfer its rights to sue the survey provider to the EPC contractor, thereby leaving the contractor to carry this risk vis-à-vis the owner. Alternatively, the risk of the survey could remain with the owner and it could identify the survey (or any other information regarding the site) as 'rely upon information' for which the owner carries the risks of mistakes or omissions. Banks are sometimes prepared to accept the concept of rely upon information for solar power projects provided always that they are comfortable that it is reasonably comprehensive and therefore the risk is mitigated. Generally, banks are not comfortable with the 'cross your fingers and hope' concept and will expect some understanding of the site to have been gained before risks of the unforeseen are allocated.

2.2 Key equipment supply

As discussed earlier, a feature of the procurement of solar power projects is the large part of the capital expenditure that is tied up in the three key items of equipment – panels, inverters and transformers. The operational performance of this equipment will be crucial to the overall success of the project and the owner is going to want to know the manufacturer of the equipment and the performance specification.

There is an inevitable tension here between the contractor and the owner. The contractor will want to have the option to change suppliers if it can source equipment cheaper elsewhere. However, the owner will want to ensure that it is paying for the best-performing model which is available. To counter this, the contractor will argue that if the owner specifies a particular item of equipment then the owner should carry the risks associated with its delivery and performance.

Most problems can be resolved in this area if the equipment specified by the owner is highly regarded by the industry and is sourced from a reputable supplier. Under these circumstances one would expect the contractor to carry the risks associated with delivery and supply. If the equipment is not from a trustworthy source then clearly problems could arise. The instinctive answer is that on a debt-financed project this risk is unlikely to arise as the banks would not finance such a risk. However, if it were to proceed the contractor would be entirely reasonable in not accepting the risk for such owner-specified equipment.

Owner-supplied key equipment was discussed earlier in this chapter. This is becoming increasingly common in solar projects where large developers are using their purchasing power to secure large discounts from key suppliers. How then to address the issue of owner-supplied equipment and who should carry the delivery and performance risk?

It would seem to be contrary to the principles of EPC contracting if a large proportion of the value of the works is carved out of the responsibilities of the EPC contractor. However, generally the banks will accept such a carve-out and the 'free issue' of equipment to the contractor if:

- the contractor itself makes clear that it retains the installation and commissioning risk;

- there are satisfactory contracts (and guarantees) in place between the owner and the equipment suppliers;
- the logistical arrangements for the delivery of the equipment are in place and it is clear that it can be delivered on time and to an agreed location; and
- the risks of storage and protection of the equipment is passed to the contractor on delivery.

2.3 Commissioning, completion, the ascertainment of performance levels and liquidated damages

In common with many other contracts for the construction of assets for the generation of electricity, deciding when and how the construction works are complete and the extent to which contractual performance levels are achieved is crucial to both the owner and any funders.

A fundamental difference between a thermal power plant, such as a CCGT, and a solar project is that there is some control or influence over the fuel specification. Clearly, irradiation levels, while capable of reliable prediction based upon historic levels, are not under the control of either party and the contractual completion process needs to take account of these issues.

As a minimum, the EPC contract should anticipate a two-stage completion process to reflect the differing time periods over which performance needs to be monitored and guaranteed. First should come provisional acceptance which, when achieved, should lead to a Provisional Acceptance Certificate (PAC). For a solar project, the test to measure PAC should be of a limited duration and reflect the achievement of minimum performance levels.

Damages at PAC are two-fold. First, there are delay damages. These are calculated by any delay to a guaranteed completion date in the plant achieving the minimum acceptable levels of performance.

Once the minimum acceptable levels of performance are achieved there will be a calculation of damages by reference to the guaranteed performance levels at PAC. Note that these are likely to be different from the guaranteed performance levels at final acceptance. Once PAC is achieved, control and risk of the facility will pass from the contractor to the owner.

Following PAC there will then be a period of post-completion tests during which the performance of the plant will be measured over a longer period of time. This longer period is required to ascertain the overall performance of the plant under different conditions and at different irradiation levels. The post-completion tests should be designed to ascertain any difference between the actual performance ratio and the warranted performance ratio. At the end of the post-completion tests this difference should be calculated and any shortfall will result in the payment of post-completion performance damages.

Often a contractor seeks to incorporate some form of bonus mechanism into the EPC contract in the event that the project exceeds the warranted performance ratio. Conceptually this would appear to be a fair risk allocation, however, the owner should only agree to the payment of a bonus if it results in additional income. Without any additional income there will not be any money to pay a bonus.

Finally, the owner should always ensure that any payments under the contract are structured in such a way that there is always enough money held back from the contractor to meet any claims. Perhaps the most fundamental principle of law that applies to any EPC contract under any legal system is that 'possession is nine-tenths of the law'. As owner, it is always better to be in a position where you owe the contractor money than to be trying to recover money.

2.4 Support regime

To date, most solar power projects have been undertaken with some form of subsidy regime, be it tax breaks, feed-in tariff, contract for difference or green certificate. Many such regimes are time critical for the dispatch of energy to the grid to secure the support regime. This can be particularly true when a support regime is evolving and accreditation must be achieved by a fixed date. Obviously any delays in completion could result in the loss of support regime payments and the question is whether this should be reflected in the EPC contract. The loss of the support regime as a consequence of contractor-culpable delay is something a bank would expect to see in the contract and a risk that contractors will generally take. In terms of measuring the risk, the project will still recover some revenue in the event of a delayed completion, albeit at a lower level. The practice has developed that the calculation of these losses for failure to secure a support regime is based on the reduction in capital expenditure to secure the expected project return on brown power rather than green power. For example, if a project was projected to cost €50 million and was to secure an internal rate of return (IRR) of 8% with a support regime in place and that support regime is lost because of contractor-culpable delays, then the measure of damages is the amount of reduction in the €50 million capital expenditure until an IRR of 8% can be achieved.

This would always be subject to the overall contractual liability cap which, funders will demand, should not be less than 100% of the sums paid to the contractor.

2.5 Defect liability period and the assignment of warranties

The EPC contract should have a defect liability period during which the contractor is obligated to return and fix any defects in the works. Generally in a solar EPC contract one would expect to see that period equate to the time between PAC and final acceptance certificate (FAC). At the end of the defect liability period one would expect the contractor's obligations for defects in the works to end.

However, there is a substantial caveat to the end of liability for defects at FAC. Under circumstances where the contractor has sourced key equipment then the achievement of FAC (and with it the end of any liability for defects) should be conditional upon the assignment by the contractor of any warranties and guarantees relating to the key equipment. Such key equipment will come with the benefits of long-term guarantees and any future recovery will be made by the owner under these documents rather than the EPC contract.

3. Key equipment supply contracts

The contracts for the supply of key equipment such as panels, inverters or transformers is of great concern to the owner irrespective of whether it procures the equipment itself or accepts assignment of warranties and guarantees upon FAC.

As a general rule there is often very little room for negotiation of such contracts. However, in relation to key equipment supply agreements, it is important that the following issues are addressed:

What is the law governing the contract?

Often solar power equipment is manufactured in countries such as China, Malaysia or Vietnam. If the parties so wish, the law governing the contract can be something other than the 'house' country, in which case this can be specifically addressed.

When does title pass?

Title to equipment should pass when it is paid for and not when it is delivered to site (if this is later).

Who carries the cost of import duties and tariffs?

Much equipment is imported from countries where tariffs are imposed as a consequence of such importation. These costs can sometimes be significant and who carries the risk should be clear.

Who insures the equipment in shipment?

Unfortunately accidents happen when transporting heavy equipment. Liability for insuring such equipment in transit should be clearly addressed.

What level of performance is actually being guaranteed?

While this may appear to be a statement of the blindingly obvious, it is not uncommon to see performance guarantees being drafted in unclear and uncertain terms. Additionally, the guarantees should last for a period of time from the installation of the equipment, not the shipping of it from the factory.

How are you going to enforce any claim?

The enforcement of court judgments and arbitration awards are governed by a series of international treaties. It is important to ensure that the tribunal that will make any determination of liability on the part of an equipment supplier is located in a jurisdiction where awards can be enforced against the equipment supplier.

Spare parts delivery

Inevitably, over the 20-plus years of operation of a solar power plant things will break and replacement parts will be required. It is important to ensure that the equipment suppliers retain an obligation to provide replacement spare parts, preferably at predetermined prices.

4. Operation and maintenance contract

The importance of the operation and maintenance (O&M) contract can often be underestimated in the overall suite of contract documents. Many project owners do not have the resources (or inclination) to operate or maintain their facilities. Additionally, independent finance providers will expect that a competent and experienced provider of O&M services is in place to guarantee performance of the plant.

Accordingly, a properly formulated and structured O&M services sector has existed for some time and, in common with comments above regarding construction contracts, this section concentrates on key issues for solar power projects rather than a general review of O&M contracts.

4.1 Availability guarantee

Clearly, the O&M contractor cannot guarantee irradiation levels. However, the contractor can guarantee that the plant will be available and capable of generating power. Damages for any failure to meet the availability guarantee should be by way of a predetermined formula.

4.2 Response times

In many jurisdictions it is not possible to ensure that the O&M contractor of a solar power plant is permanently on site. This is because such a permanent presence is unnecessary and would be financially unviable. However, the response of the O&M contractors when things do go wrong becomes crucial. The O&M contract should set out the time periods for any response as these will clearly be determined by the level of the fee paid and the location of the plant.

The owner should push for a common response time irrespective of whether the problem occurs in or out of business hours. To assist the contractor, however, response times should be calculated by reference to hours when irradiation is above predetermined levels and not simply to hours – it may not matter if the response takes longer at night when no generation time, and therefore income, is being lost.

4.3 Liability cap

One of the problems with O&M contracts for solar power projects is that they tend to be of relatively low value. If the liability cap is set at a small percentage of annual income then the potential recovery of the owner from the contractor is correspondingly very limited. Great care should be taken to negotiate a cap at a level as high as possible to ensure that the remedies against the contractor are meaningful.

4.4 Support for any warranty claim

Responsibility for any fault in the key equipment is likely to be excluded as a contractor risk under the O&M contract and the claim should be made under the relevant equipment warranty. However, the O&M operator will be responsible for the site and the records. These records will be necessary to establish the equipment supplier's liability and keeping accurate records should be a requirement of the O&M contract.

5. Conclusion

Establishing a suite of accurate and complete contracts for the construction and operation of a solar power project is essential for the overall success of a project. Money raised for the capital expenditure will be spent through the project agreements and it is essential that these are correct.

On final analysis they are construction, supply and operation contracts and will have terms similar to such contracts in other industries. Additionally, however, solar power projects have unique features which require particular attention in the areas discussed above.

Securing project revenues – the route to market and monetisation of generation

Elisabeth Blunsdon
Orrick, Herrington & Sutcliffe LLP

Without a reliable, long-term revenue stream, no developer will build a solar project. Investors want a return on their investment and as noted in the chapter on financing perspectives, solar projects are capital-intensive to construct, so in a development stage project, money is needed upfront before anything can happen. As a result, the lynchpin of any project is, first, ensuring a physical route which will allow power generated by the solar facility to be exported and, secondly, finding a creditworthy entity which will take and pay for that electricity over a period of time.

This chapter focuses on these two aspects – connection and offtake arrangements. As noted in the chapter on financing perspectives, there are certain key bankability requirements which are common to all solar projects, whether funded by equity or third-party debt. This section therefore also deals in further detail with the important issues of a route to market and a legally binding offtake contract.

For completeness, it should be noted that lenders will require a direct agreement to be entered into with the power purchase agreement (PPA) offtaker and, if suitable regulatory protections are not in place, the grid connection provider. Under these direct agreements if there is a default by the generator, the lenders will have the right to step into the generator's place to cure the default and any rights of the counterparty to terminate will be suspended while the lenders exercise such cure rights.

Connection is, in its simplest form, a conducting wire linking a generator (in this case, the solar facility) to its customer. Without it, a project literally cannot function and will not be able to secure any investment. To ensure bankability the connection needs to be available for at least the asset life of the solar facility, and the cost of using any equipment which is not part of the solar facility needs to be understood, and ideally be predictable. In terms of offtake arrangements, the key is a stable, secure revenue stream. Hence the creditworthiness of whichever entity is paying the generator is a key factor, together with the term of the offtake arrangements. As described in the "Financing perspectives" chapter, investors want to know that the money which they have put into the project at the start will be paid back over time by the project vehicle where the financing is non-recourse.

Depending on the lenders' appetite for risk, and the characteristics of the jurisdiction in which the solar facility is being developed, the income streams which are secured by the offtake arrangements vary. Where subsidies are the principal income stream, there is often a state-backed entity taking the credit risk which is of great comfort to investors. In such cases, an offtake contract for sale of the electricity will be needed, but the main concern will be to ensure that the solar facility can generate in order to gain its subsidies. The price under that offtake contract will not be the primary issue and the generator and lenders may be prepared to take a low price, or market risk, in return for being able to despatch at will. Where subsidies are not the primary source of income (or are not available at all), pricing under the electricity offtake contract becomes more of an issue. Fixed-price offtake contracts are commonly encountered in such situations, or a collar mechanism which adjusts the price against the market price so that the generator is guaranteed a minimum income, rather than a fixed price *per se*. The frequency of such adjustments varies, and may be as infrequent as annually. The longer the period of reconciliation, the greater the risk to project cashflow (and hence the higher the level of concern for investors) in periods where market prices are low. It is always worth remembering that securing a stable revenue stream is usually more attractive to investors than securing the highest price. That said, there are some markets where investors are prepared to take merchant price risk on electricity sales, although as discussed in the "Financing perspectives" chapter, this requires good forward visibility of electricity prices, and is typically more suited to equity participants that want to access upside, rather than debt providers. The level of market risk that is acceptable will also depend on the subsidy arrangements as noted above.

1. Connection

It is not possible to ignore the basic laws of physics and so any generator must be connected to an electrical load of some description if it is to operate. Without a load, electricity will simply not flow, nor will it flow without a physical connection between the generator and the load. Whether that load is a set of premises (connected to the generator by a private network of wires) or a separate distribution network (to which end users are in turn connected) there will need to be an agreement with the owner of the premises or network operator governing the terms on which the connection is made at the connection point (where the solar facility's equipment meets the third party's equipment) and the use of the system in which the connection point is located.

In a so-called private wire solar project, the solar facility will be connected to the private wire network of the offtaker, rather than to a distribution network. The offtaker of the electricity will use the electricity in its own premises via the direct, privately owned connection between the solar facility and the offtaker's premises. The private network may be as simple as a single connection between the solar facility and the offtaker's premises, or there may be other generators connected to the private network. If the site is a large industrial site, the private network can be very complex, but whatever its level of technical specification, the solar facility will need to be operated in accordance with the rules and technical requirements of that network. In almost all circumstances, the private network will also be connected to the distribution or transmission network, which will impose certain constraints on the private network and its operation. Whether or not the solar facility benefits from direct access to the external network is an extremely important issue, particularly for lenders, which is discussed in further detail below.

In a private wire project, the precise location of the delivery point under the PPA will be specific to the particular site, however, there will be a physical route which electricity will take from the solar plant to the delivery point and the connection arrangements will allocate responsibility for the electricity to be delivered, and for the wires themselves, as between the parties. The generator may need a right to use the offtaker's system if the delivery point is embedded in the offtaker's network, or the generator may deliver to a point on the edge of the network, and the offtaker will take the power to its final destination. Unlike a use of system agreement with a distribution network operator, there will not normally be any fees due for use of system while the generator is delivering power to the offtaker under the PPA. This is one of the advantages of a private wire PPA project, the other being that the offtaker can also avoid the charges that are levied by licensed suppliers when making supplies from the grid to an end user. As result the offtaker can pay the generator a price under the PPA which is higher than the wholesale market price, but still lower than the grid delivered price that it would have to pay a licensed supplier in the open market.

In the European Union all owners of private distribution networks are required to grant third-party access to their networks on reasonable terms as a result of the *Citiworks* case, however it is always preferable to have contractual arrangements in place which can be enforced directly against the offtaker without the need to seek enforcement through the regulator. Where the PPA terminates due to the default of one or other of the parties, access to the offtaker's system is needed even where the generator is at fault. Offtakers can find this difficult to accept but they need to understand that the PPA is a contract for delivery of electricity, not a power plant, and even if the PPA terminates for the generator's default, as long as the generator has paid the relevant damages for its breach, the project should not then be penalised by being stranded. Without provisions to this effect, the project will be unbankable, and probably unsaleable. Clearly the key is the negotiation of a suitable methodology for calculating damages on termination which ensures that the breach of the PPA is compensated, but the project can continue to operate afterwards.

Where the solar facility is connected to a distribution network, a connection agreement will be required from the local distribution network operator (DNO). DNOs are invariably regulated and therefore the procedure for obtaining a connection agreement is usually set out in the relevant statute or code. The first step in obtaining a connection is to procure connection capacity, usually by means of a connection offer. The generator applies for a connection and receives an offer stating the amount of capacity available, the works that need to be done to construct the connection and the costs of those works. In many markets capacity at local distribution network level for intermittent generation is constrained so there may be issues around availability. Assuming that these can be overcome, connection works may in some instances be undertaken by independent contractors (rather than the DNO) but there will usually be a specific set of works (particularly those involving connection to high voltage equipment) that can only be undertaken by the DNO. Once works are complete and the DNO is satisfied that the facility has been correctly commissioned and connected (a process which will usually be witnessed by DNO representatives prior to issuing a certificate of compliance with the relevant technical codes), a connection agreement will be signed which sets out the obligations of the generator, the relevant technical details and the charges that must be paid by the generator for capacity and use of system. Connection agreements are rarely negotiable: the generator's (and the lender's) protection lies in the fact that the terms and the costs are regulated.

2. Offtake arrangements

The most common type of offtake agreement (ie, the means by which the electricity produced by a project is sold), is the Power Purchase Agreement or PPA. Income under the PPA may or may not be the primary revenue stream (depending on the prevailing subsidy regime, if any), but even if the quantum of PPA income is secondary in importance to the subsidy payment stream, almost all subsidy regimes require electricity to be produced in order for subsidies to be paid, so an offtake agreement is always needed.

The PPA therefore serves two functions: first it provides income for the project, but secondly it provides a contractual means of disposing of the electricity produced by the project which allows the project to generate, and hence obtain subsidies. Due to its inherent physical characteristics, electricity needs to flow from the project to its end user, whether directly, via a network or even into a battery. The PPA, together with the connection arrangements described in this section, provides the necessary contractual framework for that to happen, and, as noted previously, the presence of these two elements is a key bankability requirement.

From a contractual point of view, PPAs are, at heart, simply a contract for the sale and purchase of electricity. As with all contracts, the terms allocate risk as between the parties, and it is always best practice to ensure that risks lie with the party best placed to manage that risk. In some markets the form of the PPA is determined by a central buyer or regulator, and there will be little room for manoeuvre; in other markets where parties are free to contract as they please, PPAs can be heavily negotiated and may take whatever form the parties wish. Whichever model is in place, certain risks are common to all jurisdictions and are discussed in more detail below.

3. The offtaker

Almost without exception, a PPA offtaker will be a corporate (or equivalent) entity. However, not every entity can be an offtaker under a PPA because the obligation to take delivery of power is subject to that offtaker having the necessary rights to access and use a wire network (either its own, or someone else's) and having the necessary regulatory and legal authorisations to take electricity. In most jurisdictions the offtaker will be either the local grid operator, or a licensed supplier. The reason that this is the most common model encountered is simply that these entities have the rights to use the grid network and to sell the electricity that they buy from the project either to end users or others in the wholesale market. The project therefore has an offtaker that can access the wholesale energy market (where one exists) and end users and has access to the network for the purposes of balancing the output of the project (as will be discussed later in this section, responsibility for balancing is a risk that is almost always passed to the offtaker).

Less common, but becoming more prevalent in the post-subsidy world, are so-called 'corporate and private wire PPAs'. Private wire PPAs involve the sale of power to a customer that is on the same site as the generator. They are also called 'within the fence PPAs' or 'embedded PPAs'. If a customer has a suitable load, sufficient space for a solar project to be built on site and either has, or is willing to construct, a private wire network to connect the project to its premises, a private wire PPA can be beneficial to both generator and offtaker. The generator benefits from entering into a long-term PPA with its direct customer, while the offtaker is able to purchase power directly from the generator, thus avoiding the often significant difference between the price of wholesale electricity and the price of grid delivered electricity supplied by a supplier. The difference arises due to the transmission and distribution costs, levies and charges which suppliers are obliged to include in the bill when they deliver to customers from the grid. As a result, under a private wire PPA, a corporate customer will, in theory, offer the generator a price which is higher than the wholesale market price, but lower than the grid delivered price, thus benefiting both parties. The downside for any generator (and its lenders) under a private wire PPA is how to manage the risk that the offtaker disappears before the PPA expires. There is a risk that the project will become stranded if the offtaker defaults, or abandons the site and there is no separate connection from the project to the grid. If not managed adequately this risk can prevent a project from obtaining investment. A separate grid connection could be installed during construction but this would significantly increase the project's costs. Therefore, it is common to see provisions relating to the consequences of termination which set out a mechanism whereby the offtake grants the necessary land rights to construct a connection to the local grid and, if the termination is due to the offtaker, pays for the construction of the connection and compensates the generator for lost PPA income during the construction period. Even where termination is due to the generator, the right to construct a connection with the assistance of the offtaker is an important feature which any lender will require. In such circumstances, the generator would pay the agreed contractual compensation under the PPA, and pay for the construction works.

A corporate PPA is the term used to describe a PPA where the offtaker is not a licensed entity but is instead a commercial company which contracts directly with the project for the purchase of the project's power. A corporate PPA can be a private wire PPA, but need not be, and typically involves a large corporate entity that contracts for power which is then delivered via the grid network to various sites. As the corporate is not itself licensed, there is also a contractual arrangement with a licensed entity which performs the licensed activities required to take delivery of electricity on behalf of the corporate. The obligations on the offtaker in the PPA are therefore stated in terms of "will perform or will procure the performance of".

Whether these arrangements with a licensed entity are bilateral (ie, the generator and the corporate offtake enter into a PPA, and the corporate offtaker and a licensed entity enter into a separate back-to-back arrangement) or tripartite (all three parties enter into a single agreement) depends on the applicable regulatory environment and the appetite of the corporate offtaker for risk. Some corporate offtakers are prepared to enter into long-term bilateral arrangements with generators for corporate and social responsibility (CSR) reasons and will take the risk of not being able to achieve a back-to-back arrangement with a licensed entity, while others will not take that risk and will seek to share it with the generator. The tension arises because, particularly where financing is involved, generators tend to favour a long-term PPA, whereas large corporates like to keep their energy supplies keen by changing suppliers at regular intervals (such intervals being much shorter than a typical PPA term). Entering into a long-term bilateral PPA therefore gives the generator what it needs but means that the corporate takes on long-term liabilities which it may not be able to pass through if it changes supplier. On the other hand, a tripartite agreement gives the corporate entity certainty that it will be able to protect itself against any supplier risk, but the generator, and its lenders, may not feel comfortable with the level of uncertainty under a tripartite arrangement.

There is another variant on the corporate PPA known as the 'virtual' or 'synthetic' PPA. These are financially settled arrangements, whereby physical delivery is made to a supplier or the local grid by the generator, and the counterparty pays a fixed or market price to the generator while taking its power requirements from its own supplier. There is no physical delivery of power, or any attempt to 'trace' the power from generator to end user. While being a simpler mechanism, at least on the face of it, than the tripartite or bilateral PPAs described earlier, this lack of traceability can be unattractive to corporate offtakers. A potentially bigger issue is that, as a financially settled contract, in all likelihood financial regulatory rules will apply. As most project developers, and indeed corporate offtakers, are keen to stay outside the jurisdiction of the financial regulatory bodies, this model has proved rather less popular in some jurisdictions than others.

4. Term

Traditionally, any project that is financed using third-party debt would require a PPA with a term equal to the tenor of the facility, plus somewhere between 18 months and three years as a buffer. Hence PPAs of 10 to 15, and even 20 years are not uncommon. Given that reliable price curves may not be available (power price curves beyond 24–36 months are notoriously unreliable, at least in Europe), the decision surrounding the term of a PPA is tightly linked to the chosen pricing mechanism. The generator's lenders invariably prefer either a fixed price, or a collar on any market price exposure to ensure that debt repayments can be made and other financial commitments met. Offtakers, and indeed generators, may wish to take more market price risk if they perceive a potential upside but they must still take a view on the level of certainty that they require looking forward. If, however, the principal source of income for the project is some form of subsidy payment, it may be that the lenders will take some market price risk, and use the projected value of the subsidy over the life of the project as the base case under the financial models. See Section 8 below for further discussion on pricing.

The start date of the PPA will depend on the project's specific circumstances. Where a project is already operational, the start date will be agreed between the parties, depending only on any metering or other arrangements that may need to be completed before the offtaker can take the project's output. Where, however, the project is under development the PPA will provide that the start date is conditional on certain milestones being reached.

If the offtaker has agreed to offtake the electricity produced during commissioning, the offtaker's obligation will usually commence on the date notified to the offtaker by the generator as the start of commissioning. Typically the generator will be required to provide notice of the start of commissioning window several months before the process of commissioning starts, and that window will narrow down to a specific day as the required notice period gets shorter. During commissioning the price paid for power is usually low, typically no more than the system spill price. The generator's main concern is to ensure that power is offtaken so that the project can generate and undertake the various tests and processes that are required to commission the solar facility. On the generator's part, there is no obligation to make detailed notifications or to perform to any particular service level. It is not unheard of for there to be a different offtaker for the commissioning power. The contractual mechanism remains as described above.

The term of a PPA which is in development will usually start on the commercial operations date or COD. This is the date on which the project is complete under the construction contract as evidenced by the issue of the Taking-Over Certificate. On a solar project, this is relatively straightforward and is determined by the number of panels installed and the power being generated (compared with the position for thermal generation plants where there are usually performance tests to be run between taking over and completion). There will generally be several conditions to COD, including achieving accreditation for any renewable benefits, obtaining connection etc. However, in practice, these all occur as part of the construction programme, leaving the Taking-Over Certificate as the final condition to be met.

There is usually a funnelling mechanism which narrows down the date on which COD will occur, similar to that described above for commissioning. At COD the full set of PPA obligations kicks in, and the contract price becomes payable by the offtaker. Liquidated damages are often paid to the offtaker if either commissioning or COD is delayed beyond an agreed date. The offtaker may also have the option to terminate the PPA and will seek to receive damages if the delays extend beyond the longstop date. When negotiating these arrangements it is important to ensure that any liability on the generator for delay in commissioning or COD under the PPA is, to the greatest extent possible, passed through to the construction contractor under the construction contract.

5. Delivery point

The delivery point under any PPA needs to be carefully defined in order to ensure that issues of title and risk in the electricity produced by the project are addressed. Further, where electricity is being delivered across a network, the party delivering the electricity needs to have the necessary rights to use the network and will, at the same time, have liability to pay use of system charges in relation to that network. Generally, under a solar PPA, the delivery point is downstream (heading away from the facility) of the final inverter but just before the connection to any distribution network. A single line diagram is always helpful in identifying where the delivery point is located and should be annexed to the PPA for clarity.

6. Fundamental obligations

As is the case for all PPAs, whether relating to a solar project or not, the fundamental obligation on the generator is to sell power to the offtaker, and the fundamental obligation on the offtaker is to take power from the generator and pay for it. Given the generally small size of solar projects, it is normal to see the offtaker committing to sell all its output from a project to the offtaker, in return for the offtaker agreeing to take whatever is delivered whenever it is delivered. This arrangement is sometimes referred to as 'take-if-tendered' or 'as produced'. This is a common model for most renewable generation and contrasts with the situation under a typical large-scale thermal PPA where the offtaker will usually have the right to specify when it requires power to be delivered.

The *quid pro quo* for the offtaker taking all power, as and when it is produced, is that the generator will receive a flat price rather than a fluctuating price reflecting short-term volatility in the market.

7. Transfer of benefits

Depending on the jurisdiction, renewable benefits may be transferred under the PPA together with generated power. There is a wide variation as to whether power and benefits have to be transferred together, whether they are usually transferred together as a matter of market practice, and whether the price paid under the PPA is a blended figure for power and benefits or whether power and benefits are invoiced and paid for separately.

Where benefits are transferred under the PPA there will be a provision setting out what each party is required to do (and by when) to ensure an effective transfer of the benefits. The PPA will also specify at what point 'title' to any benefits passes. Many benefits are electronic certificates maintained in central registries, and are transferred between accounts with risk and title transferring on the transfer between accounts being completed.

Benefits are invariably issued against a specific set of criteria which are to be found in the relevant scheme rules and legislation. Logically, therefore, it is possible that electricity generated may not meet those criteria, leading to a revocation of particular benefits already issued, or a benefit may be revoked due to an administrative or clerical error. Given the nature of solar PV technology, once a project is accredited, the possibility for a revocation of a particular benefit on technical grounds is unlikely (most issues tend to be related to the original accreditation itself) but there is still the potential for revocation of benefits due to administrative/clerical issues. Solar PPAs therefore contain similar wording to other renewable PPAs concerning the revocation of benefits. The matrix of possible reasons for revocation is drawn up based on why the benefit was revoked (which party, if any, was at fault) and whether or not a benefit will be re-issued (when the administrative error is corrected) or a replacement procured from the market (if benefits are openly traded).

If a benefit is not re-issued, or is not re-issued within any applicable time period for compliance, the party at fault may be able to substitute another benefit, if the relevant scheme allows. Otherwise compensation is payable, with the value of that compensation subject to negotiation and the rules of the scheme – some benefits have a 'face value', others are freely traded and others can be valued on the basis of benefits lost. Where a benefit is revoked and neither party is at fault, risk allocation will need to be agreed.

While on the subject of benefits, if renewable benefits are included in the PPA, there will be provision relating to the process that must be completed for the project to qualify for benefits (referred to as 'accreditation' in this chapter) and the ongoing issue of benefits. The generator will therefore be required to obtain and maintain accreditation and to ensure that benefits are issued during the term of the PPA. It is not uncommon to see also a provision requiring either the generator or both parties to act reasonably in maximising the benefits that can be achieved by the project. Generators should be wary of accepting any obligation that also requires them to maximise revenue unless they are in a position to maximise revenue from the market. If the generator has no access to the market, it cannot optimise revenue.

8. Price

Clearly the price payable under the PPA is of huge commercial importance. From a legal drafting point of view, price provisions vary hugely, from a fixed price applicable throughout the term, which would normally, but is not always, indexed, to a floating market price based on a specific index, with or without a price collar or floor price. As with all pricing mechanisms, agreeing the underlying commercial principles is the key to success.

Most solar PPAs that do not have fixed prices will be priced against a market price which is short term – typically in Europe a day ahead price – as this most closely reflects the output profile of a solar facility. As mentioned above, some PPAs are priced against the market price for their full term, which is more likely where the principle income stream is derived from the sale of renewables benefits.

Some offtakers offer the ability for the generator to fix prices for specific periods (say, for example, six months forward at a time) based on market prices from time to time. If the generator wishes to fix prices it will ask the offtaker for a price at which the offtaker is prepared to fix for the relevant period. If the generator likes the price, the offtaker will fix forward. This gives the generator a hybrid between market and fixed price and allows the generator to capture seasonal price variations if it wishes to. Note that a lender would expect input into any decision to fix prices.

9. Invoicing and payment

At first glance, the invoicing sections of any agreement may not appear to be the most exciting, or even important. However, the invoicing provisions provide an immediate overview of all the money flows that can occur under a contract and are therefore a useful place to start when reviewing a PPA. It is important to draft the invoice clause properly, meaning that all possible payments that could be made between the parties, whether for volumes delivered, liquidated damages, rebates from the distribution network operator or any other specific payment, are listed together with a catch all "and any other amounts due under this Agreement". The exercise of compiling the list of invoice items is in itself helpful as it tends to flush out any issues that have been omitted or forgotten.

The reasons for this level of granularity are twofold – first, from a legal perspective, the payment provisions invariably state that the payment obligation is to pay the amounts stated in the invoice. If the full list of items is not stated in the invoice there may be arguments about what needs to be paid. Secondly, as a practical matter, no commercial entity will pay any amount without an invoice to pay against, and again, if the invoice is not comprehensive, any amounts not in the invoice will simply not be paid. One option is to include a *proforma* invoice as an annex to the PPA. This also assists greatly in implementing the PPA once power starts to be sold.

Often the generator does not have access to much of the data needed to prepare the invoices under the PPA, whereas the offtaker does, particularly the metered data for inputs into the system. As a result, the offtaker may be required to prepare a monthly statement within a set number of days of the end of each month once it has received the relevant metered data from the system operator. Depending on the jurisdiction and the applicable tax rules, the offtaker may issue a self-billed invoice with the monthly statement or, if this is not possible, the offtaker will send the monthly statement to the generator who will then issue an invoice. This will add an additional number of days into the payment cycle as the generator will need additional time to process the monthly statement and issue the invoice.

Payment terms are often not as heavily negotiated as might be expected as many markets have an existing payment cycle in place driven by the wholesale markets settlement timetable and it is often easier for the PPA parties simply to slot into that cycle.

The PPA will state the time for payment, usually by reference to the date of delivery of the invoice, and will contain provisions allowing for invoices, or amounts therein, to be disputed and for the charging of interest on late payments. The PPA will also provide for set-off of payments owed by each party to the other, so that a single net sum is paid by the party owing the larger amount. This is an important provision for lenders as it reduces risk in the event of insolvency.

10. Metering and data

Clearly, metering is an important aspect of any PPA and Solar PPAs are no exception. Metered data is not only needed for system settlement purposes, it also determines the volume for which the seller will be paid under the PPA and, depending on the subsidy regime in place, may also be determinative of the amount of renewable benefits which the project can claim. In many jurisdictions the relevant industry code will determine which party is responsible for installing, operating and maintaining metering systems. There will invariably be a set of meter standards, often international in nature, with which all metering equipment must comply. Audit provisions are included as standard, allowing the party that is not responsible for the operation of the meter to audit meter records and to inspect the meters on reasonable notice. It is also standard to see provisions which allow either party to call for a meter to be tested if that party believes that the meter has been recording erroneous data. If the meter is found to have been reading incorrectly, there is a mechanism to correct any faulty historic measurements and adjust PPA revenues accordingly.

Most projects of a size large enough to warrant a PPA will have a Supervisory Control and Data Acquisition (SCADA) system installed. As the offtaker is responsible for balancing, it is standard to see a provision which requires the seller to allow the offtaker access to the real time data from the SCADA feed. This gives the offtaker access to information about conditions at site (including weather conditions) and the generation being produced, which in turn allows the offtaker to manage the imbalance risk in real time.

11. Forecasts and balancing

By its nature, solar generation is intermittent, and even the largest solar generating facilities are small compared with more traditional thermal generation assets. The combination of intermittency and small size means that balancing risk is invariably best managed by the offtaker, and most lenders will insist on this. Balancing risk in this context means the requirement of system operators that the amount of electricity which is to be put onto a system in any given trading period should be notified in advance of that period and should be matched by an equal amount taken off the system. If the amount notified and the amount delivered do not match, there is a penalty due to the system operator to reflect its costs of managing the imbalance. Solar generation's intermittent nature makes predicting output tricky, even at relatively short notice, and the nature of generators as special purpose vehicles mean that they lack a portfolio of generation and demand that can be used to absorb balancing volumes from trading period to trading period. Hence the offtaker takes the balancing risk under the PPA and reflects the management of the risk in the price payable under the PPA.

Despite the intermittency inherent in weather conditions, the seller will, as a minimum, be required to provide forecasts of availability to the offtaker. These are given at least annually (and often more frequently) and inform the offtaker of periods of planned maintenance. Any updates to the planned maintenance regime must be notified to the offtaker as they occur, and in any event before the planned maintenance starts. Other outages or constraints are notified as soon as reasonably practicable after the seller becomes aware of them, and where there is a forced outage the seller must keep the offtaker informed as to likely cause, duration and when the outage ends.

The seller should not guarantee or warrant the level of output that will be produced in advance, as there is no way of knowing whether there will be sunshine or not. It can, however, warrant availability against its planned maintenance programme. It is sometimes agreed that the seller should achieve a certain level of availability (meaning that the facility is capable of generating, even if it is not actually producing anything) and is penalised if it fails to reach the required level, but this is still more common in wind PPAs than in solar PPAs.

12. Credit support

The importance of the creditworthiness of the offtaker has already been discussed in several places. Given the key importance of the revenue stream under the PPA in the financing structure of any project (even where the project is funded entirely by equity), credit support will always be a consideration. In an ideal world, the offtaker will have an investment grade rating which can be relied on. However, this is rare, even where the offtaker is a large corporate entity or a licensed supplier, as any investment grade credit rating will generally be group wide and will not sit with the entity which is the counterparty to the PPA. In these circumstances, a parent company guarantee securing the offtaker's PPA obligations is the preferred solution. If there is no suitably rated entity to give a parent company guarantee, a bank guarantee or letter of credit is provided. It is common to include wording covering what happens if an offtaker or credit support provider loses the required rating during the term of the PPA. In this situation, suitable credit support has to be provided within a short period of time, and failure to do so will give the generator a right to terminate. Similar terms also apply to the replacement of credit support on expiry or default. Sometimes an offtaker will require a generator to provide credit support to cover the mark to market cost of replacing lost generation if the generator defaults, but this is usually strenuously resisted by financiers.

13. Default of the offtaker or grid operator

Given that the preceding sections of this chapter have set out the pivotal importance of the grid connection and the offtake arrangements, it is worth briefly turning to what happens if either (or worse still, both) the grid operator or the power offtaker defaults.

In regulated markets, such as those found in Western Europe, the default of a grid operator is specifically dealt with in regulations, and there is invariably some sort of 'last resort' mechanism in place whereby operation of the network continues, regardless of any issues affecting the operator. As a result, the grid will almost always remain operational, meaning that while there may be some extra administration involved, power will continue to flow. The default of a licensed supplier that is an offtaker under a PPA has precedent in Europe (TXU and Enron being notable examples) and will involve invoking the terms of the PPA and pursuing damages. In a liquid market, while this can cause cashflow issues, it is usually possible to find a replacement offtaker – albeit that it may not be possible to obtain the same price. In certain subsidy regimes, such as the UK CfD regime, there is a backstop PPA mechanism in place to ensure that there is always a route to market for CfD generators, but the price has a hefty discount applied to the market price.

The consequences of losing the offtaker under a private wire PPA are described above. In all cases of offtaker default, the generator will seek to invoke the credit support instruments provided by the offtaker. It may also be appropriate to consider termination of the PPA under its terms, which will usually allow termination for an unremedied material breach or insolvency. However, any decision on termination will involve lenders. In practice, termination scenarios tend to develop over time, and the lenders will be fully aware of a deteriorating situation as a result of monitoring the various financial performance tests included as part of the lending arrangements. If termination becomes a possibility, the lenders will have various rights to direct what happens under the lending arrangements, and the generator will not be able to exercise any termination right without the lender's consent.

In markets which are not regulated, or are not liquid, and particularly where state-owned entities are involved, rights to terminate will invariably be present but options to mitigate the grid operator's or offtaker's default may be significantly limited. In such circumstances the adequacy and enforceability of any credit support or performance guarantees is paramount. A state performance guarantee under which the state assumes liability for performance would be seen as the gold standard, but lenders and generators would need to be comfortable at the outset that such a guarantee would be enforceable. A payment guarantee of all losses – including equity returns – incurred as a result of the offtaker's default backed by the state is also helpful. A detailed discussion of choice of law and dispute resolution provisions is outside the scope of this chapter, but it is worth remarking that provisions relating to choice of law, dispute resolution and enforceability, in certain jurisdictions can become key bankability issues.

Financing perspectives

William Evans
Simon Middleton
Elgar Middleton Infrastructure & Energy Finance LLP

1. Introduction

Large solar power installations, either as standalone sites or as part of a portfolio, are capital intensive to construct. This requires the use of significant funding to pay for the development and installation costs. Thankfully, once built, these sites generate stable long-term cash flows which can be used to repay this capital.

Financing of solar installations is generally in the form of corporate loans, equity or third-party debt. The requirements vary by type of funding solution, which is in turn reflected in the pricing and seniority of repayment. There are, however, a number of generic factors common to all funding options that must be met to enable finance to be successfully raised.

The amount of finance required can also dictate the best funding option available. For example, third-party debt financing of less than £20 million is uneconomical due to the upfront costs incurred and will be of limited interest to potential funders. While this is not a problem for large ground-mounted schemes it has resulted in portfolios of rooftop and small-scale ground-mounted installations being created prior to debt being arranged.

One funding route often utilised is for a solar installation to be built using a corporate loan (ie, on the balance sheet of the company doing the construction). Once constructed, the company then elects to either refinance their corporate loan with third-party debt, or to sell the solar installation. If a sale occurs, then the buyer will in turn decide to either finance their acquisition costs with equity, or also raise third-party debt to part fund their acquisition.

Pricing for all forms of solar finance compressed over the past five years as solar became increasingly recognised as a distinct sector within the renewable energy space. The increased familiarity, coupled with low default rates, has enabled finance providers to assign a lower risk weighting which in turn generates a lower return requirement.

This chapter will elaborate on the funding options available for solar installations, the key requirements that must be met for finance to be successfully raised and the site-specific variables that impact on the costs and scale of finance which is available.

2. Sources of finance

In broad terms, solar installations can be financed through three means:

- corporate loans (also referred to as on-balance sheet or HoldCo finance);

- equity; and
- third-party debt (also refered to as off-balance sheet or non-recourse finance).

2.1 Corporate loans/HoldCo loans

A corporate or HoldCo loan is a loan provided by a parent company to fund its subsidiaries. This is typical for the vast majority of funding activities undertaken and requires a company to raise debt on the market at the level of the ultimate parent company (such as at PLC level). This debt is then used internally for various means.

An illustration of the financing structure that occurs with a corporate loan is provided below

Figure 1: Financing structure for a corporate loan

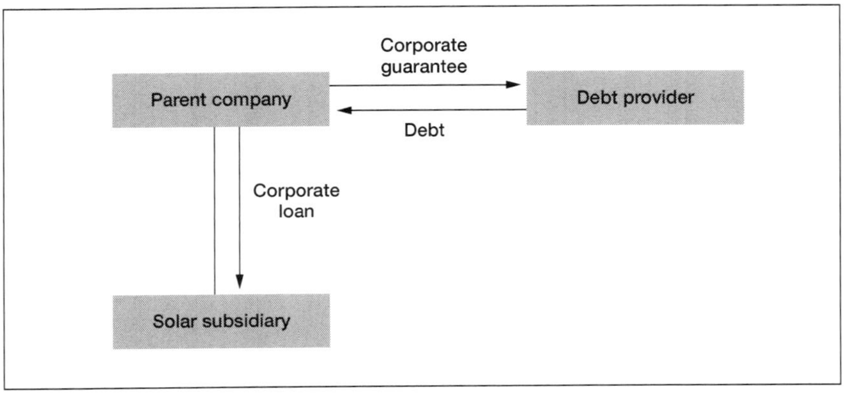

The attraction of this approach is that the parent company will have already raised the debt meaning that the loan is instantly available. However, this capital cannot be used for other means and is also registered on the parent company's accounts as a liability.

In the solar sector, a corporate loan is sometimes used as a bridging loan by a contractor so that it can complete construction prior to some arbitrary deadline (such as a decrease in the support tariff the solar installation will receive if completed after a certain date) or is used for its simplicity (ie, no third-party involvement required, less reliance on third-party due diligence reports and simpler legal documentation). Once the site is constructed the company will then look to refinance the corporate loan through either a sale or by raising third-party non-recourse debt. If a sale occurs, then the buyer will in turn decide to either finance its acquisition costs with equity, or raise third-party debt to part fund its acquisition.

The due diligence undertaken prior to the use of a corporate loan is far less extensive than that undertaken by equity and third-party debt providers. This is due to the finance being sourced in-house and secured against the parent company's balance sheet, eliminating the need for extensive enquiries. However, this approach can lead to potential problems at a later stage when third parties become involved and conduct their own due diligence. For these reasons solar installers using a corporate loan should always maintain discipline when developing a site, otherwise it can potentially become unbankable or be sold at a loss.

The use of a corporate loan is also limited by the company's available capital. This means that for larger sites the sheer capital requirement may be too great for a company to self-finance and this option may therefore not be viable. In these circumstances an installer will typically require third-party finance in the form of equity or third-party debt to finance construction.

2.2 Equity

Equity is similar to a corporate loan in that the funds are sourced from a parent company. The term 'equity' is used here in a broad sense as it covers both shareholder capital as well as a shareholder loans (sometimes referred to as subordinated debt). Although these two forms of investment are legally distinct, their cash-flows are often blended together to give a combined return for the equity investor.

An illustration of the financing structure that occurs with equity financing is provided below.

Figure 2: Financing structure for equity finance

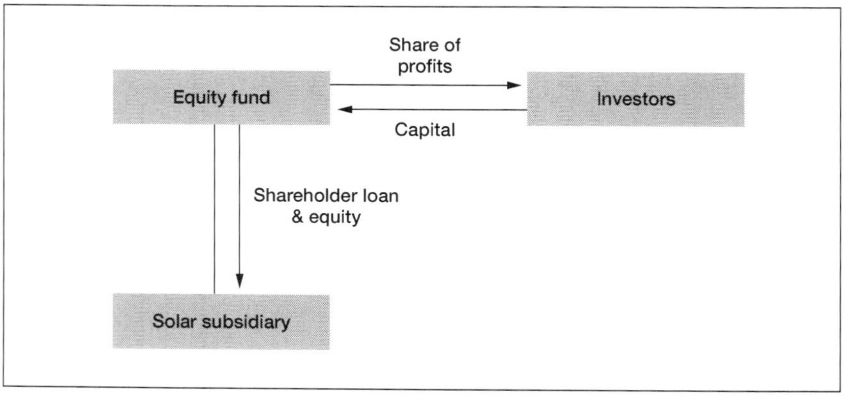

The advantages of using both equity and shareholder loans are that an investor receives four cash-flow streams:

- interest on the loan;
- loan principal repayment;
- dividends; and
- repayment of the shareholder capital.

This diversification is useful as it can enable an investor to receive payments under the shareholder loan at a time when dividends may not be payable, for example during the early operating stage of a solar installation when cash reserves are insufficient for distributions to be made.

An additional benefit is that interest on an equity loan is often tax deductible, thereby reducing the corporate tax payable by the solar company. However, the tax treatment of shareholder loan interest is subject to an ongoing review by various tax bodies globally under the so-called Base Erosion and Profit Shifting (BEPS) project and this tax treatment advantage may not be available to the same extent in future.

Equity finance typically occurs as the result of a purchase of solar assets by an equity fund. These funds may be large international investors, regional retail funds with a stock market listing or private wealth management funds. The equity funds in turn raise their money from pension funds, retail clients or high net worth individuals.

While many equity funds do not require any additional finance, others seek third-party debt in addition to provide leverage for their own equity investments. The difference in choice can be down to a fund's objectives (some, for example, are not permitted to raise debt under the terms of their own prospectus) as well as the availability of capital (once all equity is deployed a fund is unable to buy any more solar assets). The advantage of adding third-party debt is that it is normally cheaper than equity, thereby boosting the equity investor's returns, as well as freeing up equity for future investments.

Where third-party debt is raised, this will rank senior to the equity investor's shareholder loan. This means that, in each repayment period, third-party debt will be paid first, with any remaining cash flow made available for equity-related payments such as shareholder loan interest, loan repayments or dividend payments, the latter being subject to compliance with the local accounting regulations.

2.3 Third-party debt

Third-party debt or non-recourse debt is essentially debt that is provided by a funder unconnected to the solar company. It also differs from a corporate loan as the debt is invested directly into the solar company and cannot be used for any other purpose. Third-party debt is typically invested alongside equity investments (both shareholder loans and equity – as outlined above) as it will be insufficient to cover all of the solar company's costs. A third-party debt provider will also require part of the capital costs to be funded by equity to ensure the project's owner is fully incentivised to ensure successful project operation.

Third-party debt invested in solar projects is typically structured on a non-recourse basis, often referred to as 'project finance'. This means that the debt is associated with a specific project owned by a standalone company and it is this standalone company that borrows the debt. The loan is also structured as non-recourse debt, meaning that in the event of a default, the lender cannot seek recourse from any other company other than the standalone borrower. So even if all the other projects a parent company owns are performing well, a lender cannot seek to recover a loss from them, nor can it seek recourse to the parent company's balance sheet. Instead, the ultimate protection a lender has is a security package which includes a share pledge over the entire share capital in the standalone company and security over its assets. If a default occurs the lender has the option to enforce its security over the shares and become the owner of the company. As a result, the equity investor will lose the value of its investment and all the share value and outstanding shareholder loans for that specific company are written off. An illustration of the resulting financing structure is provided below:

Figure 3: Financing structure for third-party debt

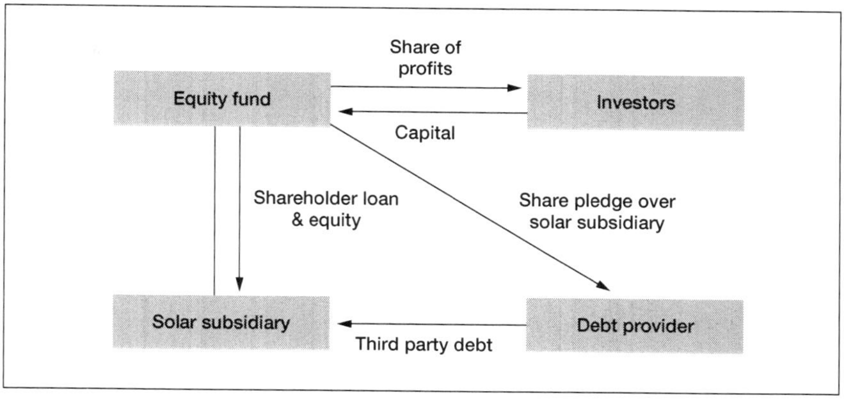

Third-party debt for solar projects comes in two forms: bank debt and bonds (either publicly listed or private placements). In most respects these two investments are very similar; however there are key differences which mean that either one may be more suitable for a specific project. These differences are outlined below:

- Tenor – While bonds in the context of renewable energy projects are typically long-term investments of 18+ years, bank debt can be of any tenor from 18 months to 19 years;
- Repayment – Bonds are more expensive to prepay early as they require a net-present-valuation of all future interest payments, known as the 'Spens redemption clause'. While bank debt can incur early repayment fees, and associated swap breakage costs, they are nearly always lower than those of a bond;
- Draw down – Bonds are typically drawn in full upfront whereas bank debt can be drawn in tranches;
- Rating – A bond often (although not always) requires a rating from a credit agency such as Moody's, Standard & Poor's or Fitch. This is not required for bank debt;
- Inflation link – Bank debt is structured as a nominal loan (ie, the repayments are fixed and not usually linked to changes in inflation) whereas bonds can be either nominal or linked to a prescribed inflation index. This can be important where a solar company receives inflation-linked revenues as it can then align its interest costs with changes to its revenues; and
- Repayment flexibility – Bond repayments are typically based on a fixed amortised profile over the tenor of the loan. Bank debt repayments are similar, but can have individual repayment periods adjusted to compensate for forecasted volatility (such as extra planned costs) to ensure the project is not at risk of a default.

Due to the predictability of solar revenues, a significant amount of third-party debt can be raised. This can be up to 85% of a solar company's required funds, with the balance provided by equity. The amount of debt varies by project and depends on key bankability requirements. These will be explored in the next section.

In addition to non-recourse financing of a single asset, borrowers will often seek project finance debt in respect of a portfolio, which may be as small as two standalone assets or an entire portfolio belonging to the same investor, in some cases as many as 50 standalone projects. In the case of a non-recourse portfolio, financing the loan is provided against a security package that captures the assets and cash flows of the entire portfolio.

The use of a portfolio has both pros and cons for the borrower. One clear positive is that if a particular solar park performs poorly, the impact is mitigated by the performance of the portfolio's other solar parks. It is therefore not unusual for lenders to run sensitivities where the largest solar park's revenues are removed to determine the portfolio's ability to still repay the debt. Some financings cater for a partial default whereby there may be an Event of Default trigger at each project level but this will only trigger a portfolio default if it has a material adverse effect on the portfolio as a whole. Cure mechanics can be introduced to allow a borrower to 'buy out' an underperforming asset in order to avoid a long-term defaulting project pushing the entire portfolio into permanent lock up or default.

The downside of financing a portfolio is that the borrower's ability to subsequently sell any of the portfolio's assets will be governed by the terms agreed in the facility agreement. This may allow a partial sale without lender consent (such as up to a stipulated percentage of the portfolio's capacity) with disposals over such thresholds, or indeed with some lenders any sale at all, requiring the lender's approval. In the event approval is required, the lender will typically rank senior to the borrower's investors in the cash waterfall and will use the sale proceeds to repay some of the outstanding senior debt. The lender may agree a pro-rata reduction in debt, typically to ensure the debt ratios are unchanged, or they may insist that 100% of the sale proceeds be applied in respect of debt service. The borrower may also have to amend the interest rate swap which can result in a break cost. Such restrictions may therefore limit the sale proceeds available for the borrower's investors.

3. Key bankability requirements

No matter whether a solar installation is funded by equity or third-party debt, there are many common key requirements that a solar installation must meet for funding to occur. These are outlined below.

3.1 Scale of funding requirement

While equity has a lower investment threshold than third-party debt, there is still a minimum amount that must be invested. This is to ensure that a project has economies of scale as well as being of interest to the funders.

In the case of equity this varies according to the investor. Private wealth funds will invest as little as £100,000, while an international fund will seek to invest at least £10 million. This range means that there is a natural home for most solar projects in the equity fund spectrum.

Third-party debt normally requires a higher hurdle of over £20 million. This is due to the due diligence costs incurred in raising third-party debt as well as the level of internal resource a bank will allocate to a project during the evaluation phase.

3.2 Evidence of lawful construction

All funders will require evidence that a solar installation has been lawfully constructed.

This will be managed through legal due diligence which will confirm that both a property lease, and the appropriate planning permission and permits are in place. For post-construction assets the due diligence will also confirm that any planning conditions have been met.

3.3 Route to market

A second matter of due diligence is that the electricity connection, be that to an electricity network or a private network, is in place and the project can therefore export its power.

This arrangement will be in the form of a legal contract (typically an easement or wayleave) with a third party who agrees to offtake all of the generated electricity under a set of agreed terms. These contracts can vary in length from monthly through to perpetuity and can be at a fixed price or linked to the prevailing market conditions. This is a fundamental part of any financing due diligence as it dictates all of the solar company's electricity sale revenue. (See the "Securing project revenues" chapter.)

3.4 Legally binding contracts

There are three key contracts that are required for solar projects: a construction contract, an operating contract and an electricity offtake agreement. All of these are subject to legal due diligence by a funder.

The legal due diligence will confirm the status of the key contracts, what rights each party has and what the payment obligations are. It will also consider what right of redress the solar company has in the event of a contractor failing to perform, such as a delay in construction completion or underperformance of the solar assets.

Where third-party debt is employed, the funder will often require a 'direct agreement' with the key contractors. This agreement enables a lender to step in and assume control of all communications with the contractor if the solar company fails to meet certain conditions by providing a legally binding contractual link between the lender and the contactor. This provides third-party debt providers with the security of knowing they can take control of assets instantly without a need to negotiate their ability to step in. Such step in rights are normally meant as temporary measures during which the contractor cannot terminate the contract and the lenders, or their appointed representative, seek to remedy the solar company's breach. Similar direct agreements are often required between the lender and the O&M provider and the landlord.

3.5 Legislation and support tariff

Historically, solar installations globally have benefited from a degree of public sector support, typically in the form of a tariff/subsidy payment linked to the amount of power produced. Such payments may be the solar company's main source of revenues, or may be on top of any revenues the solar company receives from the sale of the generated electricity on wholesale power markets.

Where this is the case, lenders will require evidence that the solar company has secured a tariff that complies with applicable legislation.

3.6 Track record of components and key parties

To ensure that a solar installation is built to a high standard a funder will usually mandate an independent technical adviser to conduct technical due diligence on both the components used and on the contractors involved.

The analysis of the components will typically consider the panels, DC/AC invertors (central or string) and mounting frames. This will assess the historical performance of these components on other sites and any failures that may have occurred. This is a crucial analysis as many banks have a prescribed set of panels they can finance; if the project used panels manufactured by suppliers not on the list then no finance will be provided.

This analysis also feeds into the financial viability of a project as different components wear out at different times. This impacts on the rate a solar panel degrades each year as well as on the period before DC/AC invertors will need replacing.

The analysis of the contractors involved will focus on their track record but also on their financial strength. Due to the tight construction profit margins there have been a number of high profile corporate insolvencies in the solar sector globally and no lender wishes to find itself having funded a part finished site or running the risk that spare parts may not be available in the medium to long term.

The technical analysis also examines the site configuration, the power rating and sizing of all components and their configuration in order to determine the likely performance ratio (PR) of the project, which is a measure of the efficiency of the plant at converting DC generated power from the panels to exported AC power into the electric grid. PRs are typically 85%(+/–5%).

Where a solar installation is selling power to a private company, the corporate strength of the purchaser will also need to be considered. A back-up option will also need to be available, such as an ability to sell to the grid network.

3.7 Financial viability

All investors will require a financial model to forecast the revenues and costs for the remaining life of the solar installation, typically the next 25–30 years. If a portfolio of solar sites is being financed then the model will combine them together into a consolidated cash flow.

Each financial model is bespoke as it uses site-specific inputs such as the contractually agreed installation and maintenance costs, the forecast irradiance and yield for the location and panel orientation, the degradation rate likely to be experienced for the panels, the forecast expenditure for ongoing replacement of inverters, and the cost to dismantle the solar park at the end of its useful life. The model will also adhere to the local jurisdiction's tax and accounting requirements.

The model will forecast the cash flows that are expected to be available to service the equity and/or third-party debt investments, as well as assessing the ability of such payments to be achieved under various stress-test scenarios. For third-party debt, the model will also calculate the repayment profile of senior debt (typically semi-annual) as well as other key metrics such as gearing, debt service cover ratios and average loan life.

As a model is typically produced by the borrower or its financial adviser, it is usually audited by a bank's adviser prior to financial close to ensure it is accurate.

3.8 Security package

Where solar projects are funded by third-party debt, an additional bankability requirement is the provision of a suitable security package for the lender. This is required as the loan is secured only against the assets of the borrower and the lender has no recourse to any other companies. While some aspects of a security package can be project, client and funder specific, there are a number of common features such as a debenture (a fixed and floating charge over all assets), a share pledge (the ability to take over the equity of the borrower) and an intercreditor agreement (governs the cash waterfall and enforcement rights between the various connected parties).

4. Debt sizing

As noted earlier, third-party debt is unable to fund all of a solar installation's costs. Although third-party debt can potentially finance up to 85% of the costs, that still requires equity to meet the outstanding cost balance. It is for this reason that a large amount of attention is spent on the key factors that influence the third-party debt amount available, especially in projects where the solar company's shareholder wishes to maximise the amount of third-party debt raised.

As third-party debt is less expensive than equity, higher gearing will lower a solar installation's weighted average cost of capital (often referred to as WACC). In other words, by increasing the senior debt amount the solar company's shareholder not only needs to invest less equity, but the return on its equity investment is also increased. The counter to this is that since they are subordinated to senior debt service payments, equity cash flows in highly geared assets are more volatile and the probability of default is increased.

To determine the amount of third-party debt that can be invested in solar installations, debt providers analyse a number of key financial variables. These include the following.

4.1 Revenue breakdown

As outlined earlier, revenues for solar installations can include tariff support, the sale of electricity to the wholesale power market via a grid connection, or the sale of electricity directly to a consumer. Furthermore, the electricity can be sold at a fixed price, or linked to the prevailing daily market rate.

Third-party debt is always seeking to minimise risk wherever possible. Their preference is therefore to fund projects where revenue volatility is minimised. This means that solar projects with high tariff support and a fixed-price electricity sales agreement will typically be able to raise a higher quantum of debt.

The downside for investors, however, is that fixing the electricity sales price comes at a cost. This is because the purchaser is wary of agreeing a price that might end up being higher than the prevailing market rate. As such, fixed-price agreements can be 10–20% lower than the market rate. So while this structure is attractive to third-party funders, the lost revenue is detrimental to equity.

If electricity is being sold directly to a consumer (such as a private company) the funder will consider the financial strength of the purchaser as well as the length of the contract. As such these contracts are typically only structured with investment grade counterparties or alternatively entities that can provide investment grade credit support.

The most common revenue arrangement is a sale onto the wholesale power market by way of a contract with an intermediary. Here, the solar installation is paid between 92 and 97% of the prevailing daily market rate at the end of each month. Contracts can be up to 20 years in length and are agreed with companies which then sell the power to retail and corporate customers. These contracts can sometimes also include the ability for the solar company to forward-fix market prices for periods of up to several months. While this helps reduce revenue volatility, the prices are agreed on an ongoing basis rather than at the outset and, as such, cannot be deemed as fixed-price revenues for the purpose of third-party debt sizing.

4.2 Power forecast

As noted above, where electricity is sold to the wholesale power market the revenues are based on the daily wholesale price. As this is in the future and is not fixed, it is inherently uncertain. To quantify these revenues as much as possible both third-party debt and equity investors will obtain a power price forecast from a reputable independent company.

A power price forecast attempts to consider the impact that global and local changes may have on the wholesale power market. For example, it will consider the changes expected in the generation mix (such as the rise in renewables and reduction in fossil fuels) as well as the future demand (such as the rise in battery-powered vehicles). These variables are then modelled and a power curve produced for the next 25 years.

As with any forecast, power price forecasts are essentially a best look at the future and, to illustrate the variability in the projection, power curves come in three forms: high, central and low. The difference between each forecast widens over time due to the likelihood of variability increasing with time.

In sizing the debt amount, third-party lenders have differing views. Some will accept the central forecast, while others require the low forecast to be used. This in turn influences the forecast revenues and the ability to service the debt.

4.3 Power generation

To determine the likely output a solar installation will deliver, both third-party debt and equity will obtain a yield report from an independent technical adviser. This will consider the site's location, the orientation of the panels, as well as the equipment used.

A site's location is fundamental as it dictates the amount of solar energy received, known as irradiance. The further from the equator, the lower the irradiance and the higher the seasonal variability. This can be determined for a site from historical ground-based weather stations and satellite data.

The orientation and mounting angle of the panels is important in ensuring that the installation captures as much solar energy as possible. Panels typically face south in the northern hemisphere and conversely north in the southern hemisphere. The further from the equator the greater the mounting angle with UK sites typically opting for a mounting angle of 25 degrees, although the impact of the availability and cost of land results in angles ranging from 8 to 30 degrees. For ground-mounted sites this can be easily achieved, however rooftop sites have less flexibility and both orientation and mounting angles are dictated by the existing roof. A downside is that higher mounting angles generate larger shadows resulting in a larger project footprint, or where the site's area is limited a smaller installed capacity. This variability can have material implications, hence the requirement for a site-specific forecast.

Seasonal variability is calculated and produced for each month. As third-party debt is typically paid every six months this variable is important as it will help assess the drop in revenues over the winter months. To mitigate this volatility a reserve account can be used to retain a portion of the summer revenues to smooth out the seasonal impact.

The technical adviser will consider the factors outlined as well as other site-specific factors including shading losses, the performance of the equipment used, panel degradation over time and the expected electrical losses. This produces an energy yield output which is used in the funder's financial analysis. This yield is typically presented as a P50 and P90 output. A P50 is a base case assumption and shows the output that will be achieved 50% of the time, while the P90 is a lower value which should be achieved 90% of the time.

4.4 Debt service cover ratios

A key consideration for a funder is the ability of the solar installation to meet its future debt payment obligations. To manage this risk, funders stipulate a debt service cover ratio (DSCR) that must be met in the base case scenario. This ratio is calculated as follows:

$$\frac{\text{cash-flow available for debt service}}{\text{third-party debt interest and scheduled repayment}}$$

Cash flow available for debt service (CFADS), is the cash available once the solar installation's operating costs have been met. These costs typically include any lease payments, operating costs, planned maintenance, transfer to or from any reserve accounts and tax. The debt cost includes both the scheduled repayment and the interest due on the outstanding balance. It is a key metric for both equity and third-party debt as it essentially informs the financier of the cash available to them.

The ratio is typically based on annual cash flows and can be both forward looking (ie, the next 12-month forecast cash and debt payments) and historical (actual cash and debt service over the past 12 months). It is typically calculated on, or just prior to, a semi-annual debt repayment date.

If the amount of third-party debt is increased, the repayments required in each future period also increase and consequently the debt service cover ratio drops. As such, the amount of third-party debt lent to the solar company is usually sized so that the borrower just meets the required cover ratio at all future repayment dates. A bank's debt service cover ratio requirement varies from 1.20 to 1.30× for companies with a majority of their revenues based on tariffs or subsides and therefore fixed in nature, to 1.50 to 1.75× for companies where revenues are uncontracted and based on sales to the wholesale power market. Consequently, solar companies with high tariffs are able to borrow more third-party debt than those with little contractual support. It also follows that equity investors can raise more third-party debt by optimising project capacity and layout such that revenues are maximised, or by applying pressure on operating costs; all of which increase the numerator in the DSCR calculation.

The DSCR is calculated throughout the repayment period of the third-party debt to enable the funder to keep track of the solar company's performance. This is important as the funder has the ability to restrict equity distributions if this ratio ever falls below a set threshold, referred to as the 'lock-up covenant', typically 1.10 to 1.15×. If the ratio falls further and breaches a DSCR level typically set at 1.05×, the funder can call an event of default and take control of the solar company.

4.5 Sensitivities

The variables outlined above are all used to establish a base case scenario for the sizing of the third-party debt. However, a funder will also require sensitivity analysis to be undertaken to determine the ability of the solar company to repay its investment under a number of downside scenarios. These are run through the financial model to show the impact of each adjustment.

Sensitivities vary by funder and by geography, however, an illustration of typical sensitivities is provided below:

Table 1: Third-party debt sensitivities

Sensitivity	Description
Base case	The base case financial model. Used as the benchmark for all sensitivity results.
P90	The electrical output is reduced to a level where there is a 90% certainty of generation being over this value.
PR – 3%	The performance ratio (the ability to convert solar energy to electricity) is reduced by 3% to determine the impact of poor efficiency.
Low power price	The low power curve is used to calculate future electricity revenues.
Low inflation	This lowers revenue growth but not debt service requirements if the debt is nominal.
Opex + 10%	All operating costs increased by 10%.
Construction delay	Completion of the solar site delayed.
Lower availability	The solar installation is disconnected from the grid for 5% of the year thereby lowering revenues.
P90 and low power price	Combined sensitivity to reflect low production and low power price.
Power price break-even	The power price is reduced until the debt repayments are met in all periods (ie, the DSCR not falling below 1.0× at any time). This is a break-even analysis to demonstrate how low prices need to fall before a default occurs.

If any of the sensitivity requirements are breached, then typically the quantum of third-party debt is reduced and the analysis conducted again. This continues until such time as the funder's sensitivity requirements are satisfied.

5. Debt cost of finance

There are several costs associated with third-party debt that will be incurred by the solar company. These typically consist of the following.

5.1 **Upfront arrangement fee**

This is a one-off payment payable at financial close (typically the date on which the third-party debt is drawn) and can be referred to as an 'arrangement fee' or 'placement fee'. It is calculated at an agreed percentage of the total debt amount and is in essence a work fee for the funder during the due diligence, structuring and negotiation phase of a financing. Funders allow for this to be paid out of the borrowed amount.

5.2 **Ongoing administration fee**

This is ongoing payment to compensate the funder for its ongoing role in managing the loan. It can be referred to as an 'agency fee' and/or a 'security trustee fee' depending on the type of third-party debt involved. Its quantum primarily depends on the complexity of the transaction and the number of funders involved in it. The fee is usually semi-annual and increases annually with inflation.

5.3 **Interest rate**

This is the total cost of interest due on the outstanding debt which, for bank debt, typically consists of three components:

- Underlying reference rate – This refers to the financial institution's cost of funding in the interbank market, for example in the United Kingdom the London Interbank Offered Rate (LIBOR) is used. Since this underlying rate is variable and would expose the solar company to market volatility and potentially increased interest rates, funders typically require this to be addressed by way of an interest rate swap – a derivative that fixes the underlying interest rate – for the debt's repayment period; plus
- Interest rate swap margin – This is the fee charged by the financial institution that arranges the interest rate swap. In most cases the debt funder has the ability to provide this instrument so the payment is made to the same institution; plus
- Loan margin – This is the funder's actual remuneration on the loan. The rate may change over time, for instance to reflect higher perceived project risk during the construction period, or be fixed. This rate is agreed at the outset.

For bonds and private placements there is usually no need for an interest rate swap. Instead, the debt is priced over an agreed reference rate, typically an underlying government bond with an appropriate average life. This reference rate is then fixed at financial close and the margin or spread added to this.

5.4 **Cash reserve accounts**

To ensure a solar installation can meet its short-term debt repayment obligations a funder will require that the next six to 12 months' forecast debt payment is placed in a debt service reserve account. While not a direct cost, it does require a portion of the amount of debt raised being locked in a cash account until the loan is repaid. This means a smaller amount of debt is available to fund actual project costs.

Some banks allow a debt service reserve account to be replaced with a debt service standby facility (DSF). This is advantageous to the solar company as this cash is no longer held in reserve thereby allowing it to be used to fund costs. Instead, a DSF exists that can be drawn at any point over the life of the loan should the borrower be in short-term financial difficulty. While an arrangement fee is due on this facility, the solar company avoids the interest and ongoing administration fees that would have otherwise been payable. The solar company will pay a commitment fee on the amount of the facility, usually equivalent to 35–40% of the loan margin. This structure is not available where a bond is provided as bond providers are typically not able to offer such facilities.

Another cash reserve typically required by both banks and bond providers is a maintenance/inverter replacement reserve account. This is sized to ensure that the solar company has sufficient cash available in the event of a large number of invertors failing at the same time or generally for capital replacement items – lifecycle costs not factored into the O&M agreement. Although not normally pre-funded at financial close, it is typically funded over the first 5–10 years from surplus cash flow prior to dividend distributions.

5.5 Influencing factors

The debt costs outlined above vary according to a number of project-specific variables, many of which have been touched on already in this chapter, but also to factors common across all projects. Examples of these generic factors include:

- Geography – Costs of debt vary by country reflecting the prevailing government-funding rates and government default risk;
- Level of support tariff – The higher the tariff the more secure the loan and hence the lower the pricing;
- Tenor of loan – Shorter tenors typically attract lower margins, although appetite for long tenors varies widely across the lender range;
- Gearing – Higher gearing can be viewed as being of higher risk which will impact a funder's return requirements;
- Size of loan – The larger the loan the better the economies of scale;
- The market's interest rate expectations – This will impact the pricing of the interest rate swap and the reference rate used for a bond; and
- Availability of global finance – Margins increased dramatically during the global financial crisis from 2007 as bank liquidity shrank. Margins have never returned to pre-credit crunch rates.

In addition to the above, to determine if there are any project-specific variables that could impact on the costs of debt, third-party debt providers will always review the findings of their external due diligence advisers. These advisers will consider the project's technical, legal, financial and insurance risks and each will prepare a report for the lender to consider. These reports, along with the funder's own experience, will then generate the perceived risk associated with the third-party debt.

5.6 Recent trends

Over recent years the costs associated with third-party debt for solar installations have contracted significantly. This has primarily been down to three key factors:

- Familiarity – The solar sector has grown significantly in recent years and this has resulted in multiple-debt financings. This has enabled debt providers to become increasingly comfortable with solar as a distinct sector and this familiarity has enabled margins to be compressed as the perceived risk of default has reduced.
- Track record – The increase in solar projects has resulted in an ever-growing database of projects for funders and their advisers to review. The evidence gathered to date indicates a very low probability of default with the majority of projects performing at, or above, base case expectations.
- Global interest rates – In recent years the global markets have assumed low rates of economic growth and consequently have assumed that interest rates will remain low for some time to come. This has benefited solar companies borrowing third-party debt as it has resulted in them being able to fix their underlying interest rate at historically low levels. Even if interest rates now climb the solar companies are unaffected as they have a fixed cost for the full term of the debt.

An illustration of the reduction in the costs of debt for solar projects is provided below. This is based on three solar portfolios in the United Kingdom that borrowed nominally priced bank debt. While each project had its unique characteristics the downward trend is clearly noticeable.

Table 2: Pricing of third-party debt for UK solar

Project	UK solar portfolio 1	UK solar portfolio 2	UK solar portfolio 3
Date of financing	December 2011	January 2016	June 2017
Fixed swap rate	1.89%	1.66%	1.31%
Swap margin	0.25%	0.22%	0.15%
Loan margin	3.00%	1.75%	1.45%
Total interest rate payable	5.14%	3.63%	2.91%

6. Equity returns

Regardless of whether equity is invested alongside third-party debt, or as the sole financing source, its return on investment requirements are measured in the same way. These returns are based on all of the payment streams that equity investors are entitled to and are a combination of the subordinated debt interest, subordinated debt repayment, dividends and repayment of equity capital. This is typically referred to as a 'blended equity' return.

When equity makes its investment it will be based on a presumed rate of inflation. However, if the actual inflation rate is higher, so too are equity payments, while the opposite is true for low inflation. To account for this, equity returns are calculated with and without inflation, resulting in 'nominal' and 'real' equity return forecasts, respectively. The majority of investors refer to the nominal return.

When calculating their return on equity, equity investors will typically consider either, or both, of the following metrics:

- running yields; and
- internal rate of return (IRR).

A running yield is typically based on annual cash payments paid to the investor and shows the percentage return this delivers. It is calculated as follows and displays the yield equity is receiving on its investment:

$$\frac{\text{annual payment to equity}}{\text{total equity investment}}$$

Running yield is important for equity investors which seek to receive consistent returns in each and every period. Such investors include stock exchange listed funds which, in turn, may have offered their own investors a regular dividend.

While a running yield considers each year separately, the IRR looks at all payments to equity over the entire investment period. Consequently, it is used by the majority of equity investors as it allows for periodic differences in equity payments and also allows investments in different geographies or different tenors to be readily compared. It also produces a single value to define the total equity return, while a running yield generates a number for each year.

To increase their investment returns, equity investors typically look to gear their investment through the use of third-party debt. As noted earlier, third-party debt is usually cheaper than equity as it ranks senior in the cash waterfall which therefore lowers a solar installation's weighted average cost of capital. Consequently debt financing increases the IRR and the average running yield.

An investor's equity return hurdles are generally not revealed as this provides competitors with an insight, thereby weakening the investor's position on future auctions. However, general market experience for European solar suggests an IRR for solar companies without any third-party debt of 5%–7%. Where third-party debt is invested alongside equity, the required IRR increases to 7%–9%, reflecting the fact that equity is now subordinated to a third-party lender in the cash waterfall. As secondary and in some cases tertiary portfolio sales take place both running yields and IRRs continue to contract as asset consolidation and growing portfolio sizes introduce more cost efficiencies to ongoing asset management.

7. Financing subsidy-free solar

To date the vast majority of solar installations globally have received a support tariff from their national or regional governments. These tariffs were provided to encourage the construction of new solar sites as part of the global effort to decarbonise electricity production and were paid according to the amount of electricity generated. As these support tariffs are generally regarded as having a similar risk profile to government debt, they are very attractive to both equity and third-party debt providers. This has enabled high gearing and low margins to be incurred as the risk of non-repayment was deemed to be very low given the successful operational track record of solar technology.

However, mindful of the fact that the cost of constructing solar installations and panel production prices has fallen significantly, many governments have cut subsidies significantly and in some cases ended them entirely. As a result, from 2018 onwards the vast majority of new ground-mounted solar installations will not receive any subsidy support in countries with an established solar industry.

This change creates a new challenge for solar installations as, despite the decrease in installation costs, new solar sites still require financing to enable their construction. Consequently, equity and debt providers are now recalibrating their lending models to be able to evaluate what quantum of finance they can provide and at what cost.

Although a subsidy may not be available, solar installations can still sell the generated power and these revenues can still be used to secure financing. The key variable is the volatility risk behind this revenue stream. Solutions include selling power on the open market, an option that typically generates the highest revenues but with the highest volatility; versus selling to a corporate entity under a fixed-price contract, which generates lower revenues but with more stability.

Where a fixed-price export agreement occurs, debt and equity providers will assess the financial strength of the buyer to ensure it is robust enough to be bankable. If it is not, then this route adds little value to the financing as a default by the buyer would have significant knock-on implications for the solar installation. To mitigate this, a solar installation should ensure it has a back-up plan, typically in the form of access to the local grid network so that it can continue to sell the electricity to another buyer.

Where electricity is sold on the open market, the solar installation will benefit by retaining the most revenue as it captures the wholesale power price, as opposed to selling at a discount to a corporate entity. However, as noted earlier in this chapter, power price forecasts are inherently volatile and consequently less third-party debt will likely be available as funders will take a pessimistic forecast, which in turn impacts on the cover ratio and sensitivity results. To offset this, a number of the more creative intermediaries which purchase solar electricity are now offering long-term power price floors at meaningful levels. These floors set a price that will be paid for the electricity even if the wholesale price is lower, thereby providing the funder with a degree of reduced credit risk and volatility

For these reasons it is expected that, as solar moves from a sector enjoying subsidies, to one relying purely on the sale of generated electricity, the funding structures will evolve accordingly. For example, evidence in the UK solar market suggests that this evolution could see third-party debt levels decrease from historical levels of up to 85% gearing to levels of *circa* 50–60%, although it should be noted that such a change will vary by country. To compensate, more equity will be required to fund the balance.

It is worth noting, however, the risks borne by equity will actually not materially change with the loss of subsidies. Where equity is invested alongside third-party debt, the revenue associated with the subsidy is typically used for third-party debt service, while payments to equity can be made only to the extent additional revenues from electricity sales materialise. Therefore, in existing projects where both subsidies and revenues from the wholesale market are received, the equity investors' return forecasts already fully rely on wholesale power revenues being at the levels initially assumed. As such, a removal of the subsidy does not directly change the risk exposure borne by equity. As a consequence, the reduction in the availability of third-party debt does mean that the availability of global equity will become increasingly important for future solar installations.

M&A and corporate structuring

Dario Bertagna
Capital Dynamics
Anthony Riley
Yolanda Yong
Orrick, Herrington & Sutcliffe LLP

1. Introduction

M&A activity in the solar market has developed significantly in the last decade across the globe. During this period there has been a shift away from M&A in respect of construction-ready projects (ie, those projects which are fully consented but not yet physically constructed) to fully operational projects and an increasingly competitive secondary market. While in the United Kingdom, this trend has been driven in large part by a gradual decline in, and ultimately the withdrawal of subsidies for new capacity; this is a trend which has been seen across a number of other jurisdictions.

A range of factors affect the structuring of a solar M&A transaction. This chapter outlines the key points for buyers and sellers to consider when embarking on an M&A transaction of a solar asset and explores, in particular, key aspects of an M&A transaction in the United Kingdom, France and the United States in order to highlight some of the differences and similarities between civil and common law jurisdictions.

2. M&A transactions and corporate structuring

The parties' interests and commercial objectives will be key considerations when structuring and negotiating any M&A transaction and this is no different for the acquisition of a solar asset. This chapter (see section 2.1) examines some of the key considerations for different types of investor in this market and how investors value solar assets in order to decide whether to engage in any M&A activity.

The steps commonly undertaken as part of the sale or acquisition process and those covered in this chapter are:

- putting in place preliminary agreements such as heads of terms or letters of intent (see section 2.3);
- carrying out due diligence and determining which areas to cover (see section 2.4);
- agreeing the corporate structure of the acquisition (see section 2.5);
- financing the acquisition (see section 2.6);
- preparing the relevant acquisition agreement (see section 2.7); and
- disclosure (see section 2.8).

2.1 Commercial considerations

This section will explore different types of investor, how investors value a solar asset or portfolio in order to secure the price on which a buyer and seller are willing to enter into commercial negotiations in connection with a potential M&A transaction, and the key risks an investor will take into consideration when valuing the asset or portfolio.

(a) Investment strategy

An investor's investment strategy and appetite for risk are some of the key elements considered when bidding for solar projects. In broad terms, two main strategies and therefore two main investors in solar assets can be identified: short-term owners focusing on generating a capital gain by adding value and re-selling the asset for a profit; and long-term owners focusing on income generation and dividend distribution.

Short-term owners of solar assets: Short-term owners typically consist of project developers, high net-worth individuals, private equity and venture capital funds, construction and engineering companies, and components manufacturers looking to extract profits by selling equipment (eg, module manufacturers financing the construction of a project and then selling it when operational).

This investor class normally enters early in the solar value chain. They either develop solar projects from the beginning or they acquire semi-developed sites with the objective of building and selling them once operational. In contrast to acquiring an operating solar farm with a verifiable performance track record, the nature of these transactions is riskier and therefore it demands higher risk-adjusted returns. This, in turn, is reflected in a premium to the cost of capital of this investor class.

The focus of the short-term holders is on creating a capital gain by de-risking a project and then selling it for a profit. As such, when bidding for development or construction-ready solar projects the process involves assessing and managing short-term risks.

Key considerations include, but are not limited to, the following:

- assessing the planning approval or permitting process;
- securing a grid connection;
- measuring and validating the projected yield of solar resources;
- procuring construction contracts (including panel supply and inverter arrangements);
- managing the construction timetable;
- obtaining a PPA or other routes to market the power;
- ensuring that a project is eligible for subsidies (if any), including, in the United States, the investment tax credit (or ITC); and
- procuring or arranging operations and maintenance (O&M) and other operating contracts.

Long-term owners: Long-term owners of operating assets are more risk averse and, accordingly, their return expectation is typically lower. This category includes infrastructure funds, pension funds, insurance companies, strategic investors (such as large independent power producers) and listed investment funds (known as 'yieldcos'). The key driver for this investor class is ensuring that a project delivers long, stable and predictable cash flows with a minimum and controlled level of volatility.

When bidding for operational solar projects, yield-hungry investors scrutinise the robustness of the income generated by the asset and adjust return expectations by factoring in uncontrollable risks. It is typical to see an inverse correlation between the percentage of fixed revenues and the return premium (the higher the portion of fixed revenues, the lower the return of a solar investment). So in the United Kingdom, for example, a project with the renewable obligation certificate (ROC) would be cheaper to acquire than a comparable feed in tariff (FiT) project due to the higher portion of fixed revenues of a FiT project.

Often bids are conditional upon independent parties' validation of key project assumptions, such as energy yield forecasts, historical production validation, grid availability and calculation of actual panel degradation. In the United States, for projects with construction debt, developers and their lenders often insist that energy yield forecasts be delivered at the stage of investor commitment and not at closing so as to present fewer obstacles for the construction loan 'take-out' at commercial operation of the project. Other key considerations include assessing the creditworthiness of project counterparties, the O&M contractor's track record, quality of key components installed and the duration of equipment warranties.

Long-term owners have historically acquired operating assets. However, in recent years due to the maturing solar market and the increased global demand for yield and the scarcity of operating assets, it is not uncommon to see long-term investors acquire construction-ready sites in order to secure a pipeline of projects. In the United States, a recent trend is that some institutional investors are investing in even earlier stages of project development given the yield-hungry nature of this investment class and increased competition for shovel-ready projects. Bids for construction-ready projects, where a buyer effectively provides a construction finance facility, tend to include conditions around the EPC contractor's performance warranties, delay penalties and milestone payment mechanisms.

(b) How to value a solar portfolio/asset

A solar asset, or a portfolio of solar assets, presents a number of unique characteristics that need to be considered when performing a valuation. Revenue support schemes (such as ROCs and FiTs in the United Kingdom and tax credits in the United States), terms of the operating contracts, exposure to power market volatility and asset or portfolio gearing are all relevant in this context.

Regardless of the investor strategy, the most common tool for valuing a solar asset is using a financial model. A financial model is used to calculate the free cash flows that a project will generate. These cash flows are then discounted by a rate to derive the Net Present Value (NPV) of the future cash streams. The NPV represents the current price that an investor is prepared to pay to acquire cash flows in the relevant asset. Once the NPV is calculated, it is common practice to benchmark this value with financial metrics of other similar transactions in the market (commonly known as 'transaction comps').

To determine the NPV, most buyers value assets using their required equity internal rate of return (IRR) and yield-oriented investors tend to compare it with certain cash yield thresholds. As a general rule of thumb, larger portfolios are sold for a premium to smaller assets (hence delivering a lower IRR) due to the diversification of risk across a larger number of assets. For example, some risks such as solar resource volatility and grid outage events, tend to decrease if a portfolio is diversified. Larger portfolios also attract some, albeit limited, economies of scale in relation to ongoing operating costs and transaction costs.

In some cases, investors assume a terminal value or an exit price at the end of the life of an asset. In fact, with the market evolving and returns reducing, there is a trend towards trying to extract additional value by extending operational projects' lives and attributing additional value to the merchant tail of projects which attract government subsidies.

In the United Kingdom, the assumption on the useful life employed in solar valuations has shifted from 20–25 years (which correlated to the lifetime of the relevant subsidy the project benefited from) to 30–35 years and in some cases even longer due to the feasibility of having a plant generating income beyond the duration of a subsidy.

Finally, with solar transactions becoming more common and valuations being publicly disclosed in annual reports or in industry databases, it is also a widely used valuation practice to compare financial metrics of other assets bought in the market. One of the most common financial metrics to value an asset is the ratio of enterprise value (EV) and EBITDA, although, for solar transactions specific metrics considering project capacity or electricity production are also used. Among them, the EV/MWp ratio is one of the most commonly used in the industry to compare solar transactions in the same market.

The next section will cover the details of financial modelling and how these are used to calculate the purchase price of an asset.

(c) Modelling and assumptions

Any financial model must have a set of inputs or assumptions, a series of calculations and an output.

Inputs (assumptions) ⟶ *Calculations* ⟶ *Outputs*

The output or objective of a financial model is to forecast the cash flows that a solar project will generate. Financial models are used for three main purposes: to calculate the purchase price that a buyer is willing to pay to secure the cash flows generated by a project; to calculate the income and yield that a project will generate in relation to its purchase price; and to calculate the actual performance of an asset post-acquisition. In this section we focus on the first two of these purposes which are used in a solar M&A transaction.

In simple terms, financial modelling of a solar asset or portfolio requires forecasting revenues and costs for the full operational life of a project. The assumptions used to forecast revenues of a solar project are split between generation and sales.

During the bidding phase, the seller provides to the buyer the project-specific assumptions used to build the financial model. Such assumptions are normally explicitly listed in an offer letter to allow the buyer and its advisers to verify them during the due diligence phase.

Generation assumptions include:

- solar plant capacity: the DC capacity of the solar plant rated in megawatt peak (MWp);
- irradiance: the number of hours of sunlight that a project receives during a year (MWh/m^2);
- performance ratio (PR): ratio of the actual and theoretical possible output of a plant (expressed as a percentage);
- plant availability: amount of time that a plant is able to produce electricity over a certain period of time (expressed as a percentage); and
- modules annual degradation (Degradation): annual loss of performance of the PV modules caused by UV exposure and weather cycles (expressed as a percentage per annum).

Generation is normally expressed in MWh and it is calculated based on the following formula:

$$Generation = MWp \times \frac{MWh}{m^2} \times PR \times Plant\ availability \times (1 - degradation)^n$$

Assuming a 5 MWp solar plant with 1,150 MWh/m^2 Irradiance, 83% PR, 99% Availability and 0.5% Degradation, the first year (n) production will be calculated as follows:

$$Generation = 5 \times 1{,}150 \times 83\% \times 99\% \times (1 - 0.5\%)^1 = 4{,}701\ MWh$$

Sales assumptions include:

- power prices;
- income derived by subsidy schemes (if any); and
- other revenues (eg, embedded benefits).

Power markets are volatile and investors tend to use third parties to forecast long-term power prices. In the United Kingdom, the main providers of power forecasts commonly used by investors and lenders are Pöyry, Baringa and Aurora. These companies provide quarterly or semi-annual updates on long-term power prices in most of the European markets.

Finally, depending on the jurisdiction, solar projects might receive additional sources of income from subsidies which are added to the revenues received from the sale of electricity.

After determining generation and revenues, the next phase of modelling consists of calculating the project operating costs.

The main annual cost items of a solar farm are as follows:

- O&M;
- land lease;
- insurance;
- community benefits/royalty;
- decommissioning/obligations;
- metering/electricity consumptions;
- transmission and congestion (if applicable);
- accounting, audit and asset management;
- business rates and local taxes;
- financial costs (if any); and
- corporate taxes (including assumptions on capital structure and shareholders' funding).

Costs are modelled considering actual contractual terms or specific views on markets (eg, the long-term view on subsidies, corporate taxes, financing costs or operating costs). Models also include macro-economic assumptions, such as a long-term view on inflation, exchange rates and interest rates.

Once revenues and costs are determined they are then projected in the long term, adjusted for inflation and used to calculate the cash flows that a project will generate. Such cash flows are discounted by a buyer's IRR and used to calculate the purchase price of a project or NPV, net of any external transaction costs.

After the NPV or purchase price is determined, buyers tend to run sensitivity scenarios (in other words, changing the parameters of one or more assumptions) to confirm the robustness and stability of the cash flows. Some of the sensitivities include:

- per cent increase/decrease in generation;
- per cent increase/decrease in power prices;
- per cent increase/decrease in annual costs;
- change in financial leverage and/or change in debt terms; and
- change in market and macro-economic assumptions.

Understanding sensitivities is important because it shows how solar project returns are linked to changes in assumptions. The more sensitive a project return is to certain assumptions, the more a buyer will try to reduce its volatility by seeking stronger assumption validations, additional protections from the seller or by implementing other risk-reducing methods (eg, by seeking insurance to protect against the relevant risk).

For yield-oriented investors, the second main objective of a financial model is to calculate income and cash yield. Income yield is the ratio of income returned to investors after depreciation, divided by the invested amount. Cash yield is the gross cash divided by the invested amount. Both of these metrics are used by investors to measure the quantity of cash that is generated on an annual basis but, unlike the NPV, these metrics do not take into account the time value of money.

2.2 Key considerations in terms of risks when acquiring solar assets – lessons from experience

Investing in solar assets has proven to be lucrative for a number of investors, however, it is not a risk-free proposition. Risks of acquiring solar assets can be categorised as development risks, construction risks, operating risks and market risks.

Development risks are those related to the obtaining of the consents and permits which are necessary to enable the project to be constructed. Only certain types of investors (usually the short-term owners as set out in section 2.1(a)) have the appetite to acquire development-stage projects and these investors usually have in-house capabilities to develop the project to ensure that the project can then be sold on 'construction-ready' where the project can be sold on at a premium to a wider range of investors.

Construction risks are those related to the construction and commissioning of a solar project. Investors get comfortable with these risks by allocating them to third parties, typically creditworthy EPC contractors, who normally offer fixed-price contracts with penalties payable upon failing to connect a plant by a certain date.

Operating risks are those related to generating electricity and cash from running a solar facility. In particular, some of these risks include forecasting the power output, managing equipment failures and managing the operating costs.

Power output: In terms of power output there are two main uncertainties that buyers consider: solar resource uncertainty; and equipment performance. Solar resources are determined by extensive analysis of weather recordings. A number of databases are available which use actual ground measurements or satellite data. However, given that it is impossible to predict accurately the amount of solar irradiance that a certain plant will receive, some statistical production estimates are introduced to predict solar resource. These are commonly known as 'P' which stands for probability. It is typical for buyers to consider in their investment case the P50 data which means that there is a 50% chance that in a given year solar irradiance will be at least a specified amount. Lenders tend to run their cases on more conservative scenarios such as P75 or P90. This is the level of solar irradiance that is likely to be exceeded 75% or 90% of the year. It is worth noting that generation of energy from solar is more predictable than wind and this is one of the reasons why more conservative investors are comfortable to invest in solar assets as opposed to wind assets. As an example, the difference between P50 and P90 for solar is about 6% and 10% respectively, while it is between 9% and 13% for wind.

Equipment and construction: Equipment and construction works on solar plants are normally backed by the manufacturer or contractor warranties which helps manage operating risks. However, over the last few years some large players in the market, in particular inverter and module components, went bankrupt making the enforcement of warranties difficult or even impossible. Assessing the financial robustness of the project counterparties is one of the key considerations of prudent buyers.

Operating costs: Managing operating costs is another key aspect of understanding the risk of investing in solar plants. Entering into long-term O&M contracts with expensive break clauses can be a prudent approach in terms of securing the long-term operability of a plant. However, this approach can limit the upside potential in a market where costs have significantly decreased over the last few years.

External risks: Finally, market risks are the risks associated with external factors such as regulation, power markets and grid availability.

In most countries, solar revenues are subsidised and, as a result, understanding the robustness of the subsidy is a key consideration. In some European countries, including Italy and Spain, incentives granted to solar plants have been retroactively reduced. In the United Kingdom, incentives such as LECs and FiTs were withdrawn and ROCs were phased out for certain projects at very short notice which affected investor confidence in the market. This meant that buyers expecting to receive certain revenues had the profitability of their investments reduced.

In the United States, the primary method by which solar assets are subsidised is the Investment Tax Credit (ITC)—apart from some limited 'community solar' incentive programmes in the mid-western United States which act similarly to a FiT scheme for smaller solar projects. The ITC is available for solar technologies and provides a one-time US federal tax credit of 30% of the project's qualifying capital costs. Accelerated depreciation benefits are also available to solar projects. As sponsors generally do not have sufficient taxable income to use these benefits, the financing of these projects depends on tax equity investors providing 'tax equity' in return for the use of the project's tax benefits and some of its cash flows. These tax equity investors and their investments are distinct parts of the 'capital stack' which equity investors will need to evaluate when considering an investment in a solar project.

Another indirect subsidy in the United States are state-driven renewable portfolio standards (RPS). Over the last decade, states' RPS have driven significant growth in the solar sector, by requiring regulated utilities to supply a certain percentage of their energy from renewable resources. Although RPS programmes vary widely by state, renewable energy is typically tracked by renewable energy certificates (RECs) which typically represent one MW of energy. Depending on the nature of the specific RPS programme, RECs can either be bundled with the energy that is produced by the generator or can be sold separately. REC markets and the value of solar RECs vary quite widely. In some markets, including New England, RECs can be quite valuable, whereas others, such as in the Texas ERCOT market, RECs are not nearly as valuable. See the United States: California chapter for a good description of California's RPS.

Power markets and grid availability risks are somewhat related. As mentioned previously, buyers tend to use third-party consultants to forecast power prices. However, with more renewables being installed across the globe and an overall decrease in underlying oil and gas prices, electricity prices have generally decreased or, more accurately, increased less than forecast. The link between renewables and power price is significant and interlinked. A whole chapter could be written about this topic, however in simple terms, the more renewables that are installed, the cheaper the marginal price setting the wholesale electricity price will be. This is because the cost of generating a new MWh of electricity from renewables is virtually zero (solar irradiance is free) and it is primarily linked to the generator cost of capital. As generation from renewables and in particular wind and solar is intermittent, the grid suffers instability on sunny and windy days where the production and the demand of electricity are unbalanced. This might lead to grid outages, which in turn can reduce the profitability of a solar farm connected to a congested grid which has created further risks when acquiring solar assets.

2.3 Setting out the parties' intentions

It is common for the parties to a transaction to enter into and sign heads of terms, a letter of intent or a memorandum of understanding (referred to as 'preliminary agreements') which set out the key commercial terms of the sale and acquisition and a period within which the parties agree to exclusively negotiate with one another. The commercial terms set out in such preliminary agreements are generally not legally binding, save for certain provisions, but substantial deviation from the terms without good reason is unlikely to be well received during the course of negotiations. Terms such as the exclusivity period, governing law and jurisdiction, and the terms of confidentiality are usually legally binding and enforceable provisions in preliminary agreements and should provide enforceable rights and remedies if breached. Some parties choose to enter into separate confidentiality or non-disclosure agreements in advance of signing the letter of intent or memorandum of understanding in order to share data for the buyer to be able to enter into the preliminary agreements and agree to commercial terms such as price.

The parties' entry into extensive preliminary agreements in civil code jurisdictions such as France is less common due to the parties' general duty of good faith (which does not exist in common law jurisdictions such as the United Kingdom and United States unless that duty is expressly agreed to) which may give rise to pre-contractual and legally enforceable relations.

Where preliminary agreements are entered into, the signing of such documents by the buyer and seller may commonly signal the start of an exclusivity period. During the exclusivity period, the seller agrees to give the buyer exclusive rights to the transaction, to allow it to conduct its due diligence. During the exclusivity period, the seller agrees not to negotiate with, or to sell the assets to, another buyer. The term of the exclusivity period typically depends on the size of the portfolio being sold. While the main terms of preliminary agreements are not usually legally binding, the commercial terms and assumptions of the characteristics of the project(s) set out therein are usually referred to repeatedly throughout the negotiation stage by the parties and are used to fix assumptions used in the financial model as referred to in section 2.1 (c).

Auction sales are common in jurisdictions which have an active secondary market. Over the last two years in the United States and the United Kingdom, auctions for portfolios of both operational and construction-ready assets have been increasingly common as the markets mature. In auction sales, the seller sets the commercial terms against which prospective buyers submit their bids with the intention of limiting the scope of negotiation of the M&A deal between the buyer and the seller. The requirements of the bid vary from sale to sale but they will usually require the bidder to confirm price, assumptions, agree to valuation and also usually prepare mark-ups of key transaction documents such as shareholders' agreements (if relevant) and sale and purchase agreements. In the United States and the United Kingdom, these auctions have been very competitive and generally reflect a 'seller's market'. Prospective buyers of solar assets have had to make fairly aggressive bids, both in terms of price and on terms and conditions of the sale and purchase agreement. Once the seller selects the preferred bidder, the buyer will aim to obtain an exclusivity period from the seller for a period of one to two months usually with a period for renewal after the bidder carries out its due diligence exercise. This period can be longer depending on the size of the portfolio.

2.4 Due diligence

Due diligence is the information-gathering process carried out by the buyer to find out as much as possible about the target company and its assets, referred to herein as the project company (or ProjectCo). Due diligence is used to assess the risk of purchasing the ProjectCo and to guide the buyer in negotiating the terms of the transaction including, where relevant, the price payable. The scope and type of due diligence varies between transactions but a full due diligence exercise tends to include legal, financial and technical due diligence (and other areas such as environmental matters, if relevant) on the ProjectCo and its business with specialist advisers inputting into the process including insurance advisers, lawyers, technical advisers and financial advisers.

(a) Type of due diligence

At the outset, the buyer should ascertain whether it will need to conduct full due diligence itself or whether the seller is offering any form of vendor due diligence report produced by its own legal advisers and/or technical advisers. However, the buyer would need to consider the appropriateness and scope of the report for the purpose of relying on it and the cost of doing so.

Where there is debt already in place, a seller may seek to procure that the buyer can rely on existing lender due diligence reports in order to expedite the transaction and to save on costs. In the United States, in addition to reports related to the projected energy yield and certain engineering aspects of the project, investors may also seek to obtain an environmental report and related studies for the project, which may include wetlands delineations and reports, environmental site assessments, protected species and habitat reports and cultural resource studies.

The cost of being able to rely on any existing reports and the cost of updating them should be agreed up front as part of the commercial terms of the transaction.

The scope of due diligence may also have an impact on disclosure and the level of disclosure accepted by the parties (see section 2.8 for further information on disclosure). If the buyer is conducting full due diligence itself, in the United Kingdom and in civil law jurisdictions such as France, it would be common for the buyer to accept general disclosure of a well-organised and clearly labelled dataroom in which the documents it has reviewed have been hosted. In the United States, in the context of a private acquisition of projects between sophisticated parties (ie, for transactions that are not subject to registration under the relevant securities laws), detailed disclosure is not typically prepared by the seller, but rather the buyer must review the relevant diligence materials provided by seller in the dataroom and then seek to cover perceived 'gaps' in the relevant purchase agreement with representations and warranties. Depending on the negotiation leverage of the buyer, the buyer may also seek to have the seller provide a representation related to the quality of the diligence provided (eg, that the documents delivered by the seller to the buyer do not contain any untrue statement of material fact, or omit to state a material fact necessary to avoid making the statements in the agreement misleading). Given the sellers' market in the United States, this representation is included less frequently in recent transactions.

If the buyer is relying on a form of vendor due diligence, the buyer would be more likely to reject general disclosure of the dataroom or agree a limited scope of disclosure restricted to certain areas or matters which have been reviewed by the buyer.

In the United Kingdom, where Certificates of Title relating to real estate and planning assets owned by the relevant ProjectCo or one of its subsidiaries exist, which is typically the case where there has been a recent third-party financing, these can be updated and/or recycled for the benefit of a buyer in connection with an acquisition. If the seller is providing a Certificate of Title, no disclosure of any property and planning documents is usually accepted as the Certificate of Title is the buyer's reliance on the adequacy of the project's title to land and planning matters. Even where the buyer is relying on vendor due diligence reports and/or a Certificate of Title, the buyer's legal advisers will be involved in the negotiation of the scope and also the updates made to such documents.

In the United States, real property title diligence in the form of a title report from a title company is often obtained which will be supplemented by buyer legal counsel's review and subsequent title curative efforts by the buyer and its counsel. Title insurance is often procured by the buyer at the closing of the transaction.

The scope of due diligence will have an impact on both disclosure and also the terms of the warranties in the agreed acquisition agreement.

(b) Construction rights/operational projects

One of the main factors which informs the due diligence of a solar asset is the stage of development of the project.

In a construction-stage or shovel-ready project as they are often termed, the buyer's due diligence would typically be focused on the following matters: consents, grid rights, land rights and qualification for subsidies. For operational solar projects, the buyer's due diligence will additionally be focused on the offtake and operation and maintenance of the solar asset.

Consents/planning: There will be varying consents required depending on the jurisdiction of the solar asset, therefore local lawyers and advisers should be consulted prior to an acquisition to ensure that all necessary consents have been obtained by the seller for the construction of the project if that is the commercial agreement.

In the United Kingdom, the main consent required at the construction or operational stage will be planning permission for the development, construction and operation of a solar farm. A buyer will look to ensure that the term of planning permission is no shorter than the term of the lease and for a minimum period of 25 years to cover both the time period to construct and also decommission the plant. A buyer of a shovel-ready project will want to ensure that sufficient time has passed since the grant of planning permission by the local planning authority to ensure that the statutory challenge period (known as the judicial review period) has lapsed. A buyer would also look to ensure that all pre-commencement planning conditions (if any) contained in the planning permission have been discharged. The buyer will check that there are no developments planned in the surrounding area which may have an impact on the future operation of the project (eg, developments which may cause shading of the modules). All projects which have a generation capacity of more than 50 MW are considered to be nationally significant infrastructure projects and are governed under the Planning Act 2008 and are subject to a separate application process governed by the Planning Inspectorate. At the operational stage, additionally a buyer will seek to ensure the project has remained in compliance with the necessary consents. For example, in the United Kingdom, a buyer will look to see that there has been compliance by the ProjectCo with each planning condition and that any pre-commencement planning conditions have been discharged. The buyer will additionally look to ensure that there have been no breaches to the planning permission or risk of enforcement action from the Local Planning Authority. The buyer will also seek to ensure that the 'as-built' plant complies with the requirements of the planning permission and where there is deviation will require formal amendment to the planning permission granted by the relevant local planning authority.

In France, two types of consents will be required: operational consents and real estate authorisations. In respect of the real estate authorisations, the requirements will vary depending on the type of structures (ground-based or rooftop-mounted), and the plant height and capacity. Environmental concerns are also taken into consideration prior to granting the relevant permits, depending on the size and location of the project. In addition, depending on its power capacity, a photovoltaic installation may be subject to an authorisation to operate. Since Decree No.2016-687 dated 27 May 2016, only photovoltaic installations with an installed capacity of more than 50 MW require the Department of Energy to issue an authorisation to operate. Smaller installations are deemed authorised. During the due diligence process, the buyer will check whether the relevant permits and authorisations have been obtained (depending on the type and particular environment of the project) and ensure that they are not currently the subject of any dispute.

In the United States, project buyers will often seek estoppels from key project document counterparties, including the project offtaker (or PPA counterparty), the construction contractors (and, if applicable, the panel suppliers), and relevant landowners. These estoppels will all seek to have the counterparty confirm that there are no defaults or material breaches related to the project document; no fundamental dispute is outstanding; for both construction contractors and the offtaker estoppel, that commercial operation has occurred; and that the party is not aware of receipt of any material notices from government agencies or third parties related to the project.

In addition to these documents from project document counterparties, various permits and consents from governmental authorities may be required depending on the size and location of the project. Such permits may be local or state-specific, such as permits related to conditional use of the project or use of water. In addition, there are a significant number of state or local permits that are driven by federal environmental policy, as well as federal-specific permits and consents (including those related to energy regulatory and reliability requirements). For a couple of examples of permitting regimes governing solar projects, see the chapters on Californian and Texan solar projects.

In addition to these documents from project document counterparties, various consents from governmental authorities may be required depending on the size and location of the project. If the project buyer is not US-based, another significant consideration for inbound investors is whether to make a voluntary filing with the interagency Committee on Foreign Investment in the United States (CFIUS) for review and approval of a contemplated acquisition. The CFIUS is authorised to review transactions where a foreign entity acquires control of a US company. Depending on the nature of the transaction, energy regulatory filings on a state or federal level (FERC – Federal Energy Regulatory Commission) will likely be necessary to complete the transaction.

Finally, if the transaction is of a certain size, compliance with relevant competition laws may also be a consideration and prior approval or filings may be required from the relevant authority.

Grid rights: Fundamental to the development of a solar project are the rights to connect into the grid and export electricity to the grid. The means to do so may vary from country to country depending on the electricity grid infrastructure.

In the United Kingdom, there is a national electricity system called the National Grid. Projects will need to have a right to export electricity and to connect into the National Grid which at the development or construction-ready stage will usually be evidenced by either a grid connection offer from the relevant Distribution Network Operator (DNO) at the construction stage, or a grid connection agreement at the operational stage. However, sometimes the solar project will not be near to a point of connection to the National Grid and further grid infrastructure will need to be installed by either the DNO itself or an accredited contractor called an Independent Connections Provider (ICP contractor) and at the operational stage, parties will want to ensure that these works have been adopted by the DNO. Alternatively, there could be rights provided between private counterparties to export and purchase electricity.

In France, as with any power generation facility, solar plants have to be connected to the public grid to sell power. The export capacity of the solar asset determines who the competent authority is to grant requisite grid connection rights. For solar assets with an export capacity of 12 MW or less, the competent authority to grant a right of connection to the grid is Enedis (the entity which manages the French public electricity distribution network). If the export capacity of the solar asset is more than 12 MW the French transmission system operator (Réseau de Transport d'Electricité (RTE)) must grant the grid connection rights. At the operational stage, a buyer would want to be sure that the competent authority has issued certificates to confirm that the plant has been built by the ProjectCo in compliance with grid requirements issued by either Enedis or RTE (depending on the export capacity of the plant); that the certificates were issued by CONSUEL (Comité National pour la Sécurité des Usagers de l'Électricité) certifying that the project complies with applicable regulatory requirements of the grid; and that for projects with an installed capacity of more than 100 kWp, there is a certificate from an independent state institution to confirm that the installations conform to the information provided by the ProjectCo in its request for a PPA or premium offtake contract. (See the France chapter for a more detailed discussion of grid access issues in France.)

The United States comprises a mixture of different types of markets. Due diligence will involve an evaluation of the relevant interconnection agreement with the system operator and/or transmission provider and the project's compliance with that agreement and the interconnection protocols relevant for the market that the project is operating in. In organised wholesale markets, such as MISO, CAISO and ERCOT, independent power producers are able to enter into the interconnection queue to sell power on a wholesale market or to utilities on a wholesale basis. With very few exceptions, such as ESPs in the CAISO market (see the "United States: California" chapter) or retail electric providers in ERCOT (see the "United States: Texas" chapter), projects in organised markets are seeking to sell wholesale to utilities or in the organised markets themselves. In sophisticated power markets, such as ERCOT, a variety of instruments other than ordinary power purchase agreements, such as hedging arrangements, insurance products and corporate PPAs, have arisen in recent years (see the "United States: Texas" chapter and the chapter on power offtaking arrangements for an in-depth review of these products).

The interconnection process may take a significant amount of time depending on the project's queue position and the nature of the market. Generally, interconnecting a project is an iterative process, requiring filing of an initial application and then participating in various grid and supply/demand studies. Most connecting utilities will also determine whether system upgrades are needed, which can be expensive depending on the market and the amount of congestion on the system. Assuming these studies and upgrade determinations are acceptable to the parties, a standard-form interconnection agreement is typically used to interconnect the solar facility to the relevant grid. As mentioned, most solar projects in organised markets sell to the wholesale market, but for solar projects that sell to retail customers, transmission services may also need to be obtained.

In regulated, vertically integrated markets, there are very few opportunities for power producers to sell power on a wholesale basis other than by developers of renewable energy projects invoking their rights under the Public Utility Regulatory Policy Act of 1978 (PURPA) to interconnect their projects with transmission systems owned by utilities in these markets, and sell power to electric utilities in those states. Qualifying cogeneration and small power production facilities may require utilities to purchase power offered by 'qualifying small power production facilities' and 'qualifying cogeneration facilities' (collectively, QFs) at avoided cost. Under the FERC regulations, small power QFs, including solar QFs, that do not exceed 30 MW, are exempt from:

- most parts of the Federal Power Act (FPA), except Sections 205 and 206 (relating to rates of wholesale sales of power);
- state regulation of electric utilities; and
- the recordkeeping and reporting requirements under the Public Utility Holding Company Act of 2005.

Land rights: A buyer will need to ensure it has the requisite rights to the land to which the planning consent and also the grid connection rights relate. For a development or construction-ready phase project in the United Kingdom and in the United States, the relevant land rights can take the form of either an option for lease (which is to be exercised by the seller or buyer before, on or after completion of the acquisition) or a lease. At the operational stage, the land rights required will be a lease. Solar projects tend not to involve the acquisition of freehold land. Therefore, consideration should be given as to the adequacy of the lease (or option agreement if applicable), that the terms set out therein are bankable and do not contain any onerous conditions or commercial terms. Additionally to an option for lease and/or lease for the development of a project, a buyer will also need to ensure that the project has the necessary rights to lay the cables for connection of the project to the electricity grid. See the "Real estate" chapter and the UK and France chapters for a more detailed discussion on land rights.

Qualification for subsidies: A buyer will want to be sure that the project meets the necessary criteria to qualify for any available subsidy in the generation of electricity from renewable sources.

In the United Kingdom, due diligence will be carried out to ascertain whether a project qualifies for a renewable subsidy and, where a project is operational, to ensure that the project has the benefit of such subsidy. From a due diligence perspective, it is important to check that the project qualifies for the relevant subsidy and, if necessary, meets the criteria for a grace period for that subsidy. A buyer will usually want to see evidence of accreditation for the relevant subsidy and to ensure that it has the documentation to support accreditation for future reference in case OFGEM decides to audit the project, which is not uncommon for grace period ROC projects.

In France, no subsidies apply. State promotion of the renewable energy sector in France is implemented through guaranteed purchase of the energy produced. For a significant proportion of new capacity (especially for the majority of installations with a capacity of more than 1 MW from 2017), development of renewable energy facilities will be subject to a call for tender process securing the energy purchase price for a certain period of time. Due diligence of solar assets in France covers verification of the power purchase mechanism (either by way of power purchase agreements or pursuant to the tender process) and checking that the developer has adhered to the relevant process for obtaining and securing the purchase price. See the France chapter for a fuller description of the French support regime.

In the United States, as mentioned above, apart from some state-specific subsidy programmes, the primary subsidy programme is the federal investment tax credit (ITC). In order for projects to qualify for the full amount of the ITC (30% of eligible basis), such projects must demonstrate that they have started construction by the end of 2019 and commission the project before the end of 2023. For projects which commence construction after the end of 2019, the tax credit steps down to 26% and 22% respectively.

One thing a cash-equity buyer of solar projects must consider is how to structure its investment around the tax equity investor. For the early investor in solar projects, such an investor should have a view as to how tax equity can be committed for the project. For investors in shovel-ready projects, timing and structuring considerations related to the tax equity's commitment and ultimate financing at commercial operation should be considered. Finally, for investors in operating projects, coordination and evaluation of existing tax equity financing will be a critical component of the relevant diligence and documentation process. Tax equity investments in the United States are quite complex, and the ability of a cash-equity buyer to execute successful investments in this area requires a strong understanding of the specialised legal and accounting structures relevant to these structures.

Co-investing in projects employing tax equity can also pose some issues for the typical institutional investor. Most tax equity investments are structured as 'partnership flips' and provide a preferred return (in the form of the ITC, depreciation and project cash flow) until a predetermined internal rate of return is reached by the tax equity investor. Then, the structure 'flips' and a preferred return (typically 95% of cash distributions and related allocations) is redirected to the sponsor – which may then have the option to buy-out the tax equity investor at fair market value (of the 5% interest). This structure has the effect of deferring cash investment to the sponsor entity. Also, transfer limitations imposed by tax laws and typical partnership agreements also make these investments less liquid than many institutional investors may prefer.

Project documents: Project contracts can be put in place at the time of acquisition of the projects or thereafter by the buyer and developers often sell projects with either or both an engineering, procurement and construction contract (an EPC Contract) in place with the developer or an affiliate of the developer and/or an operations and maintenance contract (an O&M Contract) in place as an additional revenue stream for the developer/seller. If the buyer is acquiring a project with an EPC Contract in place, it will want to ensure the project will be or has been built in accordance with market industry standards, that the testing regime is robust and to understand what contractual protections there are against the contractor should there be an issue with the design, performance or operation of the plant. For solar projects, EPC contracts are usually turnkey contracts for the design, procurement, construction, testing and remedying of defects of the plant. At the point that the plant has been built and tested, the ProjectCo will issue a provisional acceptance certificate (PAC) or taking over certificate to demonstrate completion of the design and build of the plant. There is then usually a two-year testing period of the performance and/or availability of the plant. This testing period will run simultaneously with the defects rectification period which is the period whereby the EPC contractor has a contractual obligation and right to remedy defects in the plant. The passing of this testing period and confirmation that the plant is defect-free is usually marked with the issue of a final acceptance certificate (FAC). The buyer will also want to ensure that it has the benefit of the manufacturer's warranties which are usually either assigned at PAC and then managed by the EPC contractor or O&M contractor or assigned at FAC in accordance with the EPC contract.

Electricity supply agreements and meter operating agreements: The ProjectCo will enter into a contract with a utility company for the provision of electricity on a utility's standard terms. The ProjectCo would also enter into a meter operator contract for the provision of the meters which measure the metered electricity input and output of the project. However, these agreements are generally low value, standard form and not usually negotiated.

Offtake: The offtake arrangements are unlikely to be in place for a shovel-ready or construction-stage project, however, for an operational phase project the offtake arrangements will be of key concern. The creditworthiness of the offtaker, the duration of the offtake arrangements and commercial terms will be an important part of the due diligence and financial model. The availability of a relevant subsidy may also impact on the terms of such offtake arrangements.

In the United Kingdom, if the project is an ROC project, the offtake agreement is likely also to be for the sale of ROCs at a buyout price which will be slightly less than the market trade value of the ROCs. Projects benefiting from a FiT may have a FiT statement of terms instead of a typical power purchase agreement with the relevant offtaker. In a post-subsidy world, there are more projects which have the benefit of a corporate PPA where a contractual counterparty who is not the DNO will purchase the offtake generated by the ProjectCo for an agreed tariff.

In France, as part of the due diligence carried out on regulatory aspects, the buyer will want to ensure that the power purchase agreement is validly entered into and that the price conditions correspond to what the seller declared. Terms and conditions of power purchase agreements are usually standard. Due to recent announcements by the French government, there are expected to be more and more projects granted by way of public tender and therefore a buyer will want to ensure that the public tender has complied with the regulations in place.

Long-term maintenance/lifecycle maintenance: The buyer of a construction-stage or operational project will want to be clear as to how the plant is being operated and that will typically be by way of a separate O&M contract entered into with the same entity or affiliate group entity of the EPC contractor for at least the period until final acceptance. The buyer may then want to enter into longer-term O&M arrangements with its own providers or asset managers. Key provisions to review in the context of an operational project will be the ability to terminate the O&M contract and the cost of doing so. The market price per MW of installed capacity for operations and maintenance of solar assets has greatly diminished in line with the decreasing capex costs of solar plants in recent years.

(c) ***Corporate and finance due diligence***

Whether a project is at a development, construction-ready or operational phase, the ProjectCo and any subsidiaries will need to be investigated from a corporate and finance perspective. The company's incorporation details, constitution, shareholders' registers and other constitutional registers will need to be obtained and reviewed. Depending on the jurisdiction of incorporation of the ProjectCo, that information may be held on public registers or will need to be obtained from the seller.

If the buyer is acquiring a minority interest in ProjectCo, the shareholders' agreement (if any) and rights enshrined in the ProjectCo's corporate constitution should be reviewed together with any previous sale documentation which might expose the ProjectCo or subsidiaries to ongoing liabilities pursuant to such sale documentation. The rights of other shareholders and their ability to impact the acquisition need to be received and fully understood.

Furthermore, if there is third-party debt financing in place prior to the acquisition of the ProjectCo, the buyer will need to decide on whether it wants to repay that debt or leave it in place. If the buyer were to agree to acquire the ProjectCo with the existing debt remaining in place, the seller would need to ensure that the acquisition would not trigger a default under the relevant facility agreements, particularly as a result of the change of ownership of the ProjectCo. In these circumstances, the buyer should require the seller to obtain an express waiver or consent from the ProjectCo's lenders prior to the completion of the sale. Alternatively, a buyer may require any such existing ProjectCo debt to be refinanced in full and any security granted by the target group or over the shares of the ProjectCo in support of such facilities to be released and discharged in full prior to completion of the sale. If the buyer has any existing group financing or has arranged acquisition financing, it may need to ensure that such refinancing is effected prior to completion. If the ProjectCo's debt is to be refinanced as a condition to completion of the acquisition, the buyer will then need to ensure that alternative financing is available to meet the ongoing needs of the ProjectCo and its target group (if relevant) following the acquisition.

The due diligence phase of an M&A transaction is of fundamental importance to the buyer and usually the most labour-intensive part of the transaction.

2.5 Corporate structure

The corporate structure of a solar M&A transaction tends to be fairly straightforward. In the United Kingdom, most projects are held within a limited liability company which has been capitalised through a combination of equity, shareholder loan and, more commonly, in the case of operating assets, third-party debt.

In France, a ProjectCo is usually incorporated either as a limited liability company (SARL) or a simplified joint-stock company (SAS), as both forms are companies with limited liability and allow a structure with a sole shareholder. From a tax perspective, it is usually preferred to have the ProjectCo incorporated in the form of SAS as the registration taxes upon transfer of a SARL's shares are higher than those applied to the transfer of the shares of a SAS.

In the United States, Delaware limited liability companies (LLCs) are typically utilised because they allow the parties to the transaction maximum flexibility as to how to organise and structure their investments. Typically LLCs are pass-through entities (although single member LLCs are typically disregarded for tax purposes), however, they can be taxed as corporations if elected by the shareholders.

2.6 Acquisition financing

A key factor which will influence the structure and timeline of an M&A transaction will be whether or not any acquisition financing is required by the buyer for purposes of funding all or part the consideration payable for the acquisition. If the buyer needs to arrange acquisition financing, the M&A timeline will need to allow sufficient room for the negotiation of the financing terms and satisfaction of the lenders' funding conditions. In particular, lenders will have certain due diligence requirements in respect of the acquisition that would need to be satisfied as a condition for their internal credit approvals (including as to relevant authorisations and contracts for the construction and/or operation of the ProjectCo, and the corporate status and structure of the ProjectCo and its group). The lenders may require their own advisers to undertake separate due diligence, but will often agree to base their due diligence on the due diligence undertaken by the buyer (subject to the buyer obtaining suitable reliance from its advisers for the lenders, either by way of reliance letters or reports being addressed directly to the lenders).

A seller will usually want to satisfy itself that a buyer is able to complete on a purchase before it will give the buyer deal exclusivity or enter into a sale and purchase agreement. If acquisition financing is required, therefore, the buyer will need to ensure that it is able to satisfy all conditions precedent for acquisition funding in advance of the date that payment of the consideration needs to be made under the sale purchase agreement; and, in a competitive bid process, the buyer will need to ensure that funding is available for completion of the acquisition before it tenders a bid.

There are various acquisition financing structures that may be appropriate for a buyer in these cases: for example, a buyer might be able to obtain short-term bridge financing from its lenders supported by corporate guarantees and/or secured with existing assets of the buyer's group which could then be refinanced with long-term acquisition debt following successful completion of the acquisition; or the buyer could arrange for its lenders to commit to funding on a 'certain funds' basis which would require all of the funding conditions (except for a very limited set of conditions within the buyer's control) to be satisfied in advance of the buyer submitting its bid. The appropriate structure (and associated costs) of the financing package would need to be tailored depending on the particular circumstances and timeline of the acquisition.

2.7 Negotiating sale purchase agreements

Whether the buyer makes an acquisition by way of an asset or share purchase, the acquisition agreement will be a key document which will record the transfer of ownership of the assets (shares in a share sale (see section 2.7 (a)) or specific assets in an asset sale (see section 2.7 (b)) from the seller to the buyer. The acquisition agreement will contain key provisions relating to the consideration payable (see section 2.7 (c)), the warranties to be provided by the seller to the buyer in relation to the assets being purchased (see section 2.7 (e)), the limitation of liability of the seller which will include not only time and financial limitation (see section 2.7 (f)), but also limitations in relation to any matters properly disclosed by the seller during the disclosure process (see section 2.8).

(a) Share purchase agreements

Solar projects are usually owned by newly incorporated special purpose vehicles and therefore buyers have less to investigate from a corporate perspective. Consequently, buyers are often more inclined to acquire assets by way of share transfer as opposed to an asset transfer.

In the United Kingdom, the ProjectCo is normally a private limited company (Limited or Ltd). It is usually agreed between the parties that the buyer pays for the tax payable on the consideration for the shares which usually attracts a significantly lower tax rate than the tax payable on the transfer of the assets (eg, land or project contracts) owned by the ProjectCo.

In France, depending on the type of corporate vehicle being acquired, the stamp duty payable would differ but usually the stamp/registration tax is less expensive for the transfer of shares in a SAS than a transfer of shares in a SARL. This point often becomes an issue in solar power transactions as many developers will incorporate entities in the form of a SARL rather than a SAS. It is possible to change the form of the corporate vehicle prior to completion, however, in most cases the ProjectCo is prevented from changing its legal status because of the net equity being lower than the share capital (as the ProjectCo usually has development and construction debt which its revenues have not been able to repay). In such cases, the French Commercial Code prohibits the ProjectCo from changing its legal status for as long as the net equity has not been replenished which may bring cash or valuation issues. Therefore, negotiations will take place in relation to which party will bear the cost of either the registration taxes or the cost to change the ProjectCo's legal status from a SARL to a SAS.

In the United States, Delaware LLCs are typically utilised to hold solar assets as newly formed special purpose entities. As sponsors and buyers typically enter into a Membership Interest Purchase Agreements (or MIPAs) for the acquisition of membership interests in the limited liability company (or LLC) holding the project rights, the corporate structure of operating projects will be subject to existing tax equity investments and an underlying tax equity partnership. Essentially a cash buyer will only be interested in purchasing the cash-equity shares in the project which can be held by an upstream cash equity partnership or can be represented by different classes of shares in the tax equity partnership.

(b) *Asset purchase agreements*

While some buyers may choose to purchase by asset purchase agreement thereby purchasing individual project rights and transferring these by way of novation into a buyer incorporated ProjectCo, in the United Kingdom, France and the United States, it is by far the exception as regards to choice of structure due to tax implications and the time implications involved in asset transfers. There are tax implications, additional complexity and costs involved in the transfer of the project rights to a new acquisition vehicle (a NewCo) and the transfer of project rights such as title in land to a NewCo will attract taxation and may require additional execution formalities. There are timing implications as the buyer must identify exactly which assets to transfer to the NewCo and it is hard to ensure effective dates of the relevant transfers correspond to the date of the acquisition due to the number of third parties involved in granting the project rights, which will include landowners, utilities, contractors and licensing or authorising bodies. While asset transfers are not uncommon in the context of the sale and transfer of development rights, the inability of the buyer to acquire the historical losses accrued in the ProjectCo during the course of construction means that this route is rarely used for the acquisition of operational solar projects.

(c) *Valuation*

The valuation mechanism in a share purchase agreement will usually comprise either a locked box or a completion accounts adjustment.

A locked box mechanism fixes the economic value of the assets at a certain date pre-acquisition. This mechanism is used frequently in the United Kingdom to value assets from a fixed point in time (eg, commissioning). The parties will value the assets based on agreed locked box accounts and the seller usually indemnifies the buyer for any leakage (which is essentially unauthorised payments including to the seller, persons connected with the seller or its affiliates) which reduce the net asset value of the assets from the locked box date.

A completion accounts mechanism involves a post-acquisition adjustment to the consideration paid by the buyer based on a pre-agreed completion accounts procedure. The parties can agree to adjust the consideration after the acquisition based on an agreed methodology of completion accounts. It is important that the parties agree exactly by whom and how the completion accounts are prepared and the adjustments that need to be taken into consideration and excluded. It is also important to include a dispute resolution procedure to cater for any disagreement in respect of the preparation of the completion accounts.

(d) *Structuring the consideration and completion mechanics*

A buyer and seller will ascertain what the key criteria are for the terms of the sale which will usually govern the structure of the acquisition agreement. The actual mechanism to complete the transaction may vary depending on the issues that the due diligence process reveals.

Where certain key conditions required by the buyer remain to be satisfied, the parties may enter into a conditional acquisition agreement where the acquisition is subject to the satisfaction of certain conditions. If a conditional or split exchange and completion mechanism is agreed, a buyer will want to ensure that the asset value is protected in the period between exchange and completion, and therefore the buyer will ask the seller to provide a number of undertakings not to materially decrease the asset value and to agree to continue running the ProjectCo and operate the asset in the ordinary course. The buyer will also want to ensure that it can opt not to acquire the ProjectCo if the relevant conditions are not met by an agreed date but also if warranties or seller undertakings are breached. The seller will strongly resist the buyer's right to walk away from the deal and will want to limit the instances upon which this can occur. Consideration will also need to be given as to whether the seller should be able to disclose against the warranties given by the seller in the acquisition documentation at completion.

As an alternative to a conditional contract, the buyer could request that the consideration is only partly paid with a proportion being paid upon the seller satisfying an obligation post-completion. However, this is unlikely to be an option where the issue is of key concern or would fundamentally impact the value of the asset. Other mechanisms used are an earn-out being agreed in the acquisition documentation with the uplift being paid upon satisfactory resolution of the issue. Indemnities are also used by buyers to protect themselves from costs or losses where there is a specific known risk.

(e) *Warranties*

In most acquisitions, the buyer will seek contractual protections from the seller in a form of statements (warranties) given by the seller to the buyer about the business and/or the shares being purchased.

Title and capacity: Warranties relating to the seller's title to shares or to the assets subject to the sale purchase agreement, and the seller's ability to transfer the shares or assets free of any encumbrances or restrictions are considered fundamental terms and therefore it is commonly accepted that such warranties should be given by the seller on an uncapped and unlimited time basis subject to the common or civil law statute of limitations.

Real estate: In the United Kingdom, a buyer will satisfy itself as to the real estate matters either by seeking warranties from the seller, or having lawyers prepare a Certificate of Title, or a combination of both. If the buyer is content to rely on a Certificate of Title, usually the warranties in the acquisition agreement would be limited to those relating to property and/or planning matters which have not been adequately confirmed in the Certificate of Title, the adequacy and accuracy of the Certificate of Title itself and the information provided by the seller to its Certificate of Title provider in order to produce the Certificate of Title.

In France, it is not common to use Certificates of Title. Depending on the bargaining position of the parties, warranties in relation to the entire dataroom or only the disclosure appendices will be sought. Of course, if the dataroom is well organised and there was sufficient time to review, it would be more difficult to argue that exclusion of the entire dataroom is not acceptable.

Solar-specific warranties: A buyer is likely to insist on other warranties specific to solar projects depending on their stage of development such as warranties relating to the characteristics of the project. These warranties will relate to accreditation for subsidies (if applicable), revenue streams, the viability and integrity of the grid connection, offtake arrangements, the contractual rights of the project and the status of obligations and liabilities pursuant to those contracts.

Warranties commonly found in other commercial acquisition agreements such as those relating to tax, accounting, litigation, anti-corruption and bribery are also commonly found in a solar project acquisition agreement.

(f) *Liability caps*

Liability caps range widely between jurisdictions. In some jurisdictions such as the United Kingdom, it is common for the liability cap to be based on the entire or a percentage of the enterprise value of the asset(s). Capping liability by reference to enterprise value rather than by reference to consideration ensures that the buyer is covered not only in respect of the amount it is paying for the sale shares, but also in respect of any seller shareholder debt which it is paying on behalf of the ProjectCo. In the United States and France, the liability cap would usually be between 10 and 20% of the consideration paid for the shares, carving out breaches of certain fundamental representations (eg, relating to title and capacity). In France, it would also be common to exclude any liability which remains in the ProjectCo which relate to pre-acquisition tax liabilities. The parties usually agree a *de minimis* limit that must be reached before a breach of warranty claim can be brought under the acquisition agreement. In addition to a *de minimis* limit, it is also common to see a 'basket' or further threshold that the buyer needs to meet in order to bring a claim for breach of warranty. This needs to be carefully negotiated as a claim for a few thousand pounds for a breach of warranty could have a detrimental impact on the internal rate of return of a project. In the United States, both seller-friendly 'deductibles' and buyer-friendly tipping baskets are common and which approach is used depends on the leverage of the parties. Deductibles only impose liability if the basket cap is reached (and only for amounts in excess of that cap), whereas tipping baskets impose liability back to the entire amount if the basket cap is reached.

In addition to financial limits, there are usually time limits imposed in terms of time periods within which breach of warranty claims can be brought against the seller. There is usually a distinction drawn between claims for general breaches of warranty, tax claims and title warranty claims. In UK, US and French solar M&A deals, general breaches of warranty claims are usually limited to around 18–24 months following completion. Tax warranty claims in UK deals are usually open for a longer period of seven years after the date of completion due to the tax authority's powers to investigate tax issues for a longer period of time. In the United States, tax warranties will typically survive until 60 days following the relevant statute of limitations for bringing such claims. In French deals, the period to bring a tax warranty claim is usually limited to the period of statutory limitation (which differs depending on the type of tax).

(g) ***Role of W&I insurance***

The use of warranty and indemnity insurance in solar M&A transactions has become increasingly common in the United Kingdom and in France, particularly for larger asset and portfolio transactions. As part of the process, underwriters will review the due diligence reports and acquisition documents. The lawyers acting for the buyer will usually be required to discuss the transaction, diligence and disclosure processes to get the underwriter comfortable that an ordinary course process has been conducted and that the prospect of a warranty and indemnity policy has not led to a reduction in the negotiations and tensions between the parties that one would typically expect to see. The acquisition documents should be negotiated as usual, however, the financial liability cap of the seller is usually reduced where W&I has been negotiated as part of the deal given that the buyer's primary recourse for a breach of warranty claim will be under the W&I insurance policy. Therefore, careful review of the terms of the W&I policy should be carried by the buyer in order to ensure that areas which the W&I insurers will not underwrite (eg, certain historic tax liabilities or environmental liabilities) should, if possible, be warranted or indemnified by the seller. Where a seller requires W&I insurance to be put in place as a condition to the sale it should be made clear from the outset as the responsibility for the cost of putting a policy in place will need to be agreed between the parties and the additional time required for W&I underwriters to conduct due diligence on a transaction will need to be built into the deal timetable. In the United States, W&I insurance is not very common for privately negotiated solar project acquisitions.

2.8 Disclosure

Disclosure is the procedure by which the seller qualifies the warranted statements made in the acquisition agreement in order to limit its liability. The scope of such disclosure will usually be heavily negotiated between the seller and the buyer and the disclosure process will vary depending on the jurisdiction in which the transaction is taking place.

In an M&A transaction in the United Kingdom, the principle of *caveat emptor* (buyer beware) applies which means that, with few exceptions, the only protection that a buyer will have in respect of any defects or liabilities discovered post-acquisition will be limited to what was specifically agreed and written in the acquisition documents. Buyers will seek contractual protections in the form of warranties given by the seller about the assets or company to be acquired. The buyer will have a claim for breach of warranty against the seller if a warranty is incorrect and reduces the value of the shares or assets purchased. However, the seller's liability will be limited not only by the liability caps but also by disclosure to the extent that proper disclosure of matters is made against the warranties provided. Indeed, proper disclosure will restrict the buyer's ability to claim for breach of warranty. As a result, the disclosure process and contents of the disclosure letter are of key importance and remains heavily negotiated between the parties. Not only does it act as a potential completion or post-completion price adjustment mechanism, but it also allows the buyer to assess the assets or value of the shares to be purchased and the risk of acquiring them. Consequently, the disclosure process can be used to obtain a reduction of the consideration, seek an indemnity or for the buyer to withdraw from the transaction before entering into any binding agreement with the seller. Indemnities may be sought in respect of known risks and typically are not limited by matters disclosed.

(a) *Disclosure process*

In the United Kingdom, the standard approach to the disclosure process is to make disclosures in a separate disclosure letter given by the seller to the buyer, which will contain general disclosures (in relation to matters contained in public records or matters of which the buyer ought to be aware on the basis of pre-contract enquiries or searches made), and specific disclosure (comprising matters which, if not disclosed, would or might constitute a breach of warranty). The disclosure letter will also attach copies of the documents disclosed (referred to as the disclosure bundle). A seller will want the contents of the dataroom prepared by it and made available to the buyer generally disclosed. Where the buyer has relied on a vendor due diligence report issued by the seller's advisers and/or Certificates of Title, the buyer may refuse general disclosure of the dataroom or only accept a limited scope disclosure of the dataroom containing documents in relation to matters which fall outside of the vendor due diligence report and/or Certificates of Title and the warranties will only be qualified by the information disclosed in the adviser's vendor due diligence report and/or the Certificate of Title.

In France and in the United States, the standard approach to the disclosure process is to make disclosure in appendices to the acquisition agreement rather than in a separate disclosure letter.

(b) *The standard of disclosure*

The buyer will seek a high standard of disclosure expecting full and specific details of all relevant matters on the basis that the seller owns the ProjectCo, runs the business and is in possession of all relevant facts. A compromise is usually agreed between the parties' expected standard of disclosure and fair disclosure[1] is generally defined as providing the buyer with sufficient details to enable it reasonably to assess the situation and its impact on the ProjectCo or business.

There are also discussions on the definitions of 'seller's knowledge' for those representations that are limited to the seller's knowledge and awareness. The buyer will want to ensure that the seller's board of directors and other key management personnel are included and would also seek inclusion of the people involved in the transaction or the development of the projects that the seller should consult to understand whether or not it has awareness of the matter subject to the disclosure.

(c) *Buyer's knowledge*

The acquisition agreement will often contain a clause addressing the buyer's knowledge and the extent to which the buyer's knowledge derived from other

---

1 Example of disclosure which the UK courts have not regarded as fair: *Levinson v Farin* [1978] 2 All ER 1149; *Daniel Reeds Ltd v EM ESS Chemists Ltd* [1995] CLC 1405.

sources (including knowledge acquired during the due diligence process) may limit the seller's liability in respect of a warranty claim.

In the United Kingdom, case law has been unclear on the position of buyer's knowledge with one case suggesting that it would be wrong to allow a buyer to rely on such a provision where the buyer had actual knowledge of a particular matter (even if such matter had not been disclosed in the disclosure letter) and that the buyer's actual knowledge would be taken into account in assessing the buyer's loss arising from a breach of the warranty.[2] However, a subsequent case clarified the position and confirmed that a buyer may be able to bring a claim for a breach of warranty notwithstanding the buyer's knowledge of the particular facts of the breach subject to the stipulated type of knowledge the buyer has and what type of knowledge (actual, constructive or imputed) that may preclude a claim being brought by the buyer in respect of the warranties as set out in an acquisition agreement.[3] Notwithstanding the buyer's knowledge, the seller should ensure that adequate and specific disclosure is made to limit its liability for breach of warranty.

In the United States, case law varies quite widely on a state-by-state basis, and thus parties to the transaction will seek to qualify representations and warranties based on carefully defined knowledge qualifiers where relevant. In a seller-friendly purchase agreement, sellers will often seek to insert clauses known as 'anti-sandbagging' defences which are defences against the buyer's right to indemnification. The position taken by the buyer and seller of course depends on the leverage of either side in the negotiation. And the enforceability of such clauses will depend on the state-specific case law governing the contract. Nonetheless, buyers of solar projects in the United States should be aware of these issues and seek to allocate risks related to 'knowledge' appropriately.

This chapter was written by Anthony Riley and Yolanda Yong with contributions on US law aspects from Chris Gladbach, French law aspects from Gergana Rodriguez and commercial considerations from Dario Bertagna. The views expressed in this chapter by Dario Bertagna of Capital Dynamics Limited (together with its affiliates, Capital Dynamics) are his personal views and do not necessarily reflect the views of Capital Dynamics itself. The views expressed reflect the current views of Dario Bertagna as of the date of publication and neither Dario Bertagna nor Capital Dynamics undertake to advise on any changes in the views expressed herein.

2 *Eurocopy plc v Teesdale* [1992] BCLC 1067.
3 *Infiniteland Ltd and John Stewart Ariss v Artisan Contracting Ltd and Artisan (UK) PLC* [2004] EWHC 955 CCL.

Algeria

Foued Bourabiat
Bourabiat Associés

1. Brief history of the solar sector in the region

1.1 Algeria's urgent need to diversify its energy mix

(a) ***Algeria faces an urgent need to diversify its energy mix***

Algeria's economy is heavily dependent on the hydrocarbon sector. Fossil fuels account for almost all of domestic total primary energy consumption and to date only 2% of installed energy capacity derives from renewable energy resources.

The government is aiming to diversify the energy mix and aims to bring the share of electricity generation from renewable energy resources to 40% by 2030. Several factors explain the need to diversify the energy mix and have recourse to renewable energy.

Algeria's internal consumption of electricity has risen sharply in the last few decades. Electricity consumption per inhabitant has risen from 528 kWh per inhabitant in 1990 to almost 1,500 kWh today.[1] Algeria's internal consumption absorbs almost 50% of its production of natural gas and this trend is due to continue in the near future.

Each cubic metre of natural gas consumed in Algeria cannot be exported. Since hydrocarbons represent 60% of the budget of the country and 95% of its exports, it is in Algeria's best interest to export as much hydrocarbons as possible and to bring the share of electricity generation from renewable energy resources as high as possible to cover internal consumption.

The drop in the costs of photovoltaic (PV) technology is also a key factor and in many respects PV energy is becoming a viable alternative to fossil fuels, while at the same time industry reports seem to indicate that proven hydrocarbons reserves are declining in Algeria.

1.2 Algeria's renewable energy programme

Until now, Algeria's power sector and renewable energies landscape has been dominated by Algerian state-owned entities, such as Sonatrach and Sonelgaz.

In the mid-term, Algeria aims to invest around US$1 billion in the development of renewable energies.

A first attempt to increase energy efficiency and promote renewable energy

1 Source: World Bank.

sources was introduced more than 15 years ago through Law 02-01 of 5 February 2002 on electricity and gas distribution through pipelines (the Framework Law) which was based on Law 99-09 of 28 July 1999 concerning energy management. The impact of the Framework Law was, however, very limited.

More recently, the Algerian Development Programme for New and Renewable Energies and Energy Efficiency (the Programme) was introduced. The Programme seeks to promote a production capacity of 22,000 MW for the needs of the national market over the period 2015–2030, of which more than 4,500 MW should be achieved by 2020.

Overall, it is contemplated that 40% of the country's electricity needs will be generated from renewable energy sources by 2030, of which 37% will be from PV and 3% from wind power.

Within the framework of the Programme, renewable energy projects can be operated under two different regimes:

- the General Regime (as defined in section 4.1 below) introduced by the Framework Law; or
- the Tender Regime (as defined in section 4.2 below) regulated by Executive Decree 17-98 of 26 February 2017 defining the tendering procedure for the production of renewable energies or cogeneration and their integration into the national electrical energy supply system (the Tender Decree). It is contemplated that a volume of around 4,050 MW would be awarded through various tenders.

The Tender Decree describes the tendering procedure. However, some details still have to be clarified. For instance, the identity of the awarding authority remains unknown.

Additionally, the shareholding structure of the project company to be incorporated after the tender is awarded has not been disclosed. For instance, it remains unknown whether Sonatrach (which is a solvent company contrary to Sonelgaz) or its subsidiary will own all of the 51% bloc of share capital reserved for Algerian stakeholders or whether local private investors will hold a minority stake (eg, 10%).[2]

2. Overview of market participants

2.1 Ministry of Energy

The Ministry of Energy was created by Decree 71-199 establishing the organisation of the central administration of the Ministry of Industry and Energy dated 15 July 1971. The current Minister of Energy is Mr Mustapha Guitouni.

As part of the Government's general policy, the Minister of Energy is responsible for drawing up policies and strategies for research, production and development of hydrocarbons and all types of energy resources and the development of the related industries.

2 As described in section 3 below.

2.2 Ministry of Environment and Renewable Energy

In May 2017, the newly appointed government created the Ministry of Environment and Renewable Energy (MEER) which has the mission to ensure energy security and promote sustainable sources of energy. The current Minister of Environment and Renewable Energy is Mrs Fatma-Zohra Zerouati.

The objective of the MEER is to increase the share of clean energy in the country's energy mix, in line with the targets set by the Programme. However, the remit of the MEER and its practical role and involvement in the context of the Programme remain unclear.

2.3 Ministry of Industry

Within the Government's general policy, the Minister of Industry proposes national policies in the fields of industrial policy, industrial competitiveness, promotion of support services and management of the state's shareholdings in the public industrial sector. The minister monitors their implementation and compliance with the laws and regulations in force.

The Ministry of Industry should be heavily involved in the Programme and, more specifically, in relation to the industrial component of the tender regime.[3]

The current Minister of Industry is Mr Youcef Yousfi.

2.4 The Electricity and Gas Regulation Commission

The Electricity and Gas Regulation Commission (CREG) was created by the Framework Law and established on 24 January 2005. It is managed by a steering committee composed of a president and three members appointed by presidential decree, upon the recommendation of the Minister of Energy.

CREG's role is to ensure the competitive and transparent functioning of electricity and gas markets in Algeria, in the interests of consumers and operators. CREG has three main missions:

- Controlling the public utilities of electricity and gas distribution by pipelines;
- Advising public authorities on the organisation and functioning of the electricity and gas markets; and
- Monitoring compliance with laws and regulations relating to the electricity and gas markets.

2.5 The National Agency for the Development of Investments (*Agence Nationale De Développement De L'investissement*)

Created in the framework of first generation reforms in Algeria during the 1990s, the National Agency of Investment Development (ANDI) is a governmental institution which aims to promote and support investments. ANDI supports national and foreign investors through various subsidies and incentives.

2.6 Sonelgaz

Sonelgaz, the national company for electricity and gas, is a state-owned company in

3 As described in section 4.2 below.

charge of electricity and natural gas distribution in Algeria. It was established in 1969, replacing the previous company Electricité et Gaz Algérie (EGA). Sonelgaz was granted a monopoly over the distribution and the sale of natural gas within the country as well as the production, distribution, import and export of electricity.

Sonelgaz has recently been reorganised. The new organisation, approved in February 2017, has established Sonelgaz as an industrial group composed of 16 subsidiaries. Its core businesses include the generation, transmission and distribution of electricity as well as the transmission and distribution of gas through pipelines.

2.7 GRTE

GRTE (*Gestionnaire du Réseau de Transport de l'Electricité*), is a subsidiary of Sonelgaz which was created on 1 January 2004, in accordance with the Framework Law. GRTE's mission is to operate the electricity transmission network.

GRTE is in charge of the operation, maintenance and development of the electricity transmission network and should therefore guarantee adequate capacity in relation to transit and reserve needs.

In accordance with the Framework Law, GRTE has a monopoly over the transmission of electricity. The remuneration for the transit of energy is fixed by CREG and is the same throughout Algeria.

2.8 CEEG

CEEG (*Compagnie de l'Engineering de l'Electricité et du Gaz*) is a subsidiary of Sonelgaz created in 2009. CEEG is in charge of the management (studies, management, supervision of works, acceptance and commissioning) of energy infrastructure projects, specifically electricity and gas production facilities, gas and high voltage transmission networks.

Although unclear at this stage, it is anticipated that the Minister of Energy will delegate the tendering proceedings relating to solar projects to CEEG.[4]

3. General legal framework for investments in Algeria

Algerian legislation ensures a balance between freedom of trade and the necessary regulation of the market based on universal legal principles and standards, such as free competition and free movement of goods.

3.1 Corporate issues

The main constraint on foreign investors is the foreign ownership restriction rule. According to Article 66 of the Finance Law for 2016, foreign investments in Algeria can be made only within the framework of a partnership in which one or more national shareholders resident in Algeria must hold at least 51% of the shareholding (known as the 51/49 Rule).

As a result, the share capital of any project company or producer wishing to invest in a PV installation should be owned in majority by Algerian shareholders. Foreign investors should therefore find a local partner(s) in Algeria to realise its

---

4 As described in section 4.2 below.

investment. In the context of the Tender Regime (as defined below), the project company would certainly have Sonatrach[5] or one of its subsidiaries as shareholder. It is anticipated that approximately 41% of the project company would be held by Sonelgaz[6] and 10% by Algerian private investors. This needs to be confirmed by the Algerian authorities.

The current economic crisis obliges the Algerian government to adopt additional measures to attract foreign investments. In that respect, Algeria adopted Law 16-09 of 3 August 2016 (the New Investment Law) to improve the investment climate and increase economic growth.

While the 51/49 Rule enacted in 2009 is still in place, the New Investment Law has eased a number of the previous restrictions. Notably, the New Investment Law confirms that dividends and sale proceeds of foreign investments in Algeria may, under certain equity investment conditions, be repatriated outside Algeria. In addition, there is no longer an obligation to present a positive foreign exchange balance for the benefit of the Algerian government.

In contrast, the pre-emption right of the Algerian state was maintained in relation to share transfers from or to foreign investors regarding the investment.

3.2 Financing and bankability issues

Since 2016, a finance law has allowed foreign debt financing of direct foreign investments in Algeria, subject to strict controls. Previously this was totally prohibited.

However, only strategic investments can be financed by foreign debt. In spite of the uncertainty as to the exact scope of this provision, it is likely that the projects implemented in the context of the Programme will be eligible for a foreign financing.

The Algerian authorities would need to waive a number of practical hurdles which could prevent foreign financial institutions from financing projects in Algeria. By way of example, Algerian foreign exchange control regulations could render the enforcement of security interests (eg, mortgages or pledges) difficult or uncertain.

Indeed, in case of enforcement of a security interest by a foreign financial institution, the latter may not be able to repatriate the proceeds of enforcement in a foreign currency to a bank account located outside Algeria. In other words, foreign lenders would recover proceeds of enforcement in Algerian dinars but would not be able to transfer such proceeds outside Algeria.

Moreover, in case of enforcement of a share pledge by a foreign financial institution, the latter would be able to acquire only 49% of the project company due to foreign ownership restrictions and, more specifically, the 51/49 Rule. This is likely to be a serious concern for foreign lenders that typically require a pledge of over 100% of the shares of the project company.

In light of the above issues, we believe that the Algerian authorities will probably have to provide guarantees to secure foreign debt financing. It is likely that such guarantees will be provided by Sonatrach, one of the few solvent state-owned entities in Algeria.

---

5 Which is fully owned by the Algerian state.
6 Which is fully owned by the Algerian state.

3.3 Other issues

It should be noted that Algeria has entered into double-taxation treaties with around 40 countries.

Moreover, the protection of foreign investments is an established principle in Algeria. Foreign investors cannot be subject to expropriation, unless provided for by the law. In any event, any expropriation for public interest reasons duly authorised in compliance with the law requires a fair compensation of the foreign investor.

Algeria is not a member of the World Trade Organisation but has signed a number of free trade agreements. However, Algeria and the World Trade Organisation are discussing a potential integration of the latter.

It should be noted that international arbitration clauses are valid. In that respect, Algeria ratified the New York Convention dated 10 June 1958 on the recognition and enforcement of foreign arbitral awards and the Convention dated 14 October 1966 for dispute resolution by the International Centre for Settlement of Investment Disputes (ICSID).

Furthermore, except for mandatory legal requirements (eg, notarised documents such as the articles of association should be drafted in Arabic), the legal documentation can be drafted in English and the parties are free to choose the governing law of the contract.

4. The process for operating a renewable energy project

The development of solar PV projects in Algeria sits within the general framework for renewable energy project development and can be operated under two separate regimes: the general regime introduced by the Framework Law (the General Regime) and the tender regime regulated by the Tender Decree (the Tender Regime).

Both regimes are considered further below, although it should be noted that their interaction and respective scopes remain somewhat uncertain.

4.1 The general regime

(a) Step 1: Construction phase

Issuance of the guarantee of origin certificate: The producer of electricity from renewable energy sources or cogeneration, wishing to benefit from investment incentives, must submit an application to CREG in order to obtain a guarantee of origin certificate.

The guarantee of origin certificate is the document which certifies that the electricity produced is of renewable origin or originates from a cogeneration system. It is issued within one month and 10 days following the submission of an application to CREG accompanied by the relevant technical file.

The guarantee of origin certificate includes, in particular, the identification of the electricity production plant and the location of the installation, the capacity of the plant and the nature of the energy sources produced.

Installations which have received a guarantee of origin are subject to a conformity check which is carried out after the installation has been built, and then periodically during the lifetime of the installation once it has been commissioned. The purpose is to ensure that the initial characteristics are maintained over time and to prove that the electricity injected into the grid is of renewable origin or from a cogeneration system.

Authorisation to operate: The Framework Law provides that an operating licence issued by CREG is required in order to operate a new power generation facility. The request for authorisation is made to CREG and must contain:

- a duly completed application form;
- technical explanations concerning the integration of the new installation in the grid;
- a study on the impact of the integration of the new installation in the grid;
- a report on technical, economic and financial capacities and experience of the applicant;
- a report with the main characteristics of the installation;
- a report on the safety and security of the networks, installations and associated equipment;
- an environmental impact study;
- a commitment by the applicant that it will comply with the specifications setting out the rights and obligations of the project company;
- a map with the location of the installation; and
- any other relevant information.

CREG will adopt a decision on the application within four months. The decision to authorise operation of the installation is published by CREG and the beneficiary has a period of 12 months from the date of its issuance to start the construction of the installation. Failure to fulfil this obligation may lead to the operating licence being withdrawn.

Power offtake arrangements: The Framework Law contains guaranteed purchase tariffs for producers of electricity from renewable sources for each kWh produced and injected into the grid.

The benefit of feed-in tariffs: A guaranteed feed-in tariff may be granted to the applicant company subject to connecting its facility to the transmission system or the electricity distribution system, and submission of an application to CREG at the same time as the application for the authorisation to operate.

CREG issues an opinion within a maximum period of two months and then forwards the application to the Minister of Energy, who decides on the application within 15 days.

The decision to grant the guaranteed feed-in tariff is notified by the Minister of Energy to the applicant, who will have 12 months from the date of issuance of the decision to begin the construction of the installation. In the absence of such a decision, the benefit of the guaranteed feed-in tariff is withdrawn.

The obligation to purchase electricity produced: The Framework Law contains a 'take or pay' obligation for Sonelgaz or its subsidiaries which are therefore subject to an obligation to purchase all the electricity generated from the generator's facilities. Sonelgaz cannot, therefore, refuse to purchase the electricity produced from renewable sources and injected into the grid or negotiate the purchase price of electricity.

Guaranteed feed-in tariffs: The feed-in tariffs are set out in Decree of 2 February 2014 on feed-in tariffs applicable to solar plants (the Decree on Feed-in Tariffs). Feed-in tariffs are guaranteed for a period of 20 years from the date of commissioning of the connection of the PV installations. In a first five-year phase, the company benefits from a fixed power price which is calculated on the basis of a reference 1,500 hours of operation at full load.

By way of example, the fixed purchase price in Phase 1 is currently set as shown below:

- 0.145 US$/kWh for ground-based PV installations with a capacity of 1 to 5 MWC; and
- 0.116 US$/kWh for ground-based PV installations with a capacity exceeding 5MWC.

In a second phase, and for the remaining duration of the contract, the fixed power price can be adjusted, depending on the actual capacity of the solar plant. The new tariff is applicable from the sixth to the twentieth year.

In order to monitor the actual capacity of the solar plant, the company is required to install energy metering equipment, and also data and software equipment enabling the determination of the real energy potential of the site where the installation is located.

Guaranteed feed-in tariffs are reviewed periodically to take into account the evolution of the costs of the different technologies available for generating electricity from renewable energy sources and cogeneration. In any event, the new power prices do not have a retroactive effect and do not apply to power purchase agreements already signed.

Power purchase agreement features: Upon obtaining the decision to grant a guaranteed feed-in tariff, the producer will enter into a power purchase agreement (PPA) with Sonelgaz.

According to the PPA prepared by CREG and to be signed by Sonelgaz, the main features are as follows:

- A separate signing and entry into force of the PPA. Subject to providing several technical documents, the PPA will enter into force only from the commissioning date of the plant. Furthermore, the commissioning date of the plant is defined as the commissioning date of the data system measurement in accordance with the installation certification requirements.
- The PPA is entered into for a period of 20 years from the commissioning date of the solar plant.
- A guaranteed feed-in tariff set forth in the notification of the decision. Beyond a certain threshold (a threshold determined as the number of operating hours multiplied by the installed peak capacity of the plant), the

electricity produced will be purchased at a price "equal to the average price of conventional electricity on the Algerian market". Such price fixing may represent an issue since there is no defined method to price electricity in the Algerian market.

- 'Take or pay' arrangements. Sonelgaz commits to take all the electricity produced and injected by the solar plant and pay for it, except in the case of *force majeure*, and impossibilityto take the production.

In that respect, *force majeure* is defined in accordance with the Algerian civil code as 'any act or event unforeseeable, irresistible and independent from the will of the parties'.

Moreover, 'impossibility to take the production', is defined as scheduled unavailability of the grid, when it has been scheduled and agreed between the parties, or unscheduled unavailability of the grid, when it does not exceed a total of 24 hours per year. Beyond this 24-hour limit, the electricity produced but not injected shall be paid for on the basis of the guaranteed feed-in tariff.

- Billing and the relevant payments will be executed on a monthly basis, and Sonelgaz will be required to pay within 30 days from receipt of the relevant invoice. The payments will be made in Algerian dinars to a bank account opened with an Algerian bank.
- The producer will have to enter into an insurance policy granted by an Algerian insurance company. Under Article 207 of Law 95-07 dated 25 January 1995 on insurance (the Insurance Law), "Algerian residents, as well as assets and risks located or registered in Algeria can only be insured by an insurance company having a licence in Algeria".

Moreover, the Insurance Law provides that a civil liability policy, product liability policy, air transport policy, and sea transport policy are mandatory. Failure to comply with such a mandatory provision may lead to a fine of up to DZD100,000 (approximately US$1,000).

In addition, any owner of assets located in Algeria and any entity operating commercial or industrial activities in Algeria should insure such assets and activity against natural disasters. It should be noted that failure to comply with such provision may lead to an exclusion from state indemnities for compensation for damage to property, and a fine of an amount equivalent to the premium increased by 20%.

There is no mechanism in the current version of the PPA for revision of the guaranteed feed-in tariff in case of inflation or devaluation of the Algerian dinar. In light of the current financial situation, this constitutes a source of major concern for foreign investors and lenders.

(b) Step 2: Post-construction phase

GRTE is required to connect the producers of electricity from renewable sources to the grid under the same financial conditions as those of the ordinary regime set out in the Framework Law. Currently, the tariff for the transmission of electricity is set at the equivalent of US$0.006 per kWh.

The request for access to the electricity transmission and distribution network is submitted to GRTE. In order to be connected to the grid, the producer must:

- provide evidence that it holds an authorisation to operate issued by CREG;
- submit a request for connection to the grid;
- provide GRTE with all the information necessary for connection studies; and
- pay the connection costs to the grid if the distance between the existing grid and the installation is more than 50 kilometres.

GRTE processes the connection request and must provide a response to the applicant on possible connection solutions within four months from the date of the request. Following agreement on the connection solutions, GRTE enters into a grid connection agreement with the applicant at least two months before the planned end date of the grid connection works.

GRTE has an obligation to keep the producer informed of the progress of the connection works. However, GRTE is not legally bound by any time limit for the actual connection to the grid.

4.2 The tender regime

The Algerian government intends to launch a call for tenders for the development of 4,050 MW solar PV projects (the Projects).

(a) *General framework of the tender regime*

In February 2017, the Tender Decree which defines the tendering procedure for the production of renewable or cogenerated energies, and their integration in the national grid, set out two types of invitations to tender (ITs): *ITs to Investors*, which apply to larger projects such as the Projects, and *ITs by Auction*, for projects generating between 10 and 20 GWh.

Pursuant to the Tender Decree, ITs to Investors relate to

- the design and supply of equipment;
- the construction and development of production and generation plants from renewable energy sources; and
- the commercialisation and sale of the electricity generated therefrom.

Under the Tender Decree, the tendering process should in principle be conducted by the Ministry of Energy. However, it is expected that this task will be delegated to a public entity. It is understood that the Minister of Energy will delegate the tender process to CEEG, a subsidiary of Sonelgaz.

ITs to Investors require bidders to include the development of an 'industrial project' in Algeria as a condition to their bid. The Tender Decree defines industrial projects as investment projects in the manufacturing of equipment used in the production of electricity from renewable energy sources and/or in the provision of services.

Accordingly, the development of a solar panel manufacturing plant is understood to fulfil this condition. Otherwise, if bidders do not wish to invest themselves in the industrial project, the Tender Decree allows for their bid to be accompanied by a bid by a third-party industrial investor selected by the bidder.

In any event, the Decree further provides that the Minister of Energy and the Minister of Industry together may decide to waive the industrial project requirement. Many stakeholders have raised concerns in relation to this obligation to develop an industrial project due to the limited size of the Algerian market.

Therefore, it remains to be seen whether the Algerian government will systematically waive the requirement in relation to the Projects.

(b) The specifications of the tender

The Tender Decree further describes the general guidelines for the submission of tenders, such as the technical and financial offers. It states that the specifications for each IT will set out the conditions the investors should comply with, including the technical specifications for the installations such as the primary energy used, the power, availability and performance requirements for energy efficiency. In addition, the IT should specify the financial and economic conditions for the commercialisation of electricity production, which may not exceed 25 years per installation.

The Tender Decree provides that, when required, the offer for the development of the industrial project shall itself include a financial and commercial offer which contains a detailed financial evaluation including all capital expenditures, operating expenditures and maintenance costs, in addition to the price for equipment and conditions for adjustment thereof.

In addition, ITs to Investors must specify the location of the renewable energy installation and will automatically grant rights to the successful bidder over the land of the selected site. Conversely, within the framework of ITs by Auction, it will be up to the successful bidder to choose and acquire rights over land, including by way of concession under the conditions discussed in section 6 below.

The Tender Decree replaces the mechanism of guaranteed purchase tariffs contained in the General Regime,[7] with an allocation based on the price for kWh offered by the bidder.

(c) Rights of the retained bidder

A power purchase agreement will be entered into between the producer and Sonelgaz or any other relevant operator for a maximum period of 25 years. In this respect, the ITs by Auction documents will include a draft power purchase agreement. There is no template power purchase agreement in the Tender Decree. Nevertheless, a power purchase agreement template drafted by CREG in the context of the General Regime could in practice be used as a reference.

5. Support regime

5.1 The National Renewable Energy and Cogeneration Fund

As part of the financing of the Programme, a national fund for renewable energy and cogeneration (*Fonds national pour les énergies renouvelables et la cogénération*) (the

7 As described in section 4.1 (a) above.

Fund) was created by Decree 15-319 of 13 December 2015. The Fund may provide a wide range of support, including a contribution to financing:

- electricity generation projects;
- the purchase of electricity generation equipment; and
- additional costs incurred in connection with the production of electricity from renewable energy sources and/or cogeneration systems.

The Fund is financed through, among other things, a 1% levy on the hydrocarbon tax. There is limited information publicly available on this Fund and the amount potentially levied. It is currently unclear whether the Fund could in practice finance renewable energy projects or if the money has been used for different purposes. It should be remembered that Algeria is currently facing a financial crisis due to the drop in oil price and the possibility cannot be excluded that the Fund has been used to finance the current expenses of the Algerian State.

5.2 Support measures and specific benefits

The New Investment Law provides that investments registered with ANDI, and not appearing in the list of 'excluded' investments, will automatically benefit from the benefits provided for by the New Investment Law. Investments amounting to more than approximately US$45,000,000 are subject to the prior approval of the National Investment Council.[8]

The New Investment Law distinguishes between:

- benefits common to all eligible investments;
- additional benefits for privileged and/or job-creating activities; and
- exceptional benefits reserved for projects of particular interest to the national economy.

(a) Benefits common to all eligible investments

Under the regime of benefits common to all eligible investments, the incentives are automatically granted in relation to all investments, to the exception of those investments appearing in the list of 'excluded' investments. Renewable energy projects are not impacted by these exclusions.

During the construction phase, the following incentives are granted:

- exemption from customs duties for imported goods that are directly related to the investment and its construction;
- exemption from the value-added tax for imported goods and services or acquired locally, that are directly related to the investment and its construction; and
- exemption from the registration fees in relation to the acquisition of real estate assets acquired for the purpose of the investment.

During the operation phase, exemptions may be granted in relation to corporate tax (*impôts sur les bénéfices des sociétés*), and tax on professional activity (*taxe sur*

---

8 The CNI is composed of various ministers.

l'activité professionnelle) (for a period of three years). Since 2009, the exemptions are applicable only to investments creating up to 100 jobs.

An extension of the above benefits may be granted by ANDI if the investment allows the creation of a significant number of jobs.

(b) *Exceptional benefits for projects of particular interest to the national economy*

The New Investment Law contains a supplementary regime of benefits for projects of particular interest to the national economy, in particular when such projects use technologies that are likely to preserve the environment, protect natural resources, save energy and lead to sustainable development. Solar plants certainly have the potential to qualify for exceptional benefits.

The specific regime of incentives is subject to the conclusion of an agreement with ANDI. The agreement with ANDI is subject to the prior approval of the National Investment Council.

During the construction phase, the potential incentives are exemptions from custom duties, value added taxes and registration fees for a maximum period of five years.

During the operational phase, the potential incentives are exemptions from corporate tax, and tax on professional activity for a maximum period of 10 years.

In addition to the incentives above, other benefits may be granted by the National Investment Council.

6. Environment and real estate

6.1 Environment

Pursuant to Article 18 of 03-10 dated 19 July 2003 on the protection of the environment (the Law 03-10) specific requirements are applicable to certain installations considered to be 'classified facilities'. The operation of classified facilities is subject to various authorisations depending on the nature of the activities. However, the definition of activities subject to the classified facility regime does not specifically refer to power plants (solar or conventional). Therefore, it is reasonable to assume that solar plants are not subject to any authorisation to operate a classified facility.

Nevertheless, in order to obtain the authorisation to operate a solar plant under the General Regime, the applicant should provide an environmental impact study. It should be noted that such environmental impact study is required under the Tender Regime and should be provided together with the bid offer.

6.2 Real estate

Under the ITs to Investors regime the general location of the solar plant would be indicated in the specifications of the tendering authority. Under the ITs by Auction regime and the General Regime the location of the solar plant will be chosen by the project company.

The regime is unclear regarding the rights over the land in connection with solar plants. However, due to legal constraints on the acquisition of land by foreign or foreign-owned companies, rights to land are generally acquired through concession contracts. It is also possible that, as it has been the case in other projects in Algeria, the land could be acquired by the state-owned company as shareholder of the project company and subsequently leased to the project company (this was the case in particular in the context of the construction and operation of seawater desalination plants).

In that respect, under Algerian law, a concession is granted for a minimum period of 33 years, renewable for a maximum period of 99 years.

The concession confers on its beneficiary the following rights:

- the right to obtain a building permit;
- the right to apply for a loan for the project and the right to mortgage the conceded land;
- the right to transfer the ownership of the buildings and the real property rights resulting from the concession, upon effective completion of the investment project and its implementation duly established by the authorised bodies.

The concession can only be terminated in the event of breach of the concessionaire's obligations and by court decision.

7. The future for the solar sector in the region

Algeria is more than four-fifths desert. It has been estimated that six hours of solar energy from the world's deserts could meet the annual global energy demand. The solar potential of Algeria is enormous, since solar radiation is high and there is plenty of land for solar plants. Moreover, the price of the land and manpower is relatively low.

PV technologies offer a number of simple and economically viable applications that can be implemented both in the short and long term. The ability to generate heating or electricity from renewable sources can enhance the quality of life in Algeria, create jobs, develop technical skills, reduce the country's dependence on oil and gas, while meeting obligations of reduced greenhouse effects and global warming. Increased renewable energy production will also enable better management of fossil fuel based reserves.

Algeria is very well placed to be a major player in the potentially lucrative market of renewable energy and solar power in particular. However, recent political turmoil and changes within the executive seem to have slowed down progress around the Projects. In addition, some uncertainty remains regarding the Projects and the content of the tender. The Algerian authorities seem to have had second thoughts as to the best model and approach. That said, transition to more renewable energy use will need to start as soon as possible. In order to attract foreign investors, Algeria should show a clear political will and commitment towards renewable energy and put in place a clear legal framework. The implementing measures required under the Tender Regime should, in particular, be adopted without delay.

Brazil

João Ribeiro da Costa
Wagner Silva Rodrigues
LRNG Law Firm (in association with Tomanik Pompeu)
Urias Martiniano Garcia Neto
Tomanik Pompeu (in association with LRNG Law Firm)

1. Introduction

Electrical energy is an essential element for the development of society, as it plays a key role not only for technological progress, but also in meeting the population's basic needs. Similar to other countries, internal growth in Brazil – the largest country in South America and the fifth largest in the world,[1] with a population of over 200 million inhabitants and a GDP of around US$1.8 trillion – is directly related to its electrical energy production and consumption capacity.

In order to have sustainable growth, even countries rich in natural resources should not rely exclusively on a single source of electricity production, as happened in Brazil for many years, which depended mostly on hydro-electric power.

The need to diversify the sources of electricity production is a clear imperative in Brazil. It is very clear that when the country begins its recovery from the difficult economic period of the last few years there will be an increase in electricity consumption, which in turn will drive an increase in electricity generation, including a demand from new sources, such as solar energy.

2. Brazil's energy mix

2.1 Present status

The effort to make Brazil's energy mix more diversified is already a priority in the country. The table on the following page shows Brazil's present energy mix and its recent evolution:

Table 1 shows that Brazil's energy mix is already composed of several different sources of energy, commonly divided into two major groups:

- non-renewable sources – energy sources whose resources are limited by nature (eg, the energy produced from fossil fuels);
- renewable sources – energy sources whose use is inexhaustible, and which can be used over the course of time (eg, hydro-electric and solar energy).

Table 1 also highlights the increasing role of renewable power in Brazil's general energy mix, which has been increasing steadily in the past few years.

The need to diversify the country's energy mix to support growth, combined

1 In terms of geographical size.

Table 1: Domestic energy supply

Specification	Thousand TOE		16/15 %	Structure %	
	2015	2016		2015	2016
Non-renewable	175,093	162,975	–7.3	58.7	56.5
Oil and by-products	111,926	105,354	–5.6	37.3	36.5
Natural gas	40,971	35,569	–13.2	13.7	12.3
Mineral coal and by-products	17,625	15,920	–9.7	5.9	5.5
Uranium (u3o8) and by-products	3,855	4,211	9.2	1.3	1.5
Other non-renewables*	1,826	1,921	5.2	0.6	0.7
Renewable	123,668	125,345	1.4	41.3	43.5
Hydraulic and electricity	33,897	36,265	7.0	11.3	12.6
Firewood and charcoal	24,900	23,095	–7.2	8.3	8.0
Sugar-cane by-products	50,648	50,318	–0.7	16.9	17.5
Other renewables	14,223	15,667	10.1	4.7	5.4
Total	299,570	288,319	–3.8	100.0	100.0
Of which fossil fuels	*172,047*	*158,763*	*–7.7*	*57.4*	*55.1*

* Industrial gas from blast furnace, steel plant, coking plant, sulfur and refineries.
Source: Brazilian Energy Review – 2016 (Ministry of Mines and Energy)

with a growing awareness of the adverse environmental impacts caused by electricity from non-renewable sources and the international agreements aimed at reducing emissions of polluting gases of which Brazil is part, have forced Brazil's electricity sector to develop a scheme of incentives for electrical energy produced from renewable sources.

Solar energy fits in with this sectoral need, as it is a clean and sustainable renewable source. Solar energy also fits well with the country's meteorological characteristics, as Brazil has a high solar irradiation level (higher than that of the several European countries which depend more on this source of energy), affording the country significant potential for producing electricity from solar energy.

2.2 Low carbon generation in Brazil: the Clean Development Mechanism

As a result of several international agreements, the Brazilian federal government has been adopting measures to reduce adverse impacts on the environment, namely through the promotion of the reduction of emissions of polluting gases.

These measures include encouraging the promotion of the Clean Development Mechanism, which is the responsibility of the Interministerial Commission on Global Climate Change.

Through the Clean Development Mechanism the Brazilian government is trying to promote low carbon generation, by supporting entities that promote the reduction of emissions of polluting gases. This mechanism grants entities promoting low carbon generation with carbon credits, which can later be commercialised through the sale of such credits to parties that have emissions reduction targets to fulfil.

3. The Brazilian electricity sector: an overview

3.1 Regulatory background

The origin of Brazil's electricity sector legal framework goes back to the first half of the 20th century (with Decree 24.643 of 10 July 1934), with relevant development in the last decade of the century, together with the first round of privatisations. Among the innovations introduced in the 1990s it is important to mention the foundation of the National Agency of Electric Energy (ANEEL),[2] the implementation of the public service concessions law (Law 8,987 of 13 February 1995), the separation of activities in the distribution, generation and transmission of electricity and the approval of a set of other laws that encouraged private initiatives to invest in the energy market.

However, the current model of regulation was introduced in the first decade of the 21st century, with the launch of the so-called basis of the new model for the Brazilian Electricity Sector, supported by Laws 10,847 and 10,848 of 15 March 2004 and by Decree 5,163 of 30 July 2004.

3.2 The market players

The main players in the Brazilian electric sector are divided into:

- generators – generate electric energy from a number of sources;
- transmitters – operate the grid that connects generators to consumers;
- distributors – sell energy to consumers; and
- traders – buy and sell energy in the market.

---

2 The National Agency of Electric Energy (ANEEL) is a public agency linked to the Ministry of Mines and Energy, created to regulate and oversee the Brazilian electricity sector. ANEEL has a relevant role in the functioning of Brazil's electric market, managing and supervising the public energy policies.

Besides the above-mentioned, new types of players have been developing in the market, namely associated with the development of renewable energy technologies, in particular solar energy. An example of that is the Distributed Generation market, which is analysed later in this chapter.

3.3 The main authorities

In regulatory terms the main authorities in the Brazilian energy market are:

- National Council for Energy Policy – an advisory body of the President for the development of energy policies and guidelines;
- Ministry of Mines and Energy – the government's ministry responsible for formulating the public policies on energy;
- National Agency of Electric Energy (ANEEL) – a public autonomous agency created by Law 9.427 of 26 December 1996, linked to the Ministry of Mines and Energy. Its duties include regulating and supervising the generation, transmission, distribution and sale of electric energy, and also mediating in conflicts of interest between the electricity sector's market players; authorising energy facilities and services; ensuring fair tariffs; ensuring the quality of the service; calling for investments; stimulating competition between the operators; and assuring the universalisation of the services;
- Electric Sector Monitoring Committee – a committee responsible for monitoring and evaluating the continuity and safety of the electric energy supply throughout Brazil's national territory;
- Energy Research Company – a public company linked to the Ministry of Mines and Energy, whose purpose is to provide services in the area of studies and research aimed at subsidising the planning of the energy sector;
- Electric Energy Trading Chamber (CCEE) – a not-for-profit civil association that has the purpose of enabling the commercialisation of electric energy in the National Interconnected System. It is also responsible for promoting energy purchase and sale auctions, as well as managing the contracts signed at these auctions.
- National Electric System Operator – a not-for-profit civil association, responsible for coordinating and controlling the operation of the electric power generation and transmission facilities at the National Interconnected System and for planning the operation of isolated systems in the country.

3.4 The National Interconnected System and the Free and Regulated Contracting Environments

Established by Law 10,488/2004, the National Interconnected System is a system of electric power generation formed by the companies from the main regions of Brazil: South, Southeast, Midwest, Northeast and part of the North. Only around 1.7% of the country's electricity production capacity is outside the National Interconnected System, in small isolated systems located mainly in the Amazon region.

The contracting of energy within this system must take place in one of two environments: the Free Contracting Environment; and the Regulated Contracting Environment.

In the Regulated Contracting Environment, distributors are obliged to contract for the purchase of electric power from the generating companies, through auctions promoted by the Ministry of Mines and Energy and ANEEL, reselling it to captive consumers. In the Regulated Contracting Environment the consumer does not have the option to buy electric power other than from distributors, except in special regimes, as in the case of Distributed Generation.

The Free Contracting Environment is the segment of the National Interconnected System that allows the free commercialisation of electricity, through bilateral contracts, called Free Setting Energy Contracts, negotiated between concessionaires, traders, free consumers and special consumers of electric energy.

The Free Setting Energy Contracts must be registered at the Electric Energy Trading Chamber. The registration process consists of notifying the amounts contracted and the deadlines settled, with no need to notify the agreed prices. The contracts are notified by the selling party and validated by the buying party, so that they can be used in the accounting and financial settlement process.

The Free Contracting Environment embodies the principle of contractual freedom, therefore the parties freely negotiate the conditions, quantity and price of electricity, and the contractual amounts are registered and validated within the scope of the Electric Energy Trading Chamber, accounted for and settled by the Chamber itself.

Access to the grid in the Free Contracting Environment and in the Regulated Contracting Environment depends on obtaining an authorisation from ANEEL or from the Ministry of Mines and Energy and requires requesting a connection to the operator of the local distribution network, in accordance with specific regulations.

4. Foreign investment in Brazil – an overview

4.1 Regulation

There are four main authorities responsible for the regulation of foreign investment in Brazil:

- the National Monetary Council, responsible for establishing the general guidelines for monetary exchange and credit policies in Brazil;
- the Central Bank of Brazil (BACEN), responsible for foreign exchange control and registration of foreign investment;
- the Brazilian Federal Revenue, that demands all foreign entities owing assets in Brazil to enrol with the Taxpayers' Registry and to have a Brazilian resident to represent them; and
- the Brazilian Security and Exchange Commission (CVM), which has an essential role in the monitoring of foreign investors' activities within the Brazilian capital markets and the suitability of shareholders of Brazilian public companies.

4.2 Foreign capital

Foreign investors may invest in Brazil by means of direct ownership of shares/quotas of Brazilian companies, by granting loans to Brazilian companies or by investing in the Brazilian financial markets. As a rule there are no requirements for having a domestic Brazilian equity stake in foreign-owned projects, but some exceptions apply, namely in projects involving nuclear power, border zones and rural properties.

Brazilian regulation considers as foreign capital machinery or equipment entering the country for the purposes of producing goods and services (eg, solar photovoltaic (PV) equipment), besides the capital brought into the country.

Foreign capital remitted to Brazil, either in foreign currency or in assets, must be registered before the Brazilian Central Bank, but there are no restrictions on the repatriation of funds or remittance of profits out of the country.

4.3 Taxes on cross-border transactions (import duties)

Imports into Brazil are subject to Import Tax (II), as well as other taxes that internal transactions are generally subject to, such as the State Sales Tax (ICMS), generally calculated at the rate of 18%, Federal Tax on Industrialised Products (IPI), calculated with variable rates according to MERCOSUL Harmonised Tariff System (TEC) code of each product and the Federal Social Contributions (PIS and COFIS), usually calculated at the rate of 9.25%.

II's rates are also variable according to TEC's code of the product. For instance, II's rate for solar PV cells is 10% and 12% for modules.

4.4 Taxation on investors' gains (foreign direct investment)

Income paid to non-residents abroad is subject, as a rule, to withholding tax at the rate of 15%, except for residents in tax havens whose rate is 25%. However, remittances of profits and dividends are generally exempt. To avoid double taxation Brazil has concluded agreements with certain other countries to avoid or reduce taxation.[3]

4.5 Industrial and intellectual property protection

Brazilian legislation provides for the protection of patents, trademarks and copyrights. The authority in charge of monitoring industrial and intellectual property in Brazil is the National Institute of Industrial Property.

Utility models, industrial designs and other inventions that may be used in Brazil, such as solar PV equipment, are patentable, which grants its owner the exclusive right to develop it. Trademarks are also protected.

5. Solar energy in Brazil

5.1 Evolution

Solar energy made its first meaningful appearance in Brazil in 2013 when the federal government included solar energy in the New Energy Auctions A-3/2013 and

---

3 Regarding the provisions of the double taxation treaties (DTT) entered into between Brazil and South Africa, Argentina, Austria, Belgium, Canada, Chile, China, South Korea, Denmark, Ecuador, Slovakia and Czech Republic, Spain, Philippines, Finland, France, Hungary, India, Israel, Italy, Japan, Luxembourg, Mexico, Norway, Netherlands, Peru, Portugal, Russia, Sweden, Trinidad and Tobago, Turkey, Ukraine and Venezuela.

A-5/2013.[4] Although these auctions stimulated interest in the market for solar energy, they did not result in any PV projects being commenced.

The scenario changed the following year. In the 6th Reserve Energy Auction (Auction 008/2015)[5], specifically aimed at solar energy, 400 projects were authorised with a total capacity of 10,790 MW. Of these, 31 projects were contracted with a capacity of 889.6 MW, at an average contract price of R215.12/MWh, with a discount of 17.89% (R262.00/MWh).

The 6th Reserve Energy Auction is regarded as an historic milestone for solar energy in the Brazilian electricity sector, as it represented the first contracting of this energy source in a regulated auction, making the generation of electricity from solar energy a practical reality in Brazil.

Since 2014, it has been the development of Distributed Generation that has seen solar energy achieve its most significant increase in Brazil, in particular under the terms of ANEEL's Normative Resolution 482 of 17 April 2012 (ANEEL REN 482/2012).

At present, solar power can be sold both in the Free Contracting Environment, by means of Free Setting Energy Contracts, and in the Regulated Contracting Environment, through the energy auctions organised by the ANEEL – new energy auctions,[6] reserve power auctions[7] and auction of alternative sources[8] – and through the Distributed Generation market.[9]

Recent years have seen a clear diversification of the sources of electricity production and the growth of solar power, which has already attracted the attention of foreign investment. Private equity companies, such as Actis and also Canadian and Chinese companies have already made significant investments in Brazilian solar projects or manufacturing facilities.

Brazilian Federal Authorities estimated that, by the end of 2017, investment in solar energy had reached R4.5 billion, including a significant element of foreign investment.

According to the Ministry of Mines and Energy, in 2018, Brazil will likely be among the top 20 solar energy producing nations. The increasing role for solar energy is foreseen in the 10-Year Energy Expansion Plan (PDE 2024), which estimates that the installed capacity for electricity produced from solar will reach 8,300 MW by 2024, which is likely to represent approximately 1% of the total.

5.2 Incentives for solar energy in Brazil

Recent governments have been consistently developing incentives to diversify

---

4 These auctions were organised to meet the distribution companies' load increase. In this case, energy from power stations that are yet to be built is sold and contracted: A-5 for power stations that are to come into commercial operation within five years, and A-3 within three years.

5 This auction was organised in order to increase security in the supply of electrical energy in the National Interconnected System.

6 New Energy Auctions are performed to attend to the increase of load of the distributors. In this case it is contracted energy from plants that will still be built.

7 Reserve Energy Auctions have the objective of increasing security in the supply of electricity in the National Interconnected System.

8 Alternative Sources Auctions are carried out to respond to market growth in the Regulated Contracting Environment and increase the share of renewable sources in the Brazilian energy matrix.

9 The Distributed Generation market functioning is explained in section 6 below.

Brazil's energy mix, including the production of electrical energy from solar sources.

There are presently several incentives for the development of solar energy projects in Brazil, the most significant of which are as follows:

- Special financing conditions offered by Brazil's National Development Bank (BNDES) for solar energy projects, including financing of up to 80% of the infrastructure necessary for the generation of solar power at interest rates at a discount to prevailing market rates;
- Issue of bonds with tax exemptions for the financing of power generation projects (Law 12.431 of 24 June 2011);
- Special Ruling for the Development of Infrastructure (REIDI), approved by Law 11.488 of 15 June 2007, establishing a number of benefits for investment in infrastructure, which are applicable to solar energy projects;
- Regulated auctions specifically for energy from solar sources (eg, Auction 008/2015 mentioned above);
- A special regime for investment in research and development projects aimed at the development of electricity generation, in particular from renewable sources such as solar energy (Law 9.991 of 24 July 1990);
- Discounts in the tariffs for the use of Transmission Systems (TUST), and Electrical Energy Distribution Systems (TUSD), under the terms of Law 9.427 of 26 December 1996;
- Sale of electrical energy for Special Consumers[10] in the Free Contracting Environment;
- Netting system for micro-generation and mini-generation on the Distributed Generation system, established by ANEEL REN 482/2012;
- Specific Tax Benefits:
    - ICMS – a state non-cumulative tax on sales of most goods payable on import, sale or transfer of goods:
        - The ICMS Agreement 101/1997 guarantees the exemption of this tax on transactions regarding some equipment and components used for solar power generation;
        - States can also grant ICMS exemption on electricity supplied by the distributor to consumers (with a electricity generation capacity less than or equal to 1 MW) in an amount corresponding to the energy injected by them into the electrical grid, according to ICMS Agreement 16/2015 and as ruled on ANEEL REN 482/2012 (Distributed Generation). These benefits have been applied by most Brazilian States;
    - IPI – a federal excise tax levied on manufactured products also applied on imported products:
- As a rule, the IPI tax rate is zero for solar PV cells, modules and converters;
    - Social Contributions on Gross Revenue (PIS and COFINS) – which are federal contributions levied on taxpayers' monthly gross revenue:

---

10 Free consumer or a group of free consumers united by joint interests, *de facto* or *de jure*, whose load is equal to or greater than 500 kW, that has acquired energy under the form established in Paragraph 5 of Article 26 of Law 9.427 of 26 December 1996.

- Law 13.169/2015 of 6 October 2017 has reduced to zero the rates of these taxes on revenues from electrical energy supplied by the distributor to micro-generators and mini-generators on the netting system (Distributed Generation), in proportion to the electrical energy injected into the electrical grid.

- IRPJ – which is the federal corporate income tax, has reduced taxation for some investments approved by the federal government through the National Policy of Regional Development. An interesting example of one of these reduced taxation provisions is the Provisional Measure 2199-14, granting a reduction of up to 75% of the IRPJ due from legal entities that have a project which includes the generation of electricity, filed and approved until 31 December 2018, for the areas controlled by the Offices of Development of the Northeast and Amazon (SUDENE and SUDAM).[11]

5.3 Real estate licences

Before installing a PV system on a property it is important to check the local requirements, regarding real estate licensing and environmental issues.

Licensing of PV systems located in rural real estate areas must be assessed on a case-by-case basis, as several different rulings might apply, due to Brazil's specific laws regarding this type of area.

For systems installed in urban areas, the need of a specific licence will depend on whether the premises already has an operational licence from the municipal authorities, as often PV equipment does not require a specific licence. Therefore, it is always important to check the local environmental authorities' rules.

6. Distributed Generation

As previously mentioned, the development of Distributed Generation systems is responsible for a clear increase in the relevance of solar energy in Brazil, in particular the several advantages enjoyed by participants of the market and the way in which it is designed. When analysing the solar energy sector in Brazil, it is therefore essential to understand this system.

6.1 The background for Distributed Generation

Decree 5.163 of 30 July 2004 defined Distributed Generation as the production of electrical energy from projects which are directly connected to the buyer's electricity distribution system, other than that produced by hydro-electric projects with an installed capacity of more than 30 MW or thermoelectric projects, including cogeneration, with an energy efficiency of less than 75%, in accordance with the regulations of the ANEEL.

The Distribution Procedures for Electrical Energy in the National Electricity System (PRODIST)[12] defines Distributed Generation as:

---

11 The Brazilian government's autonomous agencies operating under a special regime, responsible for promoting the development of specific areas.

12 The PRODIST are rulings approved by the National Agency of Electric Energy that harmonise technical activities related to the functioning of electrical energy distribution.

Electrical energy power plants, of whatever power capacity, with installations directly connected to the electricity distribution system or by means of consumers' facilities, being able to operate in parallel or in isolation and dispatched – or not – by the ONS (National Electricity System Operator).

Distributed Generation can therefore be defined as generation produced by an enterprise located close to the load centres, and connected to the local distribution system or to the consumer unit itself.

Following studies carried out by the ANEEL aimed at reducing the existing regulatory barriers to the connection of small-scale Distributed Generation units in the distribution network,[13] the agency published the previously mentioned ANEEL REN 482/2012. This regulation established the general conditions for distributed micro-generation and mini-generation to have access to the electricity distribution system, as well as the rules for participation in the netting system.

It should be noted that ANEEL REN 482/2012, in addition to adjusting the micro-generation and mini-generation concept, has specifically anticipated a role for renewable energy sources – including solar energy – which the government wishes to encourage.

ANEEL REN 482/2012 was amended by ANEEL's Normative Resolution 687 of 24 November 2015 (ANEEL REN 687/2015), which has reduced the time for connection to the distribution network, as well as the netting system rules, and most recently by ANEEL's Normative Resolution 786 of 17 October 2017 (ANEEL REN 786/2017). The amendments introduced by ANEEL REN 786/2017 are analysed in the next section.

6.2 General aspects of micro-generation and mini-generation

In order to understand the market for Distributed Generation, it is important to analyse the concepts of micro-generation and mini-generation. Micro-generation and mini-generation are regulated by ANEEL REN 482/2012, which defines them as follows:

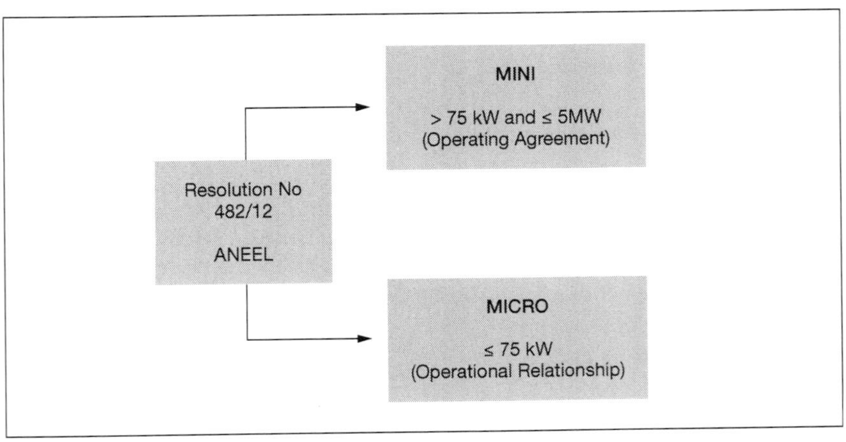

13 In the Regulated Contracting Environment, see section 3.4 above.

- Distributed micro-generation – an electricity generating plant with an installed capacity of 75 kW or less and which uses qualified co-generation, in accordance with ANEEL regulations, or renewable sources of electrical energy, connected to the distribution network through consumer units;
- Distributed mini-generation – an electricity generating plant with an installed capacity above 75 kW and not more than 5 MW and using qualified co-generation, in accordance with ANEEL regulations, or renewable sources of electrical energy, connected to the distribution network through consumer units.

It should be noted that ANEEL REN 482/2012 prohibits plants that:

- have already been registered, granted, licensed or authorised;
- have already come into commercial operation;
- have their electrical energy accounted for under the Electric Energy Trading Chamber; or
- are connected directly to an electrical energy distribution concessionaire or licence holder,

from being classified as distributed micro-generation or mini-generation, unless the project application was filed before 25 October 2017. There is also an exclusion for generating plants being split into smaller units to fit within the capacity limits for distributed micro-generation or mini-generation.

6.3 Netting system

Another key concept for understanding how the Distributed Generation market works is its netting system. The electricity netting system adopted by Brazil, under the terms of ANEEL REN 482/2012, is net metering. This system measures the flow of energy in a consumer's unit with a micro-generation or mini-generation capacity, by means of a two-way meter (the meter is able to record both the energy consumed and the energy generated at a single point of connection). Unlike the feed-in-tariff model, in which the electricity injected has a particular tariff, the Brazilian model offsets the energy consumed and the energy produced.

This netting system plays an important role in Distributed Generation, as it allows for a clear measure of the energy produced from renewable sources, in particular solar energy, due to it being the most common source of energy generation utilised in these projects.

In the netting system, the active energy injected into the distribution system by the consumer's unit is provided as a 'free loan' to the distribution company, generating a credit for a quantity of electrical energy to the consumer unit for a period of 60 months, with any future surplus energy being subject to future netting.

It is possible for a distributed micro-generation or mini-generation system to be installed in a consumer unit and, through the netting system, its surplus of energy produced to be used in another consumer's unit.

It is important to note that the netting system only promotes the netting of kWh between consumers with Distributed Generation and the distribution company, with the sale of electrical energy being prohibited.

Article 6-A of ANEEL REN 482/2012 provides that the distribution company may not include consumers in the netting system in cases where it has been ascertained, from the document proving tenure or ownership of the property where the distributed micro-generation or mini-generation is installed, that the consumer has rented or leased land, lots or properties under conditions where the value of the rent or of the lease is given in Brazilian real per unit of electricity. ANEEL REN 687/2015 states that:

equipment contracts may have clauses defining the payment of variable instalments associated with the equipment's yield and technical performance, but the value of the principal instalment should be fixed in such a way as to not characterize the sale of electricity.

Given the present legal structure, the Distributed Generation regime is having the effect of promoting the development of market participants, focusing specifically on the sale and leasing of solar power plant technology to consumers.

6.4 Practical examples of Distributed Generation

For generation next to the load, a simple example is the case of a PV unit attached to a residential home or factory. Another example is the case of projects with multiple consumer units (eg, condominiums and shopping malls) using energy independently, where each unit with individualised use represents a consumer unit and the service facilities of the common areas constitute a separate consumer unit, for which the condominium administration is responsible, with distributed mini-generation or micro-generation capacity. It is an essential requirement that the units are located on the same property or on adjoining properties.

According to the provisions of ANEEL REN 482/2012, consumers responsible for units with distributed micro-generation or mini-generation – which are members of a multiple consumer unit enterprise, which are characterised as shared generation, and which are characterised as remote self-consumption – may sign up to the netting system, allowing for projects for condominiums and shopping malls.

It is also possible to develop projects using shared generation. In this case, consumers within the same concession area of the distribution company may enter a joint venture, by means of a consortium or cooperative owning a consumer unit with distributed micro-generation or mini-generation capacity in a location different to that of the consumer units in which the surplus energy will be netted. The place where the distributed micro-generation or mini-generation is located will be considered a consumer unit, and its owner must be the consortium or the cooperative.

Another example of projects in the Distributed Generation market is the case of remote self-consumption. In this case, consumer units owned by the same legal entity (parent company and subsidiaries) with distributed micro-generation or mini-generation capacity may use that energy in a different location from the one in which the energy is generated. There is a legal limitation for these projects, as the various units must be in the same distribution concession area as that in which the excess energy will be netted.

6.5 Present status of Distributed Generation and solar energy in Brazil

According to recent figures from the National Agency of Electric Energy,[14] electricity generation by means of Distributed Generation systems has already achieved 19,260 connections (up to October 2017), with a total installed capacity of 186,218.15 kW. The source accounting for the greatest percentage is solar PV energy, with 16,592 power plants already connected.

The following graph shows ANEEL's projection for the coming years for micro-generators (installed capacity of 75 kW or less).

Figure 1: Projections for micro-generators

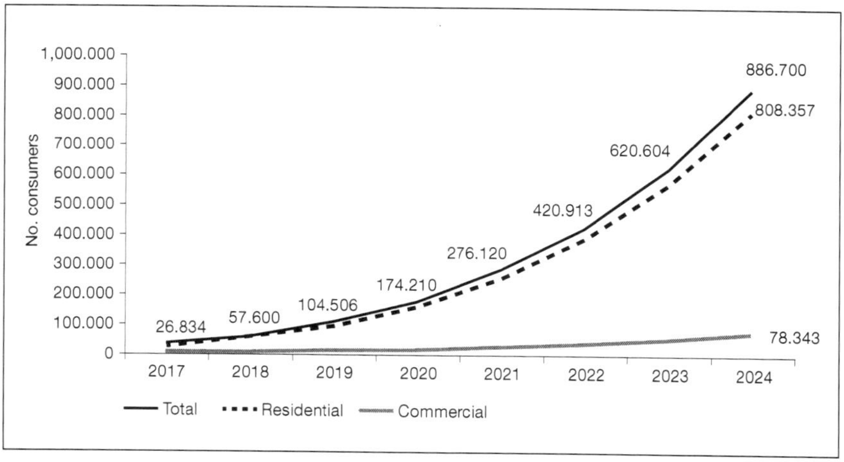

Source: Technical Note 56/2017-SRD/ANEELF

Mini-generators (installed capacity above 75 kW and not more than 5 MW) are also expected to have a significant growth, with a higher participation of commercial units.

According to Technical Note 56/2017-SRD/ANEEL, dated 24 May 2017, electricity produced by PV solar energy represents 99% of the total number of generation units and 70% of the installed capacity within the distributed generation market.

7. Outlook for Brazilian solar energy

Electricity from solar sources in Brazil has shown sustained growth, but it still represents a minor share of Brazil's energy mix, compared with other sources of energy, and also taking into account its generation potential due to the country's high solar irradiation levels.

However, the outlook for growth of solar PV electricity generation in Brazil is extremely positive. The existing incentives, together with the need to diversify the country's energy mix and the predictable increase in electrical energy consumption, will make solar energy generation increasingly relevant in Brazil.

14 Technical Note 56/2017-SRD/ANEELF.

This growth will create opportunities for power generators and consumers and for manufacturers of solar equipment and service providers. This will result in a need for the participation of foreign market participants, as Brazil does not currently enjoy a complete solar supply chain.

As detailed in this chapter, in the present scenario, electrical energy from solar sources plays a significant role in three segments:

- regulated energy auctions (Regulated Contracting Environment);
- the purchase and sale of energy in the Free Contracting Environment; and in particular,
- electricity micro-generation and mini-generation in the Distributed Generation system.

Several participants in the market are seeking further measures in order to develop Distributed Generation through micro-generation and mini-generation, and in particular, the abolition of limitations imposed on remote consumption (presently limited to the area of the distribution companies' concession), and the possibility of selling surpluses in the Free Contracting Environment. The Electric Energy Trading Chamber has recently completed a study on these issues which has already been forwarded to the federal government. The approval of these measures would be an important regulatory advance and a new milestone in the development of solar energy in Brazil.

China

Lynia Lau
Ting Su
Ernst & Young

1. Brief history of the solar sector in the region

Until the Tenth Five-Year Plan (2001–2005)[1] when China first announced plans to ramp up a renewable energy industry, China had virtually no solar industry. Under the plan, China's State Economic and Trade Commission, the predecessor of the National Development and Reform Commission (NDRC), the then economic-planning agency laid out a vision to industrialise renewable-energy industries. Specifically, the Tenth Five-Year Plan for the development of the New and Renewable Energy Industry aimed to scale up solar cell and module production by developing a robust solar supply chain in China by 2005 with an annual target of 15 MW solar-cell manufacturing capacity. However, China's solar-cell manufacturing industry grew much faster than the government had anticipated and total capacity reached 500 MW by 2005, 33 times the original target.

The production, research and development (R&D), and more extensive solar deployment within China were emphasised in the Eleventh Five-Year Plan (2006–2010).[2] More importantly, the government began funding R&D projects to address the bottleneck in technology issues. As a result, China's solar-module manufacturing capacity skyrocketed more than 17 times, from 500 MW in 2005 to 8,700 MW in 2010. It has become the world's largest manufacturer of solar panels since 2008, and by 2011 it had produced the majority of global solar photovoltaic (PV) power on an annualised basis.[3] Seven China-based solar-module manufacturers went public during this period.

More ambitious deployment goals with more specific R&D goals across the entire solar value chain, from materials to cells and modules to systems, to the tooling used in manufacturing, were pushed forward by China's Twelfth Five-Year Plan (2011–2015),[4] with a target of 21,000 MW of cumulative solar deployment in China by 2015. In fact, China has become the world's leading installer of solar PV in 2013 and the world's largest producer of PV power in 2015.[5] Industry projections estimate

1 State Council of People's Republic of China, "Tenth Five-Year Plan's New Energy and Renewable Energy Industries Development Plan", 2001, https://en.wikipedia.org/wiki/Five-year_plans_of_China#Tenth_Plan_(2001–2005); www.npc.gov.cn/wxzl/gongbao/2001-03/19/content_5134505.htm

2 China National Energy Administration, "Twelfth Five-Year Plan's Special Plan for Solar Electricity Generation Technology Development", 2012, http://182.92.181.100/uploadfile/2014/1016/20141016015752531.pdf.

3 "Annual Solar Photovoltaics Production by Country, 1995–2012", Earth Policy Institute, 31 July 2013.

4 China National Energy Administration, "Twelfth Five-Year Plan's Special Plan for Solar Electricity Generation Technology Development", 2012.

that by the end of 2017 China will have enough manufacturing capacity to produce 51 GW of solar PV per year, an amount more than twice as much as the 2010 global production of 24 GW.[6]

Further development of solar-energy technologies, including the goal of achieving large-scale commercial production of 'advanced crystalline-silicon' solar cells, have been prioritised by China and documented in the Thirteenth Five -Year Plan for Solar Energy Development by China's National Energy Administration in December 2016.[7] The plan stipulates that by 2020 China should have 110,000 MW of solar capacity installed in the country and the cost of solar-power generation should fall by half from its 2015 level.

After years of effort, the Chinese PV power generation industry has ushered in a new stage of development. Investment in solar power has experienced a rapid growth. For instance, investment in the first PV power plant project in Eerduosi, Neimengu, which started in 2006 and was put into operation in 2008, was Rmb 21 million. In 2009, the level of investment in the 10 MW solar power plant project by the Chinese government in Dunhuang, Gansu was Rmb 500 million. Investment in the city has increased since then and there were projects totalling approximately 1,090 MW approved by the government up to 2014. The 1 GW solar power plant project, which is to be developed in three stages in Chaidamu Basin, Qinghai, began in 2009 – the size of the first stage was 30 MW and the investment involved was Rmb 1 billion.[8] China's investment in solar power generation was estimated to be more than Rmb 250 billion during the Twelfth Five-Year Plan.

China currently hosts the world's largest number of solar power plants with a total installed capacity of close to 80 GW in 2016, nearly twice the amount of the United States, and contributed almost half the additions to global solar capacity (see Table 1 below). In early 2017 the largest solar farm in the world – the 850 MW Longyangxia Dam Solar Park with an area of 30 square kilometres – was started in China and the world's largest floating solar farm also commenced operation in Huainan, Anhui Province.

Table 1: Solar power in China

Year	Capacity (MW)	Installed/yr
2001	23.5	4.5

continued on next page

5 See www.reuters.com/article/china-solar-idUSL3N15533U; http://cleantechnica.com/2016/01/22/china-overtakes-germany-become-worlds-leading-solar-pv-country/; and http://cleantechnica.com/2016/07/07/china-installs-18-6-gw-solar-pv-2015-connected/.

6 "Annual Solar Photovoltaics Module Production in China, 2007–2013, with Projection to 2017", Earth Policy Institute, 8 July 2014; "World Solar Photovoltaics Production, 1975–2012", Earth Policy Institute, 31 July 2013

7 China National Energy Administration, "Thirteenth Five Year Plan for Solar Energy Development", December 2016, http://zfxxgk.nea.gov.cn/auto87/201612/t20161216_2358.htm.

8 See www.qhnews.com/newscenter/system/2009/01/09/002665434.shtml.

Year	Capacity (MW)	Installed/yr
2002	42	8.5
2003	52	10
2004	62	10
2005	70	8
2006	80	10
2007	100	20
2008	140	40
2009	300	160
2010	800	500
2011	3,300	2,500
2012	8,300	5,000
2013	~17,800	~9,500
2014	28,199	10,560
2015	43,180	15,130
2016	77,420	34,540
2017	130,250	52,830
Source: IEA, China National Energy Administration (2015 figures), China Energy Portal (2016[9] and 2017[10] figures).		

The Chinese solar market is dominated by several major power enterprises engaging in different sectors,[11] such as CHINT Group, Trina Solar Ltd (one of the largest Chinese vertically integrated solar PV manufacturers), JA Solar Holdings Co

9 See https://chinaenergyportal.org/en/2016-detailed-electricity-statistics/.
10 See https://chinaenergyportal.org/en/2017-electric-power-industry-summary-statistics/.
11 See www.greenworldinvestor.com/2017/02/02/12-biggest-and-best-solar-panel-companies-in-the-world/.

Ltd and JinkoSolar Holding Co Ltd engaging in manufacturing; GCL-Poly Energy Holdings Ltd and CGN Solar Energy Development Co Ltd engaging in plant investment; State Power Investment Corporation engaging in power generation; Shunfeng Clean Energy engaging in power station operation and solar power services; and Longi Group which has a presence in the entire solar supply chain.

The solar power markets have been very competitive in China and because the cost of funds is comparatively low for Chinese enterprises and local investors are keen to invest in those projects as a result of the incentives provided by the government, the level of foreign investment is relatively low.

The key terms of the power purchase agreement (PPA) and grid connection arrangements are governed by the Renewable Energy Law of the People's Republic of China issued on 26 December 2009 (REL).

Article 14 of the REL requires grid enterprises in China to give priority to the connection of renewable power generation and to buy the grid-connected power produced by renewable energy. Grid enterprises shall enter into grid connection agreements with renewable power generation enterprises that have been built in accordance with the renewable energy development and utilisation plan and have legally obtained an administrative licence, or for which filing has been made. They will then buy the power produced with renewable energy provided that it conforms to the technical standard of grid connection within the coverage of their power grid. Power enterprises are obliged to work with grid enterprises to ensure the safety of the power grid.

Grid enterprises are also obliged to strengthen the construction of the power grid, expand the scope of renewable energy distribution, develop and apply the use of technologies such as smart grid and power storage, perfect the operation and management of power grids, enhance the capability of absorbing renewable energy electric power, and provide grid-connection services for the generation of power from renewable energy.

Article 29 of REL specifies that if power grid enterprises breach Article 14 and fail to purchase in accordance with renewable power regulations in full, which results in economic loss to the renewable power generation enterprises, such power grid enterprises shall be liable to pay compensation, and the national power supervisory institutions shall order them to make a correction within a stipulated period of time. In the case of a refusal to make such a correction, a fine covering full compensation for the economic loss of the renewable power generation enterprises will be imposed.

2. Legislative and regulatory background and support regime

2.1 Legal framework, relevant laws and regulations

In China, solar power belongs to the renewable energy sector and the overall legal framework is a complex system with distinct hierarchies. There are national laws, ministerial regulations, guiding opinions, measures and procedures, local rules and regulations, self-regulation rules of the industry and internal governance rules for each of the state-owned power and grid companies.

The National Development and Reform Commission (NDRC) regulates pricing, the State Electricity Regulatory Commission (SERC) focuses on administration and market regulations and the state-owned Assets Supervision and Administration Commission (SASAC) supervises power corporations. Other specific regulations, such as environmental issues and land use rights, are managed by the corresponding governmental bodies that regulate project development, construction and operations.

The main relevant laws and regulations in the development, construction and operation of solar power projects are listed below, including some of the proposed rules and procedures, but it should be noted that this list is subject to change due to ongoing reforms.

National laws (approved by the National People's Congress)

- The Renewable Energy Law
- Energy Conservation Law
- The Environmental Impact Assessment Law (Law of the People's Republic of China on Evaluation of Environmental Effects)
- The Public Bidding Law

Central governmental regulations

State council regulations

- Regulations on Supply and Utilisation of Electricity
- Regulations on Power Grid Dispatch Administration

Market supervision regulations

- Measures of Power Markets Regulation (Trial) (SERC)
- Basic Rules of Power Markets Operation (Trial) (SERC)
- Guiding Opinion of the Construction of Regional Power Markets (SERC)
- Tentative Measures of Pilot Projects on Direct Power Purchase from Generation Enterprise by Users (SERC and NDRC)
  With:
  - Model Contract of Direct Purchase of Electricity by Major Users (Trial)
  - Model Contract of Direct Purchase of Transmission and Distribution of Electricity by Major Users (Trial)

Pricing regulations

- Tentative Measures of Price Administration of Electricity Uploaded to Grids (NDRC)
- Tentative Measures of Price Administration of Transmission and Distribution of Electricity (NDRC)
- Tentative Measures of Price Administration of Electricity Sales (NDRC)
- Tentative Rules of Price Review of Inter-Regional Transmission of Electricity (SERC)

Grid operation regulations

- Grid Operational Rules (Trial) (SERC)
- Tentative Rules of Inter-Regional and Inter-Provincial Power Dispatch Optimisation (SERC)

Regulations on state-owned enterprises

- Tentative Measures of the Administration of Generation Right Trading (SERC)
- Measures of the Entry of Newly-installed Generation Set into Commercial Operation (Trial) (SERC)
- Implementation Measures for Disclosure of Information of Electricity-Supply Companies (Trial) (SERC)

Other administrative regulations and rules

- Measures of Electricity Accountability Management (SERC)
- Measures of Electricity Demand-Side Management (SERC and NDRC)
- Measures of Standardisation of Power Industry (Trail) (Energy Bureau, NDRC)
- Measures of Environmental Protection in Power Industry (MEP)
- Land Quotas for Electricity Project Construction (Thermal Power Plant, Nuclear Power Plant, Substation, Converter Station) (MLR and Ministry of Housing and Urban-Rural Development)
- Notice on Implementation of Photovoltaic Power Generation Project Construction in 2016 of National Energy Administration (NEA)
- Notice on Establishment of a Monitoring and Evaluation System for the Market Environment to Lead the PV Industry Development (NEA)

Examples of local rules

- Tianjin Municipality: Rules of Electricity Supply and Utility
- Hunan Province: Provisions on Electricity Development

Examples of self-regulation rules of the CEC and of the state-owned power companies and grid companies

- Measures of Review of Industry Standard of Electricity Sector (CEC)
- Specification of Establishment of Industry Standard of Electricity Sector (CEC)
- Measures of Connection of New Power Plant to the Grid (Southern Grid Co)

2.2 Government policies and incentives

The central government prioritises the development and use of renewable energy, and emphasises the development of clean energy industries, including solar energy. A renewable energy development fund has been established and is funded by the central government. The central government can also grant tax benefits. In addition, financial institutions can offer preferential loans with a financial interest subsidy to renewable energy development for projects that are listed in the National Renewable Energy Industrial Development Guidance Catalogue.

Grid operators are obliged to schedule power with a preference for renewable sources of generation as follows:

- renewable power generation (RPG);
- power generation from energy-conserving, environmentally friendly and high-efficiency thermal power units.

In addition, the central government has established a clean development mechanism fund to support construction and industrial activities that are beneficial to strengthen proper responses to climate change.

Corporate investment in solar installations is met with generous incentives. In 2013, the Chinese government introduced new feed-in tariffs (FiTs), deployed at both state and provincial levels to fuel the growth of distributed solar rooftop installations. The central government currently provides 20-year subsidies of US$0.06/kWh of output from solar rooftop projects. For surplus power, Rmb 0.40/kWh is awarded from the state grid for any surplus power they generate.

The National Energy Administration aims to deregulate the electricity market for small distributed solar generation, including rooftop projects. Projects producing fewer than 6 MW of energy are exempt from the requirement to hold a power generation licence. Since 2014, payments and subsidies have been streamlined to aid solar projects undertaken by individuals and private enterprises. Industrial and commercial entities setting up solar rooftop projects are offered a payback period of seven to nine years with an 8% investment rate of return.

China also subsidises its solar manufacturing industry, which includes rebates on value added tax and corporate income tax, the provision of land and leased buildings for factories at below-market rates, and administration by provincial and local governments rather than by the central government. The size and breadth of these sub-national subsidies vary significantly depending on the locale.

Recently, the central government has been shifting the focus of its solar subsidies from manufacturing to deployment. China is now redesigning its solar-deployment subsidies, restructuring them for more economic efficiency, and learning from past policy-design mistakes both in China and in other countries that have heavily subsidised solar deployment.

3. Environmental, real estate and planning issues

Developers for solar power projects in China often get off to a late start in the preliminary work, and the development progress is slow in going through procedures for land, forestry, environmental protection, water conservation and power grid access. Typical difficulties include:

- the change of site due to the nature of the land;
- the length of time in coordinating the change of site,
- leasing period, and grassland occupancy;
- an increase in land cost in some projects;
- coordination of liaison between forestry and land departments;
- requirements of environmental protection and water protection for the fire and corrosion resistance technology for solar thermal power generation to be discussed;
- procedures and requirements for power grid access system, and guarantee of absorption.

Some local governments are not acquainted with solar thermal power projects in the processing of procedures for preliminary work, resulting in longer processing times. China does have abundant solar energy resources with more than 60% of its land area having abundant solar radiation, especially in the northwest, Tibet and Yunnan regions. Although China is rich in solar resources, the provinces that receive the most sunlight are predominantly rural and not well connected to the national power grid.

Much of the solar power generating capacity has been built in the relatively less populated western part of China, whereas the major electricity consumption areas (such as Shanghai and Beijing) are in the east, thereby causing difficulties and expense in transferring the energy between regions with curtailment rates of 30% or more reported in 2017,[12] but there is no significant issue on real estate and planning or zoning of the solar power projects.

China is a large producer of polysilicon which is used in first generation solar cells around the world. A by-product of the process is the toxic silicon tetrachloride, which is normally processed and recycled at a higher cost in the developed world, but sometimes dumped by Chinese green energy start-ups.[13] On 15 September 2011, more than 500 people from the village of Hongxiao protested over the death of a large number of fish in a nearby river. Some of the protesters stormed the factory compound of Zhejiang Jinko Solar Co Ltd, a subsidiary of Hong Kong-invested JinkoSolar Holding Co Ltd (NYSE Stock Code: JKS). An official of Haining's environmental protection bureau confirmed the factory's waste disposal had failed the pollution tests since April 2011 and the environmental watchdog had warned the factory but it had not effectively controlled its pollution.[14]

4. Power off-take arrangements

Since the Renewable Energy Law was enacted, China has significantly increased renewable energy subsidies. The original Renewable Energy Law expressly stated that the costs associated with FiTs and the reasonable costs associated with connecting renewable generators to the grid were to be shared nationwide through a surcharge on electricity end-users.

China's rapid deployment of renewable energy since the enactment of the Renewable Energy Law has been primarily funded through a national surcharge on electricity consumption. The NDRC issued the Interim Measures on Renewable Energy Electricity Prices and Cost Sharing Management in 2006, which directed the NDRC's pricing department to set a nationwide renewable energy surcharge levied on electricity users at a unified standard based on the users' consumption of electricity. China first implemented its solar FiT in 2007 and then developed a national FiT by providing a minimum price above the prevailing government-set power rate with adjustments for projects located in different regions in order to promote the economic development of underdeveloped areas. In 2013, China

12 Chris Baraniuk, "Future Energy: China leads world in solar power production", BBC News, 22 June 2017.

13 "Solar Energy Firms Leave Waste Behind in China", *Washington Post*, 9 March 2008.

14 "Protest over factory pollution in E China enters third day", *China Daily*, Xinhua, 18 September 2011: "Hangzhou – Hundreds of villagers in East China's Zhejiang Province protested for the third day on Saturday at a solar panel manufacturer, whose parent is a New York-listed firm, over concerns of its harmful wastes".

restructured its FiT to produce a minimum investment rate of return of 8% for investors in solar projects by taking into consideration the quality of the solar resource of the projects and required the country's grid operators to buy all the solar power that solar generators produced.

The policies, such as the Interim Measures on Revenue Allocation from Renewable Surcharges and the Interim Measures on the Renewable Energy Development Special Fund, defined the scope, standard and collection of the tariff surcharge subsidies for renewable energy.

Tariffs of solar PV power in China have gone through the approval tariff stage, bidding price stage and fixed price stage. The specific implementation of solar PV FiT is as follows. In 2007, the solar power price of Inner Mongolia, Shanghai and Kunming approved by NDRC was Rmb 4/kWh. In 2009, the FiT for the first batch of concession projects was Rmb 1.09/kWh, while the temporary FiT for the solar PV power plant project in Ningxia was Rmb 1.15/kWh in April 2010. In October 2010, the lowest FiT for the second batch of concession projects was Rmb 0.73/kWh, while the highest was Rmb 0.99/kWh. In July 2011, China's energy regulator announced a new circular on the FiT for solar PV power.

The Improving Policies on the Feed-in Tariff of Solar Photovoltaic Power, issued in 2011, introduced a unified FiT for solar PV power projects. The national FiT is Rmb 1.15/kWh for projects completed by 31 December 2011 and Rmb 1.0/kWh for projects approved by July 2011 but not completed before the end of the year. The NDRC will from time to time adjust the price according to certain factors such as investment cost changes and technical advances.[15] It is reported that the Beijing government has decided on the FiT rates for solar power projects for the year 2017. The new FiT rates will be higher than figures proposed in the government's previous plan. For the year 2017, FiT rates for PV projects will differ from region to region according to a datasheet acquired by PVmen, a Chinese online network. The FiT rate for PV projects in Resource Region I will be Rmb 0.6/kWh, Resource Region II will be Rmb 0.7/kWh and Resource Region III will be Rmb 0.8/kWh. Meanwhile, certified and grid-connected PV projects in Tibet will receive a FiT subsidy of Rmb 1.0/kWh.

The Interim Measures on Renewable Energy Surcharge Electricity Subsidy Funds Management issued by the NDRC stipulated that grid connection expenses paid by grid companies for the purchase of renewable power and other reasonable expenses may be included in the power transmission cost of grid companies and retrieved from the selling price.[16] Depending on transmission distances, renewable power will be collected at Rmb 0.01/kWh for distances within 50 km, Rmb 0.02/kWh for 50 to 100 km and Rmb 0.03/kWh for more than 100 km.[17]

5. The future for the solar sector in the region

Most of China's solar capacity has been installed by state-owned power companies and independent power producers. China's largest state-owned power companies, in

15 See www.sciencedirect.com/science/article/pii/S1364032113002785#bib22.
16 See www.sciencedirect.com/science/article/pii/S1364032113002785#bib25.
17 See www.sciencedirect.com/science/article/pii/S1364032113002785#bib26.

line with government policy, have grown increasingly aggressive in deploying solar projects in the country. For example, China Longyuan Power Group, a subsidiary of China Guodian Group, which is one of the country's top five state-owned power producers, has become the country's largest solar-power producer. It has been particularly active in solar deployment in western China, especially in Qinghai Province, home to some of China's best solar resources and some of its largest solar farms.

Private investment groups are becoming major solar market participants in China. One such firm, China Minsheng New Energy Investment Co, is developing in the Ningxia region of northwest China what will be, if finished according to plan, one of the biggest solar projects in the world. The solar farm is stated to have a capacity of 2,000 MW; as of mid-2016, half the project had been constructed and roughly 380 MW of its capacity had been connected to the grid, according to Minsheng, which has said it intends to spend ¥100 billion ($14.5 billion) over the next five years to deploy some 12,000 MW of solar capacity.[18]

China's solar manufacturers are a small but growing – and globally influential – driver of the country's massive development of solar power. Even though development by China's solar manufacturers represents a small slice of China's total solar deployment, it is of particular relevance both because the slice is getting bigger and because it constitutes a major new force shaping the solar industry in China and around the world. Project development is a way for these manufacturers to expand demand for their modules in an increasingly competitive market. It also typically offers them higher profit margins than manufacturing alone. Project development is therefore increasingly important to the profitability of China's top solar manufacturers – companies that are defining the solar industry's global future.

According to the US Department of Energy, a panel's lifespan ranges from 20 to 30 years, depending on the environment in which they are used. High temperatures can accelerate the ageing process for solar cells, while other negative factors – such as the weight of snow or dust storms – can cause material fatigue on the surface and internal electric circuits, gradually reducing the panel's power output. It is suggested that the country's cumulative capacity of retired panels would reach up to 70 GW by 2034. By 2050, these waste panels would add up to 20 million tons, or 2,000 times the weight of the Eiffel Tower, according to Lu Fang, secretary general of the PV division in the China Renewable Energy Society. China may have the world's largest issue with ageing solar panels in less than two decades, according to a recent industry estimate. It is believed that the Chinese government may develop and implement respective environmental-related policies to sustain the future development of the solar sector.

China's tariff policy has promoted the rapid development of solar energy but certain policies may hinder or jeopardise the future healthy development of the solar sector. There are issues in the specific implementation of tariff policy in respect of the

18 "Out of China's Dusty Northwest Corner, a Solar Behemoth Arises", Bloomberg News, 19 September 2016, www.bloomberg.com/news/articles/2016-09-20/out-of-china-s-dusty-northwest-corner-a-solar-behemoth-arises.

FiT, grid connection expenses and tariff surcharge subsidies. For instance, due to the intensive bidding competition for solar PV power generation projects, the bidding price has materially deviated from the actual cost of investment, which in the long run will affect the upstream industry chain and damage the sustainable development of the solar power industry.

The subsidies for grid connection are also insufficient to cover the investment in the grid connection and the return on investment can only barely cover the debt service costs of power transmission and transformation. This inadequate funding may affect the normal production and operation of the power grid companies. Hence, it is expected that Chinese policy-makers will strengthen the FiT policy and the implementation guidelines in order to provide a reasonable return on investment in the solar sector by taking into account issues such as environmental protection, quality control and the long-term sustainable development of the solar sector.

China's solar industry and market is often misunderstood in the West. There is a perception that the global solar industry is centralising in China and that its solar market is largely closed to foreign investment. However, researchers and scholars have found recently[19] that China's dominant solar companies are, as they grow larger, increasingly accessing international capital and spreading their operations across the globe. Senior Chinese officials and corporate executives are eager to apply a range of more efficient financing mechanisms that have been used in the West to dramatically scale up solar deployment in China. This new-found interest on the part of Chinese leaders represents a potentially valuable opportunity for the United States and other countries in the West. Another Western perception is that China does not innovate, but uses other countries' technological advances to reduce its manufacturing costs. While it is true that manufacturing processes have been China's strength, the country is intensifying its solar R&D efforts and Chinese researchers are producing increasingly notable results in solar research.

Solar power currently supplies only about 1% of global electricity, but the International Energy Agency predicts that solar PV technology could grow to 16% by the middle of this century. Others predict even greater growth. China is not only leading the world in terms of manufacturing solar equipment, but it is also the largest developer of solar energy. China is also getting increasingly competitive in R&D, an area that the United States and European countries have historically dominated. China is now the major driver of the global solar industry and it is expected that China, while still capitalising on its strengths, will work more closely with other countries in the future.

19 See https://energy.stanford.edu/news/future-solar-power-china-us-and-world.

France

George Rigo
Orrick, Herrington & Sutcliffe LLP

1. Brief history of the solar sector in France

The existing regulatory regime of the French renewable energy sector was, in large measure, put in place back in 2000 by means of the feed-in-tariff mechanism which applied to different technologies including wind, solar, tidal and biomass.

Power purchase agreements were entered into on this basis by electricity producers which provided a fixed electricity purchase price for the entire term of the contract. This incentive proved to be a deciding investment factor in this sector. The level of the applicable feed-in tariff was dependent on the technologies used and the date of connection to the grid.

Since the *Grenelle de l'Environment* and the 2015 United Nations Climate Change Conference (COP 21) – a series of political agreements regarding the environment and sustainable development – there have been several amendments to the existing regulatory framework in order to set out additional market incentives for the renewable energy sector.

The French government did not wait for the entry into force of the Paris Agreement in 2016 to enact regulations and take effective steps to encourage the development of green energy. Since 2015, under the Energy Transition for Green Growth Act (ETGGL), the French government committed to increase renewable energy consumption by up to 32% by 2030, which, if successful, would mean that renewable energy would represent 42% of the total energy produced in France.

The ETGGL and its related action plans were designed to improve the regulatory framework for green energy, including through regulatory simplification measures and the implementation of new incentives. Such policies have contributed over the last few years to a steady progression in the number of new photovoltaic installations.

In the solar sector, the French government's efforts have, in particular, taken the form of a launching of tender procedures for the construction and operation of solar facilities. Under such procedures, the electricity purchase price (or amount of the premium) is to be proposed by the tenderers. Such a mechanism is frequently used in areas in which investments are considered to be insufficient, for instance in isolated areas.

Taking a step further, the French government announced in October 2016 a detailed multiannual energy plan (MEP)[1] which notably aims at increasing the total

1 See Decree 2016-1442 dated 27 October 2016 relating to the *Programmation Pluriannuelle de l'Energie* (PPE).

solar capacity (in particular through regular tenders for building-integrated photovoltaics, rooftop photovoltaic and ground-mounted solar power plants).

2. Overview of market participants/practices

Nourished by its new incentive regimes at all levels of the value chain in the solar energy sector, the French market is well equipped with specialists all along the production line, starting from high quality research centres, industrial equipment manufacturers, installers and energy producers.

Major transactions occurred in France in 2015, notably the sale by BayWare of its 'La Coste' project (58 MW) near Bordeaux to Allianz Global Investors, and the sale of a 57.3 MW project in the Aquitaine region by GP Joule GmbH to Aquila Capital (ranked, respectively, the fourth and fifth major transactions in Europe by Solarplaza International[2]).

Also in late 2015 the largest solar farm in Europe was inaugurated in Cestas (near Bordeaux). The project, developed by Neoen, was developed over 250 hectares and has a total installed capacity of 300 MW. The plant is comprised of 25 sub-plants (each of 12 MW installed capacity) and was connected to the grid via a consortium of mainly France-based contributors, including French companies Eiffage and Schneider Electric.

The recent developments in the French market all seem oriented at consolidating the sector and the major market players are in effect all investing in companies focusing on development of solar energy projects. In September 2017, the leading petroleum group, Total, announced two investments aimed at reinforcing its presence in the renewable energy sector:

- an equity investment in Eren Renewable Energy (a total investment of €237.5 million for 23% of the share capital, with the option for Total to acquire a controlling stake after a period of five years); and
- the acquisition of GreenFlex, the French specialist in energy efficiency (completion is currently pending, subject to merger control approval).

The major French power supplier, EDF, is also focused on developing its footprint in the renewable energy sector, notably through its specialised affiliate, EDF Energies Nouvelles.

Although most developers are French, there is increased activity by foreign investors in the secondary market as they are keen to invest in good operating assets in France which is seen as a stable investment environment.

Along with the traditional development of small and medium-sized solar farms and their transfer and operation, new forms of risk and profit sharing have appeared over the last few years in the French market, adapted to larger projects. An example is the incorporation of an *Association Syndicale Libre* (ASL), established in accordance with the provisions of the French Community Planning Code (*Code de l'Urbanisme*). An ASL is a specific vehicle, organised by several entities, each owning a small-sized solar farm on the territory of a larger project. The ASL would own and run the shared

2 See www.solarassetmanagementeu.com/top-20-transaction/.

infrastructure of the project, such as substations, and each individual member of the ASL would own and run its own solar farm. The financing, installation and maintenance of the solar farms organised within an ASL are all negotiated at the level of the ASL and are binding on the individual solar farms forming part of the ASL.

Some solar sector investors might find this new form of structuring unattractive as the conditions of financing and operation of the solar farms are fixed in advance. In addition, decision-making is sometimes limited, the developer keeping a participation in the ASL and reserving certain veto rights on material decisions. However, such shared projects allow for 'turnkey' investments in large initiatives and, as such, may be considered appealing to investors which are not familiar with the renewable energy sector.

Another form of organisation was recently adapted to better meet the requirements of the renewable energy sector. The *Société Coopérative d'Intérêt Collectif* (SCIC) is a commercial company whose purpose is to serve a collective interest. The functioning of this type of company is, however, very specific as decision making is not based on the traditional link between the amount of investment and number of voting rights, but abides by the rule of 'one man, one vote'. In addition, the possible return on investment in an SCIC is limited by law with specific rules on generating and distributing profits. The use of this kind of legal structure is, therefore, very limited and mainly linked to the development of projects in local markets.

Also, as the result of recent legislation, crowdfunding has become available for financing both the equity and the debt requirements in renewable energy projects. Although this contributes to the general efforts of the government to facilitate access to financing in the renewable energy sector, such forms of financing remain, for the time being, limited both in terms of the number and amount of financings.

Traditionally, investments are structured in France within special purpose vehicles (SPVs), which are corporate entities with limited liability (most often a limited liability company (SARL) or simplified joint stock corporation (SAS)) holding the assets, authorisations and contracts necessary for the operation of a specific project. Such structuring is a practical means of isolating the risks of an individual project, as investors' liability is limited to the amount of share capital of the relevant SPV. Therefore, and except in certain specific cases (such as *de facto* management of the SPV by the investors or if the SPV is proved to be a fictitious construct), the eventual losses of a specific project lodged in an SPV would not affect the assets of other projects operated in different SPVs within the same group of companies. The cases where it is possible to pierce the corporate veil or to extend liquidation procedures beyond the SPV are rare in practice. This is not to say that there are no other structures available to invest in the solar sector in France, but the great majority of investments have been made through SPVs, notably to facilitate their disposal. From time to time, an SPV might also take the form of a joint venture between two or more parties so as to ensure respect of the regulations pertaining to the distance between two solar farms.

3. Legislative/regulatory background and support regime

The French government regularly publishes new targets for solar projects with an ambitious tendering schedule. Indeed, within the MEP, the French government has detailed its strategic priorities in terms of energy policy and focused on reaching the objectives enshrined in the ETGGL.

The objectives contained in the MEP are to allow the increase of installed capacity for electricity generation from renewable energy sources by 70% compared with their 2014 level (41 GW) to reach a total capacity of between 71 and 78 GW in 2023 (this corresponds to an increase of production of renewable electricity by 150 to 167 TWh in 2023). Accelerating the development of photovoltaic projects, particularly aimed at more competitive solutions such as ground-level solar panels, while at the same time ensuring that the locations chosen for solar facilities respect natural areas and farmland, are the main actions set out in the MEP for the solar sector. The quantitative targets are aimed at increasing installed solar photovoltaic capacity to 10.2 GW in 2018 and increase it from 18.2 to 20.2 GW by 2023. A clear programme and regular timetable for calls for tenders are planned in order to permit the launching of tri-annual tenders for ground-based power plants, with capacities from 0.9 to 1.2 GW per year and give visibility to the sector.[3]

In order to reach these ambitious objectives, the MEP provides for several concrete actions for the 2016–2017 period, including the setting up of new support mechanisms (*premium mechanism*) and the enacting of new tariff decrees for solar installations still benefiting from power purchase undertakings; these were published in May 2017.

It should be noted that, pursuant to the European Commission's Guidelines on State aid for environmental protection and energy 2014–2020, the ETGGL provides that some renewable energy facilities will no longer benefit from the feed-in-tariff mechanism but instead from a 'premium' granted in addition to the market price. This mechanism aims, on the one hand, to provide financial support for renewable energy production and, on the other hand, to prevent overcompensation at public expense. Solar energy facilities are not eligible for this new mechanism. In fact, except for projects awarded following a call for tenders, only solar energy facilities located on buildings and whose peak power does not exceed 100 kWp may now benefit from financial support from the French government (under the form of a purchase price obligation).

Besides these reforms, the authorisation regime under the French Energy Code was simplified by increasing very significantly the power thresholds of renewable energy installations under which the latter are deemed to be authorised. There is also a constant effort from the French public authorities to streamline the applicable administrative procedures to benefit from the purchase obligation with, for example, the abolition of the CODOA[4] procedure, which used to be a preliminary requirement before being able to sign a power purchase agreement.

In addition, Decree 2016-399 of 1 April 2016 was enacted in order to shorten the time necessary to obtain the connection of new electricity generating installations from renewable energy sources to the grid and the government has indicated its

3 Ministry of Ecology, Sustainable Development and Energy, *Multiannual Energy Plan – Executive summary.*
4 Certificate giving entitlement to purchase obligation.

objective is to simplify the conditions applicable to participatory financing of projects (ie, crowdfunding) for the production of renewable energy on platforms.

4. Power off-take arrangements

As a consequence of the evolution of the regulatory framework in France (see section 3 above), for a significant proportion of all new capacity (especially for the majority of installations with a capacity of more than 1 MW from 2017 onwards), development of renewable energy facilities is to be subject to a tender system.

In order to accelerate the development of renewable energy and to achieve the objectives outlined above, a simplification of the tender process has been initiated. To date, the timeline from the decision to launch a tender to the selection of the winners ranges from 18 to 27 months. The procedures were simplified to significantly reduce this timeframe. Furthermore, the specifications for these tenders have now been drafted and finalised by the minister in charge of energy.

The timetables for feasibility studies, agreeing specifications and their publication have all been shortened. Files are now submitted electronically and the specifications may provide for the creation of an automated filing system for online offers in order to further reduce the timeframe, in particular for sectors where projects are standardised.

A further improvement has been the adoption of an "open" tender planning process, to give visibility to manufacturers. Beginning in 2016, multiannual calls for tender, based on common specifications and successive application periods, have been systematised according to the following indicative timetable:

Indicative timetable for tenders for solar energy[5]

Provisional Timetable	2017				2018				2019	
	T1	T2	T3	T4	T1	T2	T3	T4	T1	T2
Solar (ground-level)		Deadline 2 (500 MW)		Deadline 3 (500 MW)		Deadline 4 (500 MW)		Deadline 5 (500 MW)		Deadline 6 (500 MW)
Solar (buildings)	Deadline 2 (150 MW)	Deadline 3 (150 MW)		Deadline 4 (150 MW)	Deadline 5 (150 MW)	Deadline 6 (150 MW)		Deadline 7 (150 MW)	Deadline 8 (150 MW)	Deadline 9 (150 MW)

The simplification of administrative procedures for the installation of new renewable electricity production facilities is an important lever for accelerating the development of renewable energy in France.

As stated above, administrative procedures to benefit from the purchase obligation have also been simplified and the obligation to obtain a CODOA beforehand has been removed.[6] In addition, the period of validity of planning

5 Ministry of Ecology, Sustainable Development and Energy, *Multiannual Energy Plan – Volet relatif à l'offre d'énergie*, p8.

6 Decree 2016-691 of 28 May 2016.

authorisations for all renewable energy generation works can now be extended several times, within a period of 10 years.[7]

For a facility that was not selected through a tendering procedure, there are nevertheless three ways to sell power:

- directly to a supplier – the terms (including the price of the power) of the contract entered into between the producer and the supplier are freely set by the parties;
- on power exchanges – the terms of the sale depend on the products offered on the exchanges; and
- to EDF-OA or the relevant local distribution system operator when the facility is eligible for the purchase obligation mechanism (ie, facilities located on buildings and whose peak power does not exceed 100 kWp) – the main characteristics of the contracts entered into for the sale of the power are set by a specific order for solar facilities; the contract template is approved by the minister in charge of energy.[8]

5. Environmental, real estate and planning issues

5.1 Planning requirements

Solar photovoltaic (PV) plants can either be ground-based or installed on rooftops. Planning requirements differ according to each type of installation.

(a) Ground-based solar PV plants

Ground based solar PV plants are considered new buildings and must comply with the planning rules applicable within the area where they are to be erected. Article R. 421-1 of the French Planning Code implements three procedures, depending on the solar plant's height and production capacity.

- For structures under 1.8 metres from ground level or for solar facilities where output does not exceed 3 kWp (kilowatt-peak), no authorisation is needed.[9] Decree 2016-687 issued on 27 May 2016 also exempts from planning authorisation solar plants receiving solar radiation of less than 50 MW.
- For solar facilities which exceed 1.8 metres high and 3 kWp or receive more than 50 MW of solar radiation, but remain under 250 kWp (kilowatt-peak), a prior declaration is needed. The prior declaration is sent to the offices of the municipality in which the construction works are to be conducted. The prior declaration is valid for three years after its notification.[10] For solar plants to be established on a registered site (ie, sites listed by the French administration that are deemed natural monuments or sites deemed of particular interest from an artistic, historic, scientific perspective or are deemed picturesque), nature reserve or protected landscape and which have a production capacity

---

7 Decision 2016-6 of 5 January 2016 on the period of validity of planning authorisations and various provisions relating to the application of soil law and associated taxation.

8 Article R. 314-2 of the Energy Code.

9 Article R. 421-2 of the Planning Code.

10 Article R. 421-9 of the Planning Code.

under 3 kWp, a prior declaration is required regardless of the size of the solar facility.

- Finally, for every solar plant where peak power production exceeds 250 kWp, a building permit is mandatory. In order to obtain a building permit for such construction (whether in protected areas or not), an environmental impact assessment and a public inquiry must be conducted.

(b) Rooftop solar PV plants

Rooftop solar PV plants can be erected on pre-existing buildings or on new constructions. When erected on to a pre-existing building, they are not subject to the same authorisations as ground-based solar facilities as they are considered to be an external modification of the building rather than a new construction. Therefore a building permit is not mandatory, however, a prior declaration is required for this type of installation as the external appearance of the building is deemed to be modified. An exception applies for solar panels erected on historical buildings which always require filing for a building permit.[11]

When a rooftop solar PV plant is part of a new building project, a building permit is mandatory for both the new building construction and the solar panels. Depending on the building project, several administrative authorities may be required to give their approval.

5.2 Real estate requirements

Under French law, real estate requirements depend on the type of solar plant involved – if ground-based, a long-term lease (eg, an emphyteutic lease *(bail emphytéotique)* or construction lease) will be preferred as the lessee will benefit from *in rem* rights (ie, rights resembling the property rights of an owner). For the erection of solar panels on a rooftop for a more limited period of time, the operator will, most often, resort to civil leases or division by volumes (see below).

All construction, including ground-based solar panels, erected under an emphyteutic lease are governed by Articles L. 145-1 to L. 145-13 of the French Rural Code. These articles allow the lessee to become the owner of any movable and immovable assets located on the leased property, whether the assets are already built or are to be built, for the duration of the lease.

The lessee is entitled to build or modify any construction on the site which is subject to the lease, which duration must necessarily be between 18 and 99 years (Article L. 145-2 of the French Rural Code). The lack of an obligation on the lessee to build is the main element which differentiates an emphyteutic lease from a construction lease, the latter providing the obligation to build. However, the two types of long-term lease grant a broader freedom of land use to the lessee. Indeed, not only is the ownership of land regarded as bare ownership, but also termination of the lease is prohibited, except in the case of a serious breach of contractual obligations, which termination is under the control of the French courts.

At the end of the lease, its term can be renewed. If it is not, the ownership of any

11 Article R. 421-16 of the Planning Code.

buildings or other assets on the site subject to the lease are automatically transferred to the lessor without any compensation to the lessee.

Rooftop solar PV panels are governed by contractual agreements of two kinds. The first possibility is the civil lease, which allows for broad contractual freedom – the parties can, for instance, decide the duration of the lease, attribute the responsibility of charges between the lessor and the lessee and set the division of the property by volumes.

A key point in civil leases is the determination of each party's right to the properties during and at the end of the lease. Civil leases only grant personal rights to the lessee but no *in rem* rights (like an emphyteutic lease described above). The consequence of such a situation is that the lessee (usually the SPV) would not be entitled to ownership rights over the assets constructed and, for instance, would be unable to grant a mortgage over such assets to secure its bank loans.

Another option for the parties would be to resort to a division by volumes. For this type of legal separation to be implemented, specific documentation needs to be prepared under the control of a French notary. The volumes (land and construction on one hand, and the solar panels on the other) are described in a specific schedule and are attributed contractually to one or more owners. Easement agreements are entered into between the owners of each volume in order to ensure access to all to its relevant volumes. The ownership of a volume grants *in rem* rights to its holder, therefore the SPV (as a holder of the volume corresponding to the solar panels) would be able to grant mortgages over such assets to secure its financing. In addition, the owner of a volume has a legal pre-emptive right over the other volume(s) in case of a sale by their owner(s).

5.3 Environment

Depending on its size and location, a photovoltaic installation may be subject to several administrative requirements under applicable environmental laws, notably:

- environmental impact assessment – ground-level installations with an output of more than 250 kWp and, on a case-by-case basis, installations on greenhouses and shade structures (*ombrières*) with an output of 250 kWp or more, are subject to environmental impact assessment;[12]
- public inquiry – ground-level installations with an output of more than 250 kWp are subject to public inquiry. Installations on greenhouses and shade structures with an output of 250 kWp or more may be subject to public inquiry if they are also subject to an environmental public assessment;[13]
- single environmental authorisation – if the solar energy facilities have an impact on water resources, it may require an authorisation pursuant to the applicable water legislation.[14] In which case, the single authorisation may also act as a blanket authorisation, declaration, registration or approval under several other regulations (should the project be subject to such legislations)[15]

12 Article R. 122-2 of the Environmental Code (category 30).
13 Article R. 123-1 of the Environmental Code.
14 Article R. 214-1 of the Environmental Code.

such as energy, classified landmarks or vegetation clearing legislation. It is a 'one-stop shop' process and has the advantage of relieving the project from having to obtain multiple authorisations, but only applies when water resources might be impacted.

In addition, ground-level installations may also need to comply with other environmental considerations depending on the zone in which they are located (within the perimeter of a prevention plan against natural risks,[16] a littoral or mountain area, a Natura 2000 zone or any other protected area under the Environmental Code).

5.4 Other regulatory requirements

Depending on its output, a solar PV plant may be subject to an authorisation to operate. Since Decree 2016-687 of 27 May 2016, only solar PV plants with an installed output of more than 50 MW will require the issuance of an authorisation to operate which is granted by the minister in charge of energy. Smaller installations are deemed authorised.

Moreover, as with any other power-generation facilities, solar plants have to be connected to the public grid to sell power. Depending on the output of the facility (less than or above 12 MW), connection to the grid has to be requested from the distribution system operator (ie, Enedis or the competent local distribution system operator) or the transmission system operator (ie, RTE).

A technical and financial proposal (*Proposition technique et financière* or PTF) stating the technical and financial terms of the grid connection is sent to the producer. The PTF includes an estimation of the time needed for the completion of the grid connection, and the date after which the system operator will offer a grid connection agreement (*convention de raccordement*) to the producer. The main terms of the grid connection agreement are determined by an order of the ministry in charge of energy.[17]

In addition to the grid connection agreement, the producer and the distribution system or transmission system operator must also to enter into a public network access agreement (*contrat d'accès au réseau public de distribution* or CARDI or CART) which defines the technical, legal and financial conditions of the injection of electricity produced on the distribution network, and an operating agreement (*convention d'exploitation*) which sets out the operating conditions of the facility. The main terms of the operating agreement are determined by an order of the ministry in charge of energy.[18]

Before being commissioned, a new solar energy plant will be required to undergo or obtain a conformity control of the connection facilities,[19] a certificate approved by counsel certifying that the plant complies with security requirements,[20] and a

15 Articles L. 181-1 *et seq* of the Environmental Code.
16 Article L. 562-1 of the Environmental Code.
17 Article D. 342-13 of the Energy Code.
18 Article D. 342-13 of the Energy Code.
19 Article D. 342-16 of the Energy Code.
20 Article D. 342-20 of the Energy Code.

conformity certificate for facilities with output over 100 kWp benefitting from the premium mechanism or the purchase obligation.[21]

6. The future for the solar sector in the region

According to the think tank *France Territoire Solaire*, 2017/2018 should mark a turning point for solar energy in France. After 2016, which was not a strong year (700 MW of solar capacity installed in France, ie a decrease of 36% compared with 2015), the trend should be reversed by means of a regular tendering programme to be implemented over the next three years and which should lead to an increase in capacity of approximately 1,500 MW per year.[22] The solar PV market prospects for France represent a total capacity, according to the medium scenario, of 15,229 MW by 2021, and new installed capacity between 2017 and 2021 of up to 8,095 MW, which represents a compound annual growth rate of 16%.

The medium scenario provided by PV FINANCING now anticipates that the 15 largest European markets will each install a minimum of 1 GW by 2021 (up from at least 500 MW in GMO 2016), with Germany being the largest adding 12.5 GW (up from 8.7 GW) and France with over 8 GW (up from 6.3 GW).[23] France and Germany are set to be the largest solar PV markets in the European Union by 2020, both broadly forecast to build over 1 GW per year by 2020. In the worst case scenario, France will reach 2 GW within the next few years.[24]

The results of the most recent tranche of the French tender process (Tranche 3) were made public on 9 January 2018. The tender process attributed 159 MWc of power to 283 projects. The average price per MWh came out at €85. On the basis of the latest tender process, it is clear that the cost of production of solar energy is decreasing. The average price of €85 MWh in the latest tender represents a 9% decrease compared to the average price in the previous tender process. In light of the progress of the output of solar panels and their price decrease, the trend should be to see further decreases in the price per MWh for future tenders by the French government.

---

21 Article R. 314-7 of the Energy Code. Until 1 January 2018, a sworn statement was sufficient to abide by this requirement (Article 7 of Decree 2016-682 dated 27 May 2016).

22 France Territoire Solaire, *Observatoire de l'énergie solaire photovoltaïque en France*, 22nd edn, July 2017.

23 EPIA – European Photovoltaic Industry Association, *Global Market Outlook for Photovoltaics 2017–2021*, p43.

24 PV FINANCING, S Dunlop and A Roesch, *EU wide solar PV business models – Guidelines for implementation*, November 2016, p9.

Germany

Niklas Ganssauge
Tilman Lorenz Petersen
CMS Germany

Solar photovoltaic (solar PV) is an increasingly important part of Germany's electricity supply. Among renewable energy technologies, it is second only to onshore wind. According to provisional data for the year 2017, 33.3% of Germany's electricity consumption was produced using renewable energy. Of that energy, approximately 18.3% was produced from solar PV. In sum, 6.1% of Germany's electricity consumption was generated by solar PV in 2017. The commitment to support solar PV has been relatively stable under different governments. The transition to renewable energies is currently supported by all relevant political parties, with differences regarding scale and timing. However, compared to the boom years of 2009–2012, in recent years growth has been controlled through the auction system.

1. A brief look back

The first programme to support solar PV was conceived in 1990 with the so-called '1,000 roofs initiatives'. The goal was to deploy rooftop solar PV. Under this scheme, 2,500 solar PV plants were built with a total installed capacity of 6 MWp. In 1999 the programme was expanded to a 100,000 rooftop solar PV-programme with the aim to install a capacity of 300 MWp which was achieved in 2003. The scale of these programmes shows the enormous technological progress that has taken place since, with currently more than 40 GWp of installed capacity.

Solar deployment finally gathered speed and gained significance with the inception of the Renewable Energy Sources Act (*Erneuerbare-Energien-Gesetz* (EEG)) which – in the early 2000s – foresaw for solar PV a guaranteed feed-in tariff of up to 57.4 cents/kWh for a period of 20 years. This was coupled with preferred grid connection and access, laying the foundations for a later boom of the industry.

As a consequence, considerable research and manufacturing capacity was established in Germany to supply the demand of solar technology. Prices, and correspondingly the feed-in tariff, remained high until suppliers of solar modules from China entered the market around 2008. Considerably cheaper supply of modules created a windfall for the developers which created an – as it turned out unsustainable – boom. Installed capacity quadrupled between 2009 and 2013 from 6.6 GW in 2009 to 31 GW in 2013. During the course of the boom, investments in solar PV increased from approximately €8 billion in 2008 to €19.58 billion in 2010. At the same time, the feed-in tariff was dramatically reduced and investments decreased significantly to between €1.5 and €1.65 billion between 2014 and 2016.

In 2014, not least due to pressure from the state aid department in the European Commission, the feed-in tariff scheme was partially amended with the introduction of a competitive bidding process for the level of support granted to a project. Solar PV was the first technology that was subject to a competitive bidding scheme in Germany, showing the way forward for the German renewables sector. This was coupled with an obligation on the solar project owner to market the electricity. The overall aim of this reform was to integrate renewable energy into the existing electricity market, a process which had failed to gain momentum previously. Another significant, recent development was the increase of self-supplied electricity from solar PV (see section 4.2 below).

2. Market participants

One key feature of the German PV market is the diversity of its participants. Homeowners, utilities, financial and strategic investors – large and small – all profit from the current support scheme.

Small rooftop solar PV is regularly installed by private homeowners contracting an engineering, procurement and construction (EPC) contractor to install the solar PV plant and to manage the grid connection. Larger projects are generally initiated by a developer. As in other regions, there are large-scale developers which develop a pipeline of projects requiring bank financing for construction and operation. Smaller developers often require a financially robust partner to fund the early development stages, in particular with regard to bid bonds, a requirement to participate in auctions. The ultimate commercial goal of project developers may vary. While some develop and build solar PV plants in order to own and operate them, other developers seek a relatively quick exit. Frequently, the exit is planned either at the stage of the project being shovel ready, after winning an auction or after building and commissioning. Professionally managed projects are mostly set up as limited commercial partnerships which can facilitate secondary M&A from a corporate perspective.

Beyond the developers, a utility-scale project requires interactions with municipalities regarding planning and other regulatory approvals. Dealing with landowners for both the plant itself and the connection infrastructure will be required, as well as communications with the distribution system operators (DSOs) or transmission system operators (TSOs) regarding the grid connection. Projects over 100 kWp need to conclude a power purchase agreement (PPA) with an offtaker.

Furthermore, solar PV is affected by all levels of government – the European Union, Germany, the federal states (*Länder*) and the municipality where the project is built. The Renewable Energy Sources Act (EEG) is regularly reviewed and reforms are prepared by the Federal Ministry of Economics and Energy. The Federal Network Agency (*Bundesnetzagentur*) implements the EEG by administering the tender and auction process and the register of solar PV plants. Although not part of the administrative state, the Clearingstelle EEG gives advice on the interpretation of the EEG as an independent body of lawyers and experts.

In Germany, the typical solar PV project is limited to 10 MW. As a consequence, market consolidation and concentration of generating capacity has, to date, been limited, but leaves room for future consolidation. Renewable energy assets are generally a popular asset class with pension funds and insurance companies since they provide a highly predictable revenue stream with limited risks. However, as a consequence of the relative project size, the acquisition of single projects by an institutional investor is uncommon. Portfolio and project pipeline transactions (including secondary M&A activities) are the common entry point for an institutional investor.

The German solar PV market is open to foreign investment and such investment is commonplace. The support regime has proven to be stable and legislative interventions which retroactively devalue investments are avoided. Certainty about future revenue, considerable home-grown industry expertise, as well as strong, stable and financially secure commercial and governmental institutions, all coincide to make Germany an attractive destination for foreign investment.

3. Regulatory background and support regime

The regulatory cornerstone of renewable energy support is the EEG. When first implemented in 2000, the EEG comprised of merely 13 paragraphs. By 2018, due to growth in the complexity and importance of the sector, the EEG increased to 104 paragraphs. The act's policy goals are climate and environmental protection, sustainable energy supply and the support of new technologies. The EEG formulates certain goals in terms of solar deployment in MW per year. Solar projects are thus built to benefit from support under the EEG. In addition, an indirect support regime exists for self-supplied electricity from solar PV, due to a number of exemptions from charges that would apply to grid-based electricity.

3.1 Benefiting from the support regime

There are three different pillars to the support regime. The first is a preferential grid connection, the second is the grid operator's obligation to give preference to offtake and to distribute electricity from renewable energy installations (see section 6.1 below). The third pillar is remuneration for the electricity produced. Currently, the EEG foresees both feed-in tariffs and a market premium regime for solar PV. Eligibility for such support requires that the project is built on eligible sites (section 3.1(a) below). The types and the amount of support is explained under section 3.1(b) below).

(a) Eligible sites

The EEG foresees three different kinds of solar PV projects that are eligible for support: solar PV projects on rooftops and similar constructions; solar PV projects on certain types of green space and farming land; and projects on construction installations.

To limit solar PV's use of green space and farming land, the EEG limits eligible sites on which solar PV can be built and receive support under the EEG. As a general rule, solar PV can be installed on sites that have been subject to an environmentally detrimental use and on rooftops. Examples include landfills, abandoned military and industrial areas and areas adjacent to train tracks and highways. To further limit the amount of ground space used for solar PV plants, the EEG foresees a seize limit of 10 MWp for most ground-based solar PV.

Recently, larger ground-mounted solar installations in excess of 60 MWp have successfully participated in an auction for support under the EEG. The limitation to 10 MW does not apply to ground-mounted solar panels which are built on top of a construction installation (*bauliche Anlage*). Therefore, finding an area that qualifies as a construction installation can be worthwhile to make use of scale advantages. The term 'construction installation' was interpreted broadly in a Federal Supreme Court ruling of 2013 (BGH VIII ZR 308/12) in which the court held that a solar PV plant in the centre of a former horse-racing track which was mostly pasture was built on a construction installation.

(b) Level of financial support for small projects determined by law

The level of financial support that solar PV plants are entitled to depends on the size of the project. Solar PV plants with an installed capacity of up to and including 750 kWh receive a statutory compensation until the end of the 20th year following the solar PV plant's commissioning. The statutory compensation – which is fixed until April 2018 – varies between 12.2 cents/kWh for small rooftop solar of less than 10 kWp and 8.84 cents/kWh for electricity produced in other eligible solar PV plants. For small projects of 100 kWp or less, the compensation takes the form of a feed-in tariff. Other projects between 100 kWp and 750 kWp are entitled to market premium as set out below. The statutory compensation is reduced on a monthly basis in circumstances where certain thresholds for deployment are exceeded. Since deployment has, so far, remained sluggish during 2017 and 2018, the feed-in tariff has been stable. This is likely to remain so in order to promote small-scale decentralised electricity generation.

(c) Level of financial support for large projects determined by auction

Larger projects with a capacity of more than 750 kWp are obliged to apply for a market premium through an auction process. In the auction, the solar PV project bids a certain value to be applied (*anzulegender Wert*) in cents/kWh. The bidders for the lowest value to be applied receive support under the EEG. The successful project is then obliged to sell the electricity generated under a PPA. In addition to the remuneration received under the PPA, the project receives a market premium (*Marktprämie*) from the DSO or TSO to which the solar PV project is connected. The market premium is the difference between the average market price and the value to be applied, which the project nominated in its auction bid.

To participate in an auction, certain requirements need to be fulfilled which, inter alia, depend on the parameters of the solar PV project making a bid. The project, furthermore, needs to post a bid bond in the amount of €5 per planned kWp. In the case of a successful bid, a further bid bond of – in most cases – €45 per planned kWp needs to be posted within 10 business days of the publication of the successful bids. Under the current regime, the Federal Network Agency invites tenders for the values to be applied on a quarter-annual basis. Each auction is for a total volume of 200 MW of installed capacity.

Recent auction outcomes show that only bids for increasingly low values stand a chance of success. The average value to be applied dropped from 9.17 cents/kWh in

April 2015 to 4.33 cents/kWh in February 2018. The lowest bid that succeeded in the auction of February 2018 was 3.86 cents/kWh. Thus, the policy goal to achieve market parity of renewable energy and conventional energy seems to be achievable in the not too distant future. Recent auctions for German offshore wind have already produced several zero subsidy bids. With ever-decreasing module prices, this is likely to be mirrored for solar PV projects.

In the German auction system to date, solar projects compete only with other solar projects for the lowest value to be applied. Since 2017, cross-border invitations to tender supplement the national invitations to tender. Therefore, the financial support for renewable energy is, to a very limited extent, open for the participation of plants located in other EU member states. In 2018, a joint auction for onshore wind and solar plants will take place. Another primer is the innovation tender, which allows project developers to place bids for combinations or mergers of different renewable energy technologies. The new auction formats are envisioned to be the beginning of the end of technologically specific auctions.

3.2 Self and local supply as a driver of solar PV deployment

In addition to the direct governmental support of solar PV deployment, self-supply among small and medium businesses and households has been an important driver of solar PV deployment in recent years. The relatively constant rise of electricity prices over the past decade can make self-supply an economically interesting alternative to grid-based supply of electricity if the right preconditions are met.

While wholesale and spot electricity market prices remained stable or even declined over the last decade, the price of grid-based electricity for households and small businesses soared over the same period from 18.9 cents/kWp in 2006 to 29.9 cents/kWh in 2017. This increase is primarily due to the increase in surcharges that are added to the electricity consumer's utility bill. The biggest cost drivers among these surcharges is the surcharge that finances the support of electricity from renewable energy which rose from 2.05 cents/kWh in 2010 to 6.79 cents/kWh in 2017 and the surcharge to finance the operation, maintenance and expansion of the grid. Although stable over recent years, taxes levied on electricity sales further increased the electricity price.

Subject to certain preconditions, exemption from or reduction of the EEG surcharge and electricity taxes are afforded to self-supplied electricity. Furthermore, self-supplied electricity is not subject to grid charges if certain requirements are met.

The exemptions and reductions from the above-mentioned charges need to be checked carefully in every case. The trend to self-supply is relatively recent and the law is subject to almost constant change. BNetzA issued a guideline (*Leitfaden*) regarding the exemptions and reductions of the EEG surcharge. These guidelines are comprehensive but not binding in their findings. Long-standing authoritative rulings are therefore scarce. Special consideration needs to be given in case the self-supply solar PV plant is financed by a bank or a third party so as to not lose the privileges afforded to self-supply.

A relatively new concept to complement self-supply is the local provision of electricity from landlords to tenants. The concept allows tenants to participate in self-supply as a contribution to energy transition. Under the Landlord-to-Tenant Electricity Act (*Mieterstromgesetz*), the landlord receives a direct financial remuneration for locally sourced electricity from solar PV systems on residential buildings.

4. Real estate, zoning and environment

Careful consideration of real estate and zoning issues is crucial. Environmental concerns are usually considered as part of the zoning and/or permitting process. Regarding real estate, long-term land use contracts with the relevant land owners are required. The requirements regarding real estate are linked with the financing bank's preferred security structure.

4.1 Real estate

The crucial real estate issue for the successful development of projects is the provision of long-term, reliable land-use contracts which are insolvency-proof and fulfil the financing bank's requirements set out below.

To secure long-term land use, lease contracts with the land owner(s) are market standard. Other forms of usage rights such as acquiring full or partial property in the real estate are rare. When drafting a long-term lease for the use of the land the solar PV project is to be built on, notice-based termination rights (*ordentliche Kündigungsrechte*) should be excluded to the extent legally possible. When drafting such lease contracts, special consideration needs to be given that the requirements to validly exclude such termination rights are met. Furthermore, the lease should include a clause that prevents the solar PV plant from becoming part of the real estate by accession.

Lease contracts are, however, not insolvency-proof under German law. Therefore, larger solar PV projects will require an *in rem* security in the land the solar PV project is to be built on. Such *in rem* rights can take the form of a full servitude or easement (*Dienstbarkeit*) or a priority notice (*Vormerkung*) for such a servitude or easement. A hallmark of German real estate law is the land register. The register affords reliable information to landowners, developers and banks. As a general rule, any right can be acquired from the person inscribed in the land register. Furthermore, security interests that are not registered cannot be held against the project that acquire a registered security interest. Title insurance or the like are therefore not required in Germany.

4.2 Permitting and zoning

While rooftop solar PV generally does not require a separate building permit, ground-based solar PV will, in most federal states, require special zoning and building permits or other forms of administrative approval. Projects over 750 kWp need to show some progress in respect of zoning in order to participate in an auction under the current support regime. Zoning is a municipal prerogative. Therefore, early interaction with the municipality where the prospective project is based is crucial for a project to succeed. In case no zoning plan exists for the prospective project area or the zoning plan does not foresee the construction of solar PV, a tailor-made, project-based zoning plan can be implemented (*vorhabenbezogener Bebauungsplan*). Generally, the municipality will require that the developer presents a lease or a lease option for the necessary real estate as a prerequisite of a project-based zoning plan. The plan must be accompanied by a project execution contract between the developer (or the project) and the municipality. In the contract, the developer promises to build the project within a certain time and to bear the cost of the development. In the zoning plan, environmental requirements, such as land-use restrictions due to biodiversity protection, will be considered.

5. Bankability issues

Most large solar PV plants include debt finance at the project level with a debt gearing around 80/20. Larger banks will have experience with solar-specific real estate and EEG compensation issues. The requirements that banks may stipulate vary depending on the project participants involved and the market situation. Regardless of such variability, the requirements set out below are common and can therefore be described as market standard. Banks will generally demand a comprehensive set of security rights comprising interests in the project assets, assignment of the principle rights the project has and the right to step into any of the principal agreements concluded by the project in order to cure default.

Regarding the security interest in the project assets, the banks will generally require security over the solar panels, cable inverters and other necessary hardware. Such fiduciary transfer of title is common in Germany. The bank will agree to (partially) transfer title back once the loan is being paid back.

Banks will take a close look at the real estate rights of the project, requiring long-term and insolvency-proof land use contracts described above for the project. Furthermore, the bank will likely demand security interests in the land the solar PV project is to be built on for its own benefit in the form of a servitude or a pre-registration.

The bank will demand the assignment of rights under the project's principal contracts, such as under the EPC and operation and maintenance (O&M) contracts and the PPA. The bank will furthermore require that warranties regarding the solar PV modules are assigned to the bank.

The final bankability issue relates to the solvency of the PPA counterparty. Requiring credit enhancement to provide security for three monthly payments under the PPA are not uncommon.

6. Power offtake arrangements

The EEG facilitates prioritised grid access for solar PV. Priority grid connection and priority feed-in rights are the first two pillars of the EEG support. Projects over 100 kWp are obliged to market their electricity and to conclude a PPA for the electricity they generate.

6.1 Prioritised grid connection and feed-in

The first EEG pillar is the right of solar PV plants as against the relevant TSO or DSO to be connected to the grid without undue delay and with priority. This includes –

in general – the obligation to provide access at the nearest possible grid connection point from the site of the project. Costs for grid connection are generally borne by the developer as far as they relate to the connection of the plant to the grid connection point. Costs incurred beyond the grid connection point for any refurbishment, extension or reinforcement of the grid are borne by the respective DSO or TSO.

The second EEG pillar is the feed-in priority afforded to renewable energy compared to conventional energy. The essence of the right is that in case of overproduction of electricity – which is frequent in Germany on sunny and on windy days – certain producers of electricity need to be restrained to match supply and demand. Before renewable energy installations are restrained, power plants using conventional fuels are required to curtail or shut down their output. In case solar PV plants are nevertheless required to curtail or shut down their output, they receive compensation of 95% for the lost feed-in. Thus, the risk of grid shortages or oversupply of electricity are borne by the energy-consuming public and not the project. This significantly reduces the risk of not being able to market the electricity generated by a project.

6.2 PPAs

Under the EEG, solar PV plants of a capacity of more than 100 kWp are required to market their electricity. This generally requires the sale of electricity to a third party. While other forms of PPAs exist, the most common type is the delivery PPA, since electricity sold under a delivery PPA qualifies for the market premium.

Under such an arrangement, the project physically exports the electricity it generates into the grid, which for electricity accounting reasons is sold to the PPA counterparty. The PPA counterparty then sells the electricity either on the spot market or to a third party and passes the proceeds – reduced by a management fee – to the project.

Unlike other PPAs, the standard PPAs in the German market do not transfer any pricing risks from the project to the PPA counterparty. Nevertheless, price risk does not affect the project, because of the support regime. In case electricity prices fall – but remain above 0 cent/kWh – the market premium rises to the extent that revenue under the PPA falls and the net income of the project remains the same. The purpose of PPAs is to integrate renewable energy into the market and to provide balancing services.

Because of its limited importance, the standard PPA is rather brief and straightforward. Market standard duration of a PPA under the EEG is two to three years. Important questions that need to be addressed are termination rights, remuneration structures, pricing and risk allocation. When agreeing on a PPA, unless another law is chosen, the specifics of German law need to be taken into consideration. Although the parties are, in principle, free to agree a contract's terms, there are some severe limitations. In particular, the rules regarding terms and conditions place in some cases onerous restrictions on the party proposing the terms. These rules also apply to b2b transactions.

7. The future for the solar sector in Germany

The further integration of solar PV will be both a challenge and a great opportunity for the sector. Recent auction results show that successful bids are close to the spot market electricity price. It remains to be seen whether the importance of the support regime under the EEG will decline and eventually be replaced by long-term PPAs which could fulfil the function of providing stable, foreseeable revenue for a solar PV project. Although solar PV deployment has significantly slowed down since the boom around 2010, reasons exist to be cautiously optimistic about solar PV's prospects. Potential market parity of electricity from solar PV and conventionally produced energy, as well as indirect subsidies of self-supplied electricity, have the potential to drive the deployment of solar PV. A big advantage of solar PV is its potential for decentralised electricity, supply and fast deployment. Differing from offshore wind, which requires utility-scale projects, solar PV has the potential to attract small and creative market participants, as well as utilities, to fulfil the demand of a changing electricity market.

India

Abhishek Saxena
Phoenix Legal

1. Introduction

Earth has enough resources to meet people's needs, but will never have enough to satisfy people's greed. (Mahatma Gandhi)

The effects of global warming and ever-rising pollution have brought an increased environmental consciousness among people globally. Various attempts have been made by governmental and non-governmental bodies to spread awareness about climate change and to control pollution. In India, use of fossil fuel-based energy resources, including in power generation, is one of the major reasons for pollution. On 2 October 2016, India ratified the Paris climate accord[1] and committed to reduce its emission intensity to 35% below 2005 levels by 2030 and achieve a 40% cumulative electricity power capacity from non-fossil fuel-based energy resources by 2030.[2] This shows India's twin focus on using renewable energy to meet its power requirements while simultaneously addressing the problem of pollution.

Among the various renewable energy resources such as solar, wind, small hydro and biomass, India, being a tropical country, has a high potential for solar energy. Most parts of India witness 300 sunny days a year,[3] which translates into an incidence of 5,000 trillion kWh per year of energy over India's land mass (with most parts receiving 4–7 kWh per sq m each day).[4] India can capitalise on this huge potential by using the latest advanced solar PV[5] and thermal technologies[6] and fulfil its commitment under the Paris climate accord.

This chapter sets out a general overview of key developments of the solar power sector in India and the Indian regulatory aspects applicable to the said sector.

2. Brief history of the solar power sector in India

2.1 Solar power sector in general

Less than a decade ago, solar projects were struggling to gain popularity and were

1 See http://unfccc.int/files/essential_background/convention/application/pdf/english_paris_agreement.pdf.
2 India's Intended Nationally Determined Contribution, see www4.unfccc.int/submissions/INDC/Published%20Documents/India/1/INDIA%20INDC%20TO%20UNFCCC.pdf.
3 The Ministry of New and Renewable Energy's Annual Report for year 2016–17, see http://mnre.gov.in/file-manager/annual-report/2016-2017/EN/index.html.
4 The Jawaharlal Nehru National Solar Mission, see www.mnre.gov.in/file-manager/UserFiles/mission_document_JNNSM.pdf.
5 Solar photovoltaic technology enables direct conversion of sunlight into energy.
6 Solar thermal technology utilises heat content of solar energy into useful application.

required to be heavily subsidised by the Indian central government. The government's focus was on coal-fired power generation to support the unprecedented economic growth being witnessed during that period. Indeed, the Indian government had planned to set up ultra-mega power projects, each having a capacity of about 4,000 MW.[7]

In contrast to this earlier emphasis, the Gujarat state government recently dropped a plan to build a 4,000 MW imported coal-based ultra-mega power project because it believes that upcoming renewable energy units could meet its power requirements.[8] This may be explained partly by the historic low tariffs offered by private bidders in the recently concluded competitive reverse auction for solar power projects in India which, surprisingly, were lower than conventional (coal-based) sources.[9] The government received a low tariff of Rs 2.44/kWh (ie, US$ 0.037/kWh) for 500 MW of capacity in one of India's largest solar parks.[10] To put this into perspective, in 2010 the tariff hovered around Rs 12.16/kWh (ie, US$ 0.17/kWh).[11] This change augurs well for the prospects of solar power plants in India.

2.2 Solar power sector – development of laws/schemes/plans

(a) National Action Plan on Climate Change

India has been a keen and active participant in multilateral negotiations dealing with global challenges posed by climate change including the United Nations Framework Convention on Climate Change. However, its commitment to a cleaner environment has been practically difficult to implement because of the imperatives of growth and the need to meet the energy requirements of its vast population. As part of its commitment to a cleaner environment, and on the premise that developed nations will provide technologies and adequate financial resources, on 30 June 2008 India announced its plan for climate change, the National Action Plan on Climate Change (NAPCC).[12]

The NAPCC, without compromising India's prospects for a high growth rate (which is essential for increasing the standard of living of Indian people and reducing vulnerability to climate change), provides Eight National Missions that form its core.

(b) Jawaharlal Nehru National Solar Mission

The central government launched its flagship programme as one of the Eight National Missions identified in the NAPCC with the name 'Jawaharlal Nehru National Solar Mission' (JNSM). Under the brand 'Solar India', JNSM set an ambitious target of deploying 20,000 MW of grid-connected solar power in India by 2022.[13]

---

7 See www.pfcindia.com/Home/VS/22.
8 See www.financialexpress.com/industry/yogi-adityanath-cancels-bids-for-3800-mw-power-in-uttar-pradesh/663124/lite/.
9 See www.qz.com/984656/solar-power-is-now-cheaper-than-coal-based-electricity-in-india-but-the-math-makes-no-sense/.
10 *Id.*
11 See www.cleantechnica.com/2017/04/17/solar-bid-prices-india-fallen-73-since-2010/ and www.thehindubusinessline.com/news/solar-tariffs-have-declined-by-73-per-cent-in-india-since-2010-mercom/article9599908.ece.
12 See www.moef.nic.in/sites/default/files/Pg01-52_2.pdf.

Achievement of this target was envisioned in three phases (the first phase up to 2013, the second phase from 2013–17 and the third phase from 2017–22).[14]

However, in June 2015, the central government approved a revision of cumulative targets under JNSM from 20,000 MW (20 GW) by 2021–22 to 100,000 MW (100 GW) by 2021–22 for grid-connected solar power projects.[15] The revised target of 100,000 MW is planned to be achieved in a seven-year period. Broadly, it consists of 40,000 MW (40 GW) of grid-connected rooftop projects and 60,000 MW (60 GW) of large and medium-sized land-based solar power projects.[16] JNSM is a policy of the central government pursuant to which several other schemes and incentives have been provided. It is discussed in further detail under section 3 of this chapter.

(c) ***International Solar Alliance***

The International Solar Alliance (ISA) was inaugurated on 30 November 2015 at the climate summit in Paris. It has been set up with the United Nations as a strategic partner and India has been chosen for its headquarters. The foundation stone for the headquarters in India was laid on 25 January 2016 jointly by the Indian Prime Minister and the French President at the National Institute of Solar Energy in Gurugram, State of Haryana. ISA creates a collaborative platform for increased deployment of solar energy technologies to enhance energy security and sustainable development. Some commentators believe[17] that the ISA needs a stronger focus and more commitment from the developed world.

(d) ***Pradhan Mantri Sahaj Bijli Har Ghar Yojana (Saubhagya Scheme)***

The central government launched the Saubhagya Scheme on 25 September 2017 to achieve universal household electrification by 31 December 2018 by providing last mile connectivity to all households in rural and urban areas in India.[18] Households in remote and inaccessible areas will be offered decentralised solar PV-based systems of 200–300 Watt peak each with battery packs, 5 LED lights, a fan and a power plug, along with free maintenance for five years.[19] Earlier, the Deendayal Upadhyaya Gram Jyoti Yojana (DUGJ Scheme) was launched in 2015 with the aim of electrifying all 18,000 unelectrified villages by May 2018. Further, the '24×7 Power for All' programme was launched in 2015 with the objective of providing 24×7 power to all consumers across India by 2019.[20] Though the central government has claimed successful implementation of the DUGJ Scheme, several reports have contradicted this claim.[21] Some commentators have attributed this to the definition of electrified

---

13 See footnote 3 above.

14 *Id.*

15 See footnote 2 above.

16 *Id.*

17 See www.bridgetoindia.com/international-solar-alliance-needs-stronger-focus-and-more-commitment-from-the-developed-world/.

18 See http://pib.nic.in/newsite/PrintRelease.aspx?relid=171103; and http://pib.nic.in/newsite/PrintRelease.aspx?relid=171101.

19 See www.powermin.nic.in/sites/default/files/webform/notices/OM_SAUBHAGYA_SIGNED_COPY.pdf.

20 See www.pib.nic.in/newsite/PrintRelease.aspx?relid=156256.

21 See www.thehindu.com/opinion/op-ed/On-paper-electrified-villages-%E2%80%94-in-reality-darkness/article14176223.ece; and www.dnaindia.com/india/report-modi-government-s-flagship-scheme-deendayal-upadhyay-gram-jyoti-yojna-remains-largely-on-paper-2236194.

village adopted by the central government, as a village is considered electrified by the government if public places in the village and 10% of its households have access to electricity.

3. Overview of market participants

3.1 Solar power developers/generators

Amendment of the Electricity (Supply) Act 1948 (which came into force on 15 October 1991) permitted the establishment of electricity projects by private sector generating companies and sale by them to the then Electricity Boards, resulting in a phenomenal growth in private sector power projects in India. Also, the Electricity Act 2003 (Electricity Act) has delicensed electricity generation (including captive generation of electricity).[22] As of 31 March 2017, of the total commissioned capacity, 47% is owned by Indian corporates, 18% by international developers, 10% by public sector undertakings (which are in effect entities owned by central and state governments) and 25% by independent power producers (backed by PE funds).[23]

3.2 Manufacture/import of solar modules

Solar module manufacturing in India has seen a downturn over recent years due to technological obsolescence and sub-scale manufacturing volumes. India's requirements are largely met by imports from other countries, particularly the United States and China. Even though the central government came up with various incentives to give a fillip to domestic manufacturing of solar modules, including imposing domestic content requirement (DCR) norms, operating cost and capital subsidy under different policies, these have failed to achieve the desired objective.[24]

3.3 Transmission

Transmission[25] of power between (Indian) states is carried out by the Power Grid Corporation of India Limited (an Indian state-owned electric utility). The intra-state transmission system is managed by the respective state transmission utilities. These transmission utilities manage the network of transmission lines to aid the distribution licensees[26] in operating a distribution system for supplying electricity to consumers in their area of supply. These power lines are operated in accordance with the standards prescribed by the authorities/regulators set up under the Electricity Act.

---

22 Section 7 of the Electricity Act. Section 2(28) of the Electricity Act defines 'generating company' as any person which owns or operates or maintains a generating station.

23 See p13 of India Solar Handbook 2017, Bridge to India, see www.bridgetoindia.com/wp-content/uploads/2017/05/BRIDGE-TO-INDIA_India-Solar-Handbook_2017-1.pdf.

24 *Id*, p.20.

25 Section 2(72) of the Electricity Act defines 'transmission lines' to mean all high pressure cables and overhead lines (not being an essential part of the distribution system of a licensee) transmitting electricity from a generating station to another generating station or a sub-station, together with any step-up and step-down transformers, switch-gear and other works necessary to and used for the control of such cables or overhead lines, and such buildings or part thereof as may be required to accommodate such transformers, switch-gear and other works.

26 Section 2(17) of the Electricity Act defines distribution licensee to mean a licensee authorised to operate and maintain a distribution system for supplying electricity to the consumers in his area of supply.

3.4 Distribution companies (DISCOMs)

DISCOMs form the distribution system[27] and are licensees which manage the system of wires between the transmission lines/generating stations and end consumers. Although few privately owned distribution companies exist, the overall financial position of this segment has not been great with many DISCOMs making losses and being debt-ridden. Therefore, the central government along with respective state governments have provided financial restructuring schemes including the Ujjwal DISCOM Assurance Yojana (UDAY) scheme to address problems faced by DISCOMs, including a reduction of their losses and improvement in overall operations.

4. Legislative and regulatory framework

4.1 Constitution of India

The Constitution of India provides for a quasi-federal structure of governance. Electricity is a subject matter on which both the central government and the respective state government can make laws.[28] However, in case of conflict between laws made by the central government and the state government, unless the President of India has provided consent to the state law, law made by the central government prevails. The Ministry of Power in the central government is responsible for the power sector (except for new and renewable energy) and for the issues relating to energy policy and coordination thereof and for administration of the Electricity Act.[29] The Ministry of New and Renewable Energy (MNRE) in the central government is responsible for new and renewable energy in India, which includes solar power.[30]

4.2 Electricity Act

The Electricity Act has constituted the Central Electricity Authority. The Central Electricity Authority, among others, advises the central government on matters relating to national electricity policy, specifies technical standards for construction of electrical plants, power lines and connections to the grid, safety requirements for construction, operation and maintenance of electrical plants and grid standards for operation and maintenance of transmission lines. The Electricity Act also provides for establishment of regulatory commissions at the central level (ie, the Central Electricity Regulatory Commission (CERC)) and at the state level (ie, the State Electricity Regulatory Commission (SERC)) in each Indian state. These commissions administer and regulate the generation, distribution and transmission of electricity. The CERC is required to adjudicate disputes[31] involving generating companies or transmission licensees, while the SERCs adjudicate[32] disputes between the licensees and generating companies in their respective states. Any appeal from the decisions

---

27 Section 2(19) of the Electricity Act defines 'distribution system' to mean the system of wires and associated facilities between the delivery points on the transmission lines or the generating station connection and the point of connection to the installation of the consumers.

28 Entry 37, List III, VII Schedule, The Constitution of India.

29 The Government of India (Allocation of Business) Rules, 1961.

30 *Id.*

31 Section 79(f) of the Electricity Act.

32 Section 86(f) of the Electricity Act.

of these commissions lies with the Appellate Tribunal for Electricity (APTEL). Appeals from the APTEL lie with the Supreme Court of India, the highest court in India.

The Electricity Act, among others, governs generation, transmission, distribution, trading and use of electricity across India.[33] The Electricity Act's object includes promotion of efficient and environmentally benign policies.[34] The Electricity Act empowers both the central and state governments to issue policy directions in the public interest. The Ministry of Power, the MNRE and the regulators (the CERC and the SERCs) have been instrumental in implementing the government's policies through statutory policies, schemes, regulations and orders.

4.3 JNSM, its schemes and targets

(a) *Administration of JNSM*

The Solar Energy Corporation of India Limited (SECI), a Government of India undertaking under the administrative control of the MNRE, helps in implementing JNSM and schemes such as viability gap funding (VGF). The VGF scheme provides gap-in-funding support to infrastructure projects that are economically justified but fall marginally short of financial viability.[35] The Indian Renewable Energy Development Agency Limited, another Government of India undertaking under the administrative control of the MNRE, is engaged in promoting, developing and extending financial assistance for setting up projects relating to new and renewable sources of energy and energy efficiency/conservation.

(b) *JNSM: Phase I Batch I and Batch II*

In order to facilitate grid-connected solar power generation, a mechanism of 'bundling' relatively expensive solar power with power from the unallocated quota generated at NTPC coal-fired stations, which is relatively cheaper, was proposed. NTPC's Vidyut Vyapar Nigam Ltd (NVVN) was designated as the nodal agency for procuring the solar power by entering into a Power Purchase Agreement or PPA. The total aggregated capacity of the grid-connected solar projects to be developed under Phase I was 1,000 MW. The projects are to be selected for deployment of both solar PV technology projects and solar thermal technology projects in a ratio of 50:50 in MW terms. Allotment of projects was based on a bidding process and the developer offering the maximum discount in tariff was short-listed to execute the PPA with NTPC. Most of the projects set up in the first phase were based on solar PV technology.[36]

(c) *JNSM: Phase II*

Currently, Phase II of the JNSM is being implemented. Phase I of JNSM (2010–13)

---

33 The Electricity Act, however, does not apply in the state of Jammu and Kashmir. See Section 1 of the Electricity Act.
34 Preamble to the Electricity Act.
35 See www.pppinindia.gov.in/faqs?inheritRedirect=true.
36 Guidelines for Selection of New Grid Connected Solar Power Projects, Batch II, Phase-I, see www.mnre.gov.in/file-manager/UserFiles/jnnsm_gridconnected_24082011.pdf.

generated huge interest in the solar power sector pursuant to which both grid-connected and off-grid projects were commissioned across the country.[37] Though the target set by Phase I was achieved, the central government has subsequently increased the cumulative installation target under JNSM from 20,000 MW to 100,000 MW.[38]

Phase II Batch I: Phase I was largely based on the bundling scheme described in section 4.3 (b) above. In Phase II Batch I of JNSM, the option of VGF has been selected. The total capacity of grid-connected solar projects to be set up on the build-own-operate model under this phase was 750 MW. SECI was designated as the nodal agency by the MNRE for implementation of the scheme with VGF support from the National Clean Energy Fund, in close association with the NVVN. SECI invited bids in two separate categories – Open Category and DCR Category, to promote local manufacturing of solar modules and equipment. The upper limit of VGF was fixed at 30% of the project cost or Rs 25 million per MW, whichever is lower. Phase II batch I has been successfully implemented by SECI with a total capacity of 750 MW.[39]

Phase II Batch II: The MNRE provided for a state-specific bundling scheme to add 3,000 MW of solar power projects to be implemented by NVVN in solar parks, to be developed through the association of central or state agencies with land provided by state governments or by private solar power developers in the respective states. MNRE facilitated the development of 25 solar parks to accelerate solar capacity addition in various states. 3,000 MW of solar PV capacity will be established based on the model of bundling solar-generated power with unallocated thermal-generated power with fixed levelised tariffs.[40]

4.4 Phase II Batch III – state-specific VGF scheme

Solar projects of 2,000 MW capacity under state-specific VGF schemes will be set up in solar parks in various states. These are to be developed through the coordinated efforts of central and state agencies. Out of the total capacity of 2,000 MW, a capacity of 250 MW will be earmarked for bidding with DCR. The upper limit for VGF was kept at 30% of the project cost or Rs 10 million/MW for open category (Rs 13.1 million/MW for projects in DCR category), whichever is lower.[41]

4.5 Phase II Batch IV

SECI has been designated as the implementing agency for the selection of

---

37 See footnote 1 above.

38 See www.mnre.gov.in/file-manager/grid-solar/100000MW-Grid-Connected-Solar-Power-Projects-by-2021-22.pdf.

39 Final Guidelines for Selection of 750 MW Grid Connected Solar Power Projects, see www.mnre.gov.in/file-manager/UserFiles/final-VGF_750MW_Guidelines_for-grid-solar-power-projects.pdf.

40 Guidelines for selection of 3,000 MW Grid-Connected Solar PV Power Projects under Batch-II 'State Specific Bundling Scheme', March 2015, the MNRE.

41 Guidelines for implementation of scheme for setting up of 2,000 MW Grid-Connected Solar PV Power Projects with VGF, see www.gov.in/file-manager/grid-solar/Scheme-2000MW-Grid-Connected-SPV-with-VGF-under-JNNSM.pdf.

grid-connected solar PV projects for a cumulative capacity of at least 5,000 MW to be set up on a Build-Own-Operate basis through the VGF mechanism.[42]

4.6 Phase II Batch V

This is a scheme for setting up 1,000 MW of grid-connected solar PV power projects by the Central Public Sector Undertaking (CPSU) (ie, entities owned by the central government).[43] The scheme targeted public sector undertakings and organisations owned by the central government such as the NTPC, the NHPC, the Indian Railways etc to set up 1,000 MW of grid-connected solar power projects. These projects were mandatorily required to procure PV modules from domestic manufacturers.

4.7 Implementation of scheme for development of solar parks and ultra-mega power projects

This scheme proposes the establishment of 25 large solar parks (of at least 500 MW) at various locations across the country targeting around 20,000 MW of solar power installed capacity. However, due to the increased overall target for installed capacity of solar power from 20,000 MW to 100,000 MW by 2022, the target was increased from 20,000 MW to 40,000 MW to be achieved by 2019–20.[44]

The solar park offers, in a one-stop-shop format, civil and grid connection infrastructure facilities to incentivise private developers to invest in the solar park. The state government will nominate the implementing agency for the solar park and also identify the land. The development of solar parks can be done by SECI, any special purpose vehicle (SPV) under the state government, a joint venture company set up between SECI and a state-designated agency, or private entrepreneurs. The implementing agency will have the responsibility of providing suitably developed, levelled land with all clearances, transmission systems, water access, road connectivity, etc to the solar project developer. A total of 34 solar power parks were approved in 21 states with a combined solar capacity of 20,000 MW.[45]

4.8 Grid-interactive rooftop and small SPV power plants programme

Out of an ambitious 100,000 MW installed capacity target by 2022, 40,000 MW are targeted for rooftop solar installations.[46] The MNRE has recently declared financial incentives for DISCOMs to support rooftop solar installations. The MNRE notification proposes financial support of up to Rs 3.75 million (US$55,000)/MW for up to 1,350 MW of rooftop solar capacity.[47] There is a huge potential for generation

---

42 Guidelines for implementation of scheme for setting up of over 5,000 MW Grid-Connected Solar PV Power Projects with VGF, see www.mnre.gov.in/file-manager/UserFiles/Guidelines-for-5000MW-Grid-Connected-power-project-under-VGF-NSM.pdf.

43 See www.mnre.gov.in/file-manager/UserFiles/Scheme-1000MW-grid-connected-SPV-batch_V-phase_II-JNNSM.pdf.

44 The scheme for enhancement of capacity from 20,000 MW to 40,000 MW for 'Development of Solar Parks and Ultra Mega Solar Power Projects' dated 21 March 2017, see www.mnre.gov.in/file-manager/UserFiles/Scheme-for-enhancement-of-capacity-to-40GW-Solar-Parks.pdf.

45 *Id.*

46 See www.mnre.gov.in/file-manager/grid-solar/100000MW-Grid-Connected-Solar-Power-Projects-by-2021-22.pdf.

47 See www.solarrooftop.gov.in/notification/Notification-08112016901.pdf.

of solar power using unutilised space on rooftops and on wasteland around buildings.[48]

The central government is implementing a programme where it is providing a subsidy of 30% of benchmark cost for the general category states and up to 70% of benchmark cost for special category states (such as states located in hilly areas or islands).[49]

The rooftop solar market in India has grown at a compound annual growth rate (CAGR) of 98% in the last four years. As of 30 September 2016, 1,020 MW of total rooftop solar capacity had been installed.[50] Growth in this market is being primarily driven by improvement in price competitiveness of rooftop solar power vis-à-vis grid power. However, grid interconnection regulations and processes remain challenging in most parts of the country, despite almost all states announcing net and/or gross metering policies for rooftop solar.

4.9 Scheme for setting up grid-connected solar PV projects by defence establishments

A capacity of 300 MW of grid connected and off-grid solar PV power projects is proposed to be set up by various defence establishments under the Ministry of Defence, with VGF support of Rs 7,500 million under JNSM.[51] The government has already given consent for the right of use of land by developers chosen by defence establishments such as the army, the navy and the Ordnance Factory Board.[52] The government has permitted these establishments to identify locations for developing solar projects anywhere in the country, including border areas.[53] These projects may be undertaken through a 'developer mode'[54] or an 'EPC mode' and must procure PV modules from domestic manufacturers.[55]

4.10 Pilot-cum-demonstration project for development of grid-connected solar PV power plants on canal banks and canal tops[56]

This scheme aims to gainfully utilise the unused space over the top of canals and unutilised land on canal banks for setting up solar PV power generation plants, up to a total capacity of 100 MW. Since these properties are owned directly or indirectly by the government (central or state), it may make it easy for the MNRE to execute the proposed projects.

---

48 See footnote 1 above.

49 *Id.*

50 See India Solar Rooftop Map 2016, Bridge to India, www.bridgetoindia.com/reports/india-solar-handbook-2016/.

51 See footnote 1 above. See also http://mnre.gov.in/file-manager/UserFiles/administrative-approval-on-Defence-300MW.pdf.

52 *Id.*

53 *Id.*

54 Under the 'developer mode', the developer is permitted to bring in investment, own the project and supply power to respective defence establishments.

55 The 'EPC Mode' involves projects being built by EPC contractors and investment being made by defence establishments.

56 See www.mnre.gov.in/file-manager/grid-solar/Administrative-Approval-reg-Pilot-cum-Demonstration-Project-for-Canal-Bank-Canal-Top-Solar-PV-Projects%20.pdf.

4.11 Off-grid solar photovoltaics

MNRE is also subsidising off-grid and decentralised applications for usage of solar PV products in remote areas where grid penetration is neither feasible nor cost effective.[57] This is consistent with JNSM's general objectives. For instance, the central government is providing capital subsidy for solar PV plants with battery storage in a micro-grid mode for unmatched community demand for electricity or in unelectrified rural areas.[58] Further, a special programme for 100,000 solar-powered water pumps was launched of which 31,472 solar-powered pumps were installed in the year 2015–16.[59] This is higher than the total number of pumps installed during last 24 years (ie, since beginning of the programme in 1991).[60]

4.12 Other regulatory measures

(a) *Renewable energy purchase obligations*

SERCs are required to specify that a certain percentage of the total electricity purchased by a distribution licensee should be electricity that is generated through renewable sources. This is known as the renewable purchase obligation (RPO).[61] For example, in the state of Rajasthan, the Rajasthan Electricity Regulatory Commission (RERC) has notified 4.75% and 6.75% as solar energy purchase obligations for the years 2017–18 and 2018–19, respectively.[62] In addition, the RERC has also fixed the RPO for non-solar renewable energy. However, some states and regulatory bodies have not been strictly enforcing RPO obligations. In this regard, the Supreme Court[63] set a precedent by ruling that fulfilment of RPO is mandatory for captive power plants and open access consumers. This is a positive development and will help in enhancing RPO compliance and further contribute to renewable energy growth in the country. Further, pursuant to the revised National Tariff Policy (Tariff Policy),[64] SERCs are required to ensure that this percentage/RPO reaches 8% of total consumption of energy, excluding hydro power, by March 2022.[65]

(b) *Renewable energy certificates*

RPOs can also be discharged by purchasing renewable energy certificates[66] (RECs) from any of the power exchanges. Entities which are not able to comply with their RPOs by purchasing power can purchase RECs and achieve compliance.

---

57 See footnote 1 above.
58 *Id.*
59 See http://pib.nic.in/newsite/printrelease.aspx?relid=161811.
60 *Id.*
61 Section 86(1)(e) of the Electricity Act.
62 Regulation 4 of the Rajasthan Electricity Regulatory Commission (Renewable Energy Obligation) Regulations, 2007.
63 *Hindustan Zinc Limited v Rajasthan Electricity Regulatory Commission* [Civil Appeal 4417 of 2015] dated 13 May 2015.
64 See www.kseboa.org/downloads/Government%20Orders/tariff_policy-resolution_dated_28012016.pdf.
65 The Tariff Policy notified by Government on 28 January 2016.
66 The Central Electricity Regulatory Commission (Terms and Conditions for Recognition and Issuance of Renewable Energy Certificate for Renewable Energy Generation) Regulations, 2010.

(c) ***Renewable generation obligations***

Tariff Policy notified by the central government, and referred above, has introduced the concept of renewable generation obligations.[67] Any generating company proposing to build a coal/lignite-based thermal power generation station after a specified date will be required to build renewable energy generating capacity or procure and supply renewable energy equivalent to such capacity, as may be prescribed by law.

(d) ***Exemption from inter-state transmission charges***

In order to further encourage renewable sources of energy, the Tariff Policy has also proposed to not levy inter-state transmission charges and losses on transmission of electricity generated from solar energy through the inter-state transmission system for a notified period.[68]

(e) ***Net-metering***

Net-metering is a relatively new concept which, pursuant to a special metering and billing agreement between the utilities and their customers, enables the connection of small, renewable energy-generating systems to the power grid.[69] When the net metering client's generator is producing more power than is being consumed by such client, the electric meter runs backwards, thereby generating credits. If the client uses more power than is being produced, the meter works normally. The client is only charged for the net power that he/she has consumed from the electricity supplied by the utility.

(f) ***Tax benefits***

Supply of solar water heating systems, solar power-based devices, solar power-generating systems and solar lantern/solar lamp and parts for their respective manufacture is subject to a goods and services tax at a concessional rate of 5%. With effect from 1 April 2017, for assets relating to solar power generating systems, machinery and plant, the Income Tax Act 1961 has provided a rate of depreciation of 40%. An additional depreciation rate of 20% in the first year of use of a new asset, being plant and machinery is also provided.

(g) ***Projects under various state policies***

Along with measures taken by the central government to promote solar power under the JNSM, various states have come up with their own policies to provide an enabling framework for growth of renewable energy in India.

(h) ***Full repatriability***

Indian exchange control regulations which, among other matters, govern foreign investments into India, permit 100% foreign investment in the solar power sector in India, either as a manufacturer of equipment, or as a project developer. In other

67 Paragraph 6.4(5) of the Tariff Policy.
68 Paragraph 6.4(6) of the Tariff Policy.
69 See www.mnre.gov.in/file-manager/akshay-urja/january-february-2012/EN/44-45.pdf.

words, a foreign person or entity is allowed to own a 100% shareholding of an Indian company set up to undertake the business of manufacture of solar equipment or project development. Further, subject to payment of applicable taxes, the profits from such businesses and capital invested is fully repatriable.

5. Major issues or risks faced by solar power developers in India

The two primary goals of the developer of a solar power project are commercial viability and bankability. Commercial viability of the project is ensured when the project is executed on a fixed timeline without compromising on technical quality and performance standards and when the project developer's expectation of a reasonable return and sustainable profits from the project are met. Bankability of the project means the ability of the project developer to secure finances/funds for the project at favourable terms.

However, these goals of the project developer have to contend with major issues of time and cost overruns, due to various inherent risks. The issues or risks prevalent in the renewable power industry in India are discussed below.

5.1 Environmental issues

Solar energy is universally understood to be environmentally friendly technology as it has "zero emissions while generating electricity or heat".[70] However, for setting up a solar power project, subject to applicable conditions, consent to establish/operate under the Water (Prevention and Control of Pollution) Act 1974, the Air (Prevention and Control of Pollution) Act 1981 and authorisation under the Hazardous Wastes (Management and Handling) Rules 1989 must be obtained. Further, the disposal of PV cells has to comply with the Hazardous and Other Wastes (Management and Transboundary Movement) Rules 2016, hence pre-authorisation from authorities must be obtained.

The recently issued Guidelines for Tariff Based Competitive Bidding Process for Procurement of Power from Grid Connected Solar PV Power Projects dated 3 August 2017[71] by the Ministry of Power, provide that in cases where the procurer of electricity has identified land, it shall also, in order to ensure timely commencement of supply of electricity, procure NOC or environmental clearance for forest clearance, as the case may be. Obtaining the aforesaid consents/NOC/clearances could be time consuming and may impact timely commissioning of the project.

5.2 Real estate and planning (zoning) issues

(a) Land acquisition

Availability of real estate is a key requirement and constraint for setting up solar power projects. At the same time, the real estate used by a solar park leaves it useless for other purposes, as the solar modules are permanently stationed on the land. The increase in the target-installed capacity for solar power parks from 20,000 MW to

70 Paragraph 2.4 of JNSM.
71 See www.mnre.gov.in/file-manager/UserFiles/Guidelines_for_Tariff_Based_Competitive_Bidding_Process.pdf.

40,000 MW may itself require around 80,000 acres of land.[72] In India, finding suitable land for a project and securing its possession is one of the major challenges faced by the project developers. Also, land ownership is highly fragmented and tracing the true ownership of land by title due diligence is difficult as records are not maintained properly by local land revenue authorities. Further, if the land falls under the category of a heritage site, a forest reserve or a wildlife sanctuary, its acquisition may face stiff resistance from local authorities, social activists and the local population.

(b) Planning (zoning) issues

Indian law categorises land use into different categories and various states have their own regulations in this regard. For example, land which has been classified as being for residential, commercial or agricultural use cannot be used for setting up a solar power project. In some instances, conversion of land use is possible after following due process. Conversion of land use requires payment of charges and acquisition, leasing or licensing of land is also subject to payment of stamp duty, a form of tax, to the relevant state government within whose jurisdiction the relevant land is located.

Since levy of land use conversion charges and stamp duty are, as per the Constitution of India, within the domain of each state government, the MNRE has put the onus on[73] states to provide the solar power industry with exemptions from the same. Certain states such as Rajasthan and Karnataka have removed land use conversion charges.

5.3 Grid unavailability risks

There is a general shortage of power transmission infrastructure in India and this factor constrains offtake of electricity for solar power plants. With the exception of solar parks for which a state is required to provide specific infrastructure for transmitting electricity, there is a shortage of proper infrastructure. There is a pressing need for transmission utilities to expedite creation of quality transmission lines to enable timely offtake of electricity. Further, the power grid in India is subject to stability problems. Solar power, being more intermittent than conventional power, poses an important challenge to grid stability.

5.4 Local issues

In some instances, the process of land acquisition may require displacement of sections of the population and issues of payment of compensation. If the land being acquired is located in environmentally fragile areas, land acquisition may become subject to opposition from various advocacy groups and organisations in the non-government space. The nature of these issues is such that local political and vested interests also tend to get involved and rally the local population against these projects. This may then result in the timeline for the relevant projects being impacted.

---

72 "Govt's Big Solar Park Push Could Run Into Land Acquisition Hurdle", *The Quint*, 27 March 2017, see www.thequint.com/news/india/solar-parks-and-land-acquisition-problems.

73 *Id.*

5.5 **Financial health of DISCOMs**

The poor financial condition of the DISCOMs, which are the consumers for most of the solar power plants, represents a significant risk factor. Payment delays or non-payment by DISCOMs may have a significant impact on the sector as a whole. However, as mentioned in section 3.4 above, the government has introduced the Uday scheme to address problems faced by these utilities, including a reduction of their losses and improvement in overall operations.

5.6 **Rise in cost of solar modules/cells and other factors**

Due to cost-efficient Chinese supplies of PV modules, India has been able to reduce the tariff for solar power to less than, or comparable to, the tariffs for conventional power sources. Modules account for about 55% of the project cost of a solar project and 80% of them are imported from China and Malaysia. If, for whatever reason, their prices rise, it could significantly impact the economics of these projects. This may then arrest the trend of continuously falling solar tariffs and thus adversely impact the future growth of the solar power sector.

In addition to the inherent risks in development of any project in India discussed above, the solar power developer also faces:

- risk of termination of a PPA due to unreasonable delay in the commissioning of the project;
- risk of loss of VGF if the project fails to generate power continuously for any year or its assets are sold or the plant is dismantled during the tenure of the project; or
- given the continuously falling tariffs for purchase of solar power by state and central government utilities, risks arising from certain utilities requiring higher tariffs in the earlier PPAs to be reduced to such lower tariffs.

5.7 **Management of issues and risks**

It is important for the project developer to identify risks, mitigate them appropriately and allocate them among the different project participants. An important method to manage risk is to choose a capable and experienced EPC contractor for construction of the project and agree to time-certain fixed-price contracts. Procurement of equipment with the right quality and specification is important to ensure that the degradation factor is managed and the project continues to produce power without any breakdown or interruption. Obtaining suitable warranties under the EPC contract also helps in managing the disruption arising from equipment breakdown and protects the developer against the financial impact of such breakdown.

Risks are also managed by having well-drafted PPAs which provide for a clear obligation on the utility to pay at a specified time, provide for a good payment security mechanism and with *force majeure* provisions that capture events that are outside the control of the developer and excuse non-performance under the PPA upon their occurrence.

5.8 **Financing related issues**

Some issues that directly impact the ability of the project developer to raise financing to construct and operate a solar plant are discussed below.

(a) ***Assignment of rights under PPAs***

Often PPAs entered into with DISCOMs require the prior consent of such DISCOMs to assign the rights under the PPA to a substitute developer in case of default by the original developer/borrower. Such consents for assignment are usually provided for as conditions subsequent in financing arrangements, and are, at times, cumbersome to obtain. Greater clarity on the mechanism for obtaining such consents and requiring these consents to be provided in a timely manner would provide comfort to lenders, and would encourage them to lend more to solar power projects.

(b) ***Lack of clarity on enforcement of mortgages***

Most states' solar project policies require DISCOMs or other government agencies to provide/arrange for land required for solar projects. Such land-related arrangements are in the nature of leases or, in rare cases, outright conveyance of the underlying land. However, in most cases, the underlying documents provide for the consent of the lessor or the seller of the land or any government agency for the granting of mortgages over the land by the solar developer. Such consents are also not easily forthcoming, and thereby act as a disincentive for lenders to provide financing for solar projects.

(c) ***Change in shareholding of a solar project developer***

PPAs generally contain provisions requiring the solar project developer to ensure that no change in its shareholding occurs for a certain period of time. While consent can be obtained from the relevant DISCOMs or government agencies for a change in shareholding, the process for obtaining such consent is generally cumbersome. This acts as a disincentive for private equity funds to invest in solar project developers by way of equity. Additionally, as lenders generally require a debt-to-equity conversion covenant in case of default, the lack of clarity on change of shareholding is a disincentive for them to provide financing for solar power projects.

(d) ***Signing of PPAs***

There have been reports in the public domain on DISCOMs refusing to sign PPAs or delaying signing PPAs with a view to seeking renegotiation of rates as discussed in greater detail in section 6 below. Given that the primary credit comfort for lenders is the income generated through contractually secure offtake arrangements, such failure to execute PPAs on agreed terms would be a disincentive for lenders to provide financing to solar power projects.

(e) ***Specific joint venture-related issues***

The above issues would also apply where a solar project is being developed by the project developer in joint venture with an Indian party or an overseas party. However, a notable specific issue that some overseas joint venture partners have faced is their inability to exit the joint venture, as PPAs in some states tend to contain a 'lock in' of shareholding covenant – ie, a covenant that the shareholding of the solar project developer will not change for a particular time period after the signing of the PPA. Such conditions may discourage overseas parties from forming joint ventures to operate solar power projects in India.

6. Power offtake arrangements

The primary driving force behind the solar power growth story is the guarantee given by the DISCOMs to purchase power for a minimum 25-year period at a fixed rate under a PPA executed with power project developers. Solar power developers generally enter into PPAs for the offtake of electricity generated from the solar power plant. The entire bankability of the project and investment made by a developer depends on the continued existence and validity of the PPA.

As mentioned above, SERCs are required to prescribe RPOs (which also include solar purchase obligations) for certain categories of consumers. The solar purchase obligations as part of the RPO prescribed by the SERCs offer assured offtake to solar power developers.

Since the PPA period influences the tariff by determining the period over which the investment is returned to the investor, longer PPAs are preferred. However, due to a steep fall in tariffs for renewable energy power plants (including solar power plants), certain DISCOMs have attempted to renegotiate or cancel signed PPAs with wind and solar power developers as they were executed at a much higher price than the current rate.[74] These actions by DISCOMs have clearly introduced a degree of uncertainty into the sector.

7. The future for the solar power sector in India

7.1 Capacity addition target missed – 2016–17

Amid all the euphoria arising from record capacity additions in renewable energy (including solar power plants) in 2016–17[75] and capacity additions in the renewable energy sector being at par with capacity additions for thermal power,[76] one cannot ignore the fact that against the targeted capacity additions for solar power for 2016–17 of 12,000 MW, India achieved only 5,526 MW, although this itself was a record as far as capacity additions in the solar power sector was concerned.[77] This is less than 50% of the target and is particularly concerning as the target for 2017–18 is even higher at 15,000 MW and any shortfall may hinder India meeting its target for installed capacity by 2022.[78] The future of the solar power sector, in particular solar power generation, will significantly depend upon the governmental policies, rules, regulations, their execution and how they address the issues and risks discussed in this chapter.

7.2 Conclusion

India has set a very ambitious target of 100 GW of installed capacity for solar power by 2022. This has already started acting as a catalyst in the power sector as is evident

---

74 See www.deccanchronicle.com/business/economy/210817/ppa-renegotiation-by-state-discoms-to-impact-re-sector-icra.html.

75 See www.thehindubusinessline.com/economy/capacity-additions-in-renewable-energy-on-par-with-thermal-power-in-201617/article9679836.ece.

76 *Id.*

77 *Id.*

78 See http://mnre.gov.in/file-manager/UserFiles/OM-year-wise-cumulative-target-for-100000MW-grid-connected-SP-project.pdf.

from different announcements made in India regarding cancellation of coal-fired power plants.[79] Some of these projects had become unviable as they were dependent on imported coal and could not compete in a scenario of abundant supply of renewable resources in India at comparable prices.[80] As of 30 June 2017, installed solar power capacity in India was 13,114.85 MW, roughly 3.98% of total installed capacity in India, while for the thermal power sector this figure was 219,449.5 MW which translates to more than 66%.[81]

While an argument can be made that solar capacity growth targets set by India are overly ambitious, in practice commissioning of new projects has gathered considerable pace. If India is able to achieve the targets, it will help reduce the country's carbon footprint significantly. This will not only have a positive impact on the environment and prospects for solar plants in India, but also help in reducing a huge outflow of foreign currency. Despite several issues that remain with the sector, the government should continue to formulate favourable policies for the sector, attract domestic and international capital and assure growth for all stakeholders.

79 See http://ieefa.org/ieefa-asia-indias-electricity-sector-transformation-happening-now/.
80 *Id.*
81 See www.cea.nic.in/reports/monthly/installedcapacity/2017/installed_capacity-09.pdf.

Italy

Carlo Montella
Orrick, Herrington & Sutcliffe LLP

1. Introduction – Italian policy on renewable energy sources and background for the development of Solar Power in Italy

1.1 Italy's approach to renewable energy sources

Italy's economic support for renewable energy investments began in earnest in the 1990s, under Law 9/1991 providing for what is considered to be the first feed-in tariff programme based on written agreements between the individual energy producer and the National Entity for Electricity (ENEL). This governed the distribution and sale of all energy generated from renewable sources that was not consumed by the producer itself (so-called 'excess energy') remunerated according to prices set by the Inter-Ministerial Committee for Prices (CIP) taking into account a number of factors, including fossil fuel prices and the operating expenses of renewable energy producers (the 'CIP-6' scheme).

Specifically, the system worked as follows: energy producers received a 15-year price based on what the costs for ENEL to produce the quantity of energy purchased would have been (*costo evitato di sistema*); on the other hand, ENEL was under an obligation to buy all the energy generated not destined for self-consumption. The mechanism also provided for a bonus to be granted to so-called 'new energy' projects (ie, for energy produced by plants built after 30 January 1991, or considered as such, for a period of eight years from the date of entry into operation in order to allow for an accelerated recovery of investment costs).

In 1999, Legislative Decree 79 (the so-called 'Bersani Decree'), implementing Directive 96/92/EC, was enacted, involving a fundamental transformation of the energy market. The Bersani Decree liberalised Italy's electricity market and implemented anti-monopoly rules that required ENEL to sell off 15,000 MW of its capacity during the year 2000 and transfer the activities of generation, distribution and sale of energy to newly incorporated companies.

Additionally, a new support mechanism for renewable energy sources was established, namely 'green certificates' (GCs). Under the scheme, electricity suppliers and importers injecting into the grid more than 100 GWh per annum had to ensure that a certain percentage of such energy came from renewable energy sources. Should they fail to do so, then they had to buy GCs to such extent (which act as 'waivers' from their obligation in this respect) from GSE (the Energy Services Manager, ie the entity tasked with the management of renewable energy sources (RES) support regimes) or renewable energy producers which were given a number of GCs depending on the amount of GWh produced. Starting from 1 January 2016, the GCs have been replaced by a feed-in premium mechanism known as GRIN.

From 2001 onwards, under EU energy policy and generation targets, Italy committed to a promotion of renewable energy generation as evident from the National Action Plans on Renewables filed with the EU Commission in June 2010 on the implementation of EU Directive 2009/28/EC. Between 2009 and 2012, the growth of RES in the electricity sector has been supported by three different support mechanisms: a feed-in tariff and a premium scheme (Conto Energia) for solar photovoltaic (PV) installations; a GC scheme for all RES other than PV; and a feed-in tariff scheme for all RES other than PV and with a capacity up to 1 MW (200 kW for wind).

In 2013, support schemes were changed in two ways. First, a new support scheme (conto termico) was introduced in the heat sector, with a price-based mechanism – costs are recovered via a levy on gas. Secondly, the GCs scheme was replaced by a sliding feed-in premium/feed-in-tariff scheme with different access procedures depending on the size of the installation.

In 2016, under Ministerial Decree 23 June 2016 (the New FER Decree), the government established new terms and conditions upon which renewable energy plants, commissioned after 1 January 2013, could benefit from dedicated incentives. Specifically, incentives are awarded based on one of four different procedures depending on:

- the type of works carried out (ie, new facility construction repowering);
- capacity of the relevant plant (with a distinction between small plants, enjoying a direct access to incentives; medium plants, subject to enrolment into a specific register and qualification under the eligibility list drafted by GSE therefrom; and large plants (ie, those with a capacity higher than 5 MW) that in order to be granted incentive tariffs must successfully qualify in a Dutch auction procedure); and
- relevant energy source available capacity quota (varying according to energy source, plant size and nature of works carried out).

Incentives may take the form of either an all-inclusive feed-in tariff, which is composed of the incentive and price for the electricity produced, if the relevant facility has a capacity not higher than 500 kW, or for all other plants, a standard feed-in tariff calculated according to the following formula: Base Tariff + Premium (potential) (– Hourly Zonal Price (of the place where energy produced is injected into the grid), in the case of new plants).

Under the new National Energy Strategy presented to the public for consultation in June 2017, the government outlined targets and goals up to 2030. The main course of action for the electricity market can be summarised as follows:

- Short term. Guaranteeing full market access to distributed generation (including renewable energy sources) and promotion of flexible sources (such as storage systems).
- Long term. Introduction of new contractual agreements (such as power purchase agreements) to ensure stability and an adequate investment

timeframe and return for fixed capital initiatives, and transition to new market dynamics of balancing supply and demand including a more active role for distribution system operators in system security.

Other envisaged reforms include:

- full implementation of a capacity market; and
- alignment of the unbalancing costs framework in light of the regulation procedure ongoing at EU level.

1.2 Italian policy in the PV sector

The most relevant measures specifically designed for the solar PV sector were enacted between 2005 and 2012, namely the 'Conto Energia Decrees', consisting of five ministerial decrees that, over time, were successful in providing for fixed 20-year incentive tariffs as of the date of entry into operation of the relevant facility. These tariffs varied according to the capacity and level of integration of the plants and applied throughout the support period. Tariffs were designed to cover investment and operating costs and decreased over time based on the entry into operation of the relevant facility. The regime is no longer in place for new facilities, due to the aggregate support costs threshold (€6.7 billion/year – the 'CAP') having been reached in July 2013.[1]

A solar PV asset is granted incentives under the version of Conto Energia that was in force and applicable at the time the solar PV asset was installed and/or registered. As at 5 November 2017, approximately 550,733 plants, representing 17.76 GW of capacity, are supported by the various versions of Conto Energia, at an annual cost of approximately €6.6 billion.

Subsequent to Law Decree 91/2014, effective from 1 January 2015, Conto Energia tariffs for existing and operating facilities with a capacity higher than 200 kW were cut by a certain percentage depending on the reduction option chosen by the operators or assigned by default in case no choice was made.

Specifically, Article 26, Paragraph 3 of Law Decree 91/2014 (as converted into Law 116/2014) provided for three reduction options:

- *Extension of the incentive period from 20 years to 24 years* and recalculation of the relevant incentive tariff according to the percentage of reduction based on the outstanding support period:[2]
- *without prejudice to the 20-year support period*, from 2015–2019 the incentive

1 The threshold was calculated as the sum of the support costs under each Conto decree. In this respect, the GSE forecasted for each plant admitted to the incentive regime under the relevant Conto decree the total expected average support costs over the operational life of the plant – that being used to determine the overall costs under the relevant Conto and to monitor the CAP status (at the aggregate level).

2

Outstanding support period (years)	Percentage of reduction of the incentive
12	25%
13	24%
14	22%
15	21%
16	20%
17	19%
18	18%
beyond 19	17%

tariff would be subject to a reduction based on the outstanding support years, followed by an increase of an amount corresponding to the reduction incurred;

- *without prejudice to the 20-year support period*, the incentive rate would be subject to a straight cut according to the following percentages:
  - 6% cut for plants having a capacity between 200 kW and 500 kW;
  - 7% cut for plants having a capacity between 500 kW and 900 kW; and
  - 8% cut for plants having a capacity higher than 900 kW.

According to the legislator, the intervention on PV tariffs was necessary in order to reduce the impact of support costs on end-users' electricity bills and to remedy certain alleged overcompensation issues linked to the fact that tariff determination over time was not fully aligned to the decreasing cost of technology and generation.

As a result, PV plants that qualified under the relevant Conto decree, continue operating under reduced tariffs for the outstanding support period, while new facilities cannot access the Conto Energia incentives and there are no alternative incentive tariffs available.

2. Participants in the energy sector

The following is an overview of the main authorities and entities active in the Italian energy sector.

2.1 Regulation

(a) *Italian Electricity, Gas and Water Authority (AEEG)*

AEEG is an independent regulatory body for energy markets and integrated water services. Established in 1995, it protects the interests of users and consumers, promotes competition and ensures efficient, cost-effective and profitable nationwide services with satisfactory quality levels in the electricity and gas sectors. Its core regulatory competences cover maintaining a reliable and transparent tariff system (reconciling the economic goals of operators with general social objectives and promoting environmental protection and the efficient use of energy), setting quality of service standards and defining a framework aimed at protecting and empowering consumers in competitive markets. It also provides specialised advice and reports to the government and parliament on the regulated sectors and formulates observations and recommendations for further policy actions.

(b) *Electricity Market Operator (GME)*

GME organises the power market (forward, physical market), the market for trading daily products, the day ahead auction market, intraday auction market, and gas and environmental markets. It also establishes the conditions for admission as well as the rules for the organisation and functioning of these markets.

(c) *Energy Services Manager (GSE)*

GSE is a state-owned company, which promotes and supports RES in Italy.

Specifically, GSE fosters sustainable development by providing support for renewable electricity (RES-E) generation and by taking actions to build awareness of environmentally efficient energy uses. GSE has also been tasked with the management of the existing RES support schemes (from admission procedures to payments and plant inspections and control)

(d) Ministry of Economic Development (MISE)

MISE oversees Italy's energy policy and has regulatory powers to implement any relevant legislation passed by the Italian Parliament from time to time. Its main areas of competence in this respect include:

- national energy budget and strategy;
- transport networks, energy infrastructures and security of supply;
- single market for electricity;
- promotion of renewable energy and energy efficiency and savings;
- technology for the reduction of greenhouse gas emissions;
- peaceful uses of nuclear energy;
- decommissioning of nuclear facilities;
- national gas system and market, downstream oil system and its market (refining, logistics, stocks, fuel distribution);
- national mining policy, issuing permits for exploration and cultivation of subsoil resources (in particular, hydrocarbons on the mainland and under the sea); and
- storage of natural gas and methanisation of southern Italy.

2.2 Transmission and distribution

(a) Terna Rete Elettrica Nazionale SpA (or Terna SpA)

The national transmission system operator (TSO) is responsible for transmission and dispatching of electricity throughout the country (through the national grid), and ensuring balance between demand and supply and overall stability and safety of the electricity network. To do this, Terna is active on the ancillary services market which is a market where Terna (as the exclusive and sole buyer) purchases the quantity of power necessary to ensure:

- proper system relief with regard to intra-zonal congestion;
- the creation of energy reserves; and
- real-time balancing.

In June 2006, the Electricity and Gas Authority passed Resolution 111/06 requiring energy producers to provide advance projections about the quantity of electricity they would inject into the grid to create a predictable supply and assist Terna to balance supply and demand. If an operator fails to comply with the energy injection schedule, a charge will be applied based on the deviation from the relevant schedule (so-called 'imbalance costs'), which vary depending on whether the operator is entitled to be on the ancillary services market, the kind of production unit and the kind of energy source involved (renewable/non-renewable).

The TSO role is separate from the transmission asset owner. Terna SpA performs the TSO role while Terna Rete Italia SpA owns and manages most (99.7%) of the transmission assets. Therefore, Terna group *de facto* operates in a regime of a natural monopoly.

Connection to the network of plants with capacity equal to or more than 10,000 kW falls under Terna SpA, while for lower capacities the local distribution companies (such as ENEL) where the plant is located are responsible.

(b) ENEL Distribuzione SpA

ENEL is the major operator in energy distribution, owning and managing the low and medium voltage networks. As such, ENEL acts also as a distribution network operator (DNO) at local level and is responsible for plant connection in case of capacities less than 10,000 kW as specified above.

2.3 Generation and supply

Under Article 1 of Legislative Decree 79/1999, production, import, export, purchase and sale of electricity are not subject to any licence or permit regime, although registration with AEEG for monitoring purposes is required. Because of this, generation and supply is quite fragmented with several individual power producers (mainly RES), although there are some utility generators (primarily hydroelectric and thermoelectric power stations) owned by former state-owned companies (eg, Enel Produzione SpA or Edison SpA) or local utilities (eg, A2A SpA).

Among suppliers, there are former state-owned companies such as Enel Energia SpA, local public companies (utilities) owned by municipalities (eg, A2A SpA active in Milan and Lombardia Region).

3. Access to the grid

Legislative Decree 79/99 provides that the national TSO and the operators of local distribution networks must ensure that power plants are connected (on the third-party access principle) on request from the relevant user.

Following the enactment of Law 244/2007, the AEEG issued the Unified Text for Active Connections (TICA) containing rules that apply to high, medium and low voltage connections to the power grid.

According to TICA, the request for interconnection must be filed with either:

- the operator of the local distribution network if the output is lower than 10,000 kW; or
- Terna SpA if the output is equal to or higher than 10,000 kW.

The procedure for low and medium voltage connection includes the following:

- the applicant files a request with the local DNO seeking the issuance of a connection proposal which will include details of the connection engineering works, which, once accepted by the applicant, will occur within 45 days of receipt;
- with the acceptance of the connection proposal, the applicant must also provide evidence of the relevant authorisation procedures for construction of

the power plant and the connection works having been commenced. The applicant is allowed to carry out all the connection works (ie, including those falling within the competence of the network operator). In this instance, the applicant is required to obtain the approval (so-called *validazione*) of the network operator of the project for construction of the interconnection facilities, and communicate to the network operator the completion of the interconnection works in order to allow the grid operator to test the interconnection facilities and accept them for subsequent take-over.

- The grid connection will be activated after the construction works for the power plant have been completed and the applicant has sent to the network operator all the documentation specified by the TICA. The applicant enters into a connection agreement (*regolamento di esercizio*) providing for the terms and conditions of the connection throughout the operational life of the plant.

High voltage connections are governed by rules somewhat different from those covering medium and low voltage connections and include the following:

- the relevant plant owner files a request for a preliminary connection estimate (STMG) detailing the connection proposal for the plant and the nature of the works to be carried out. The applicant must accept the preliminary proposal within 120 business days of receipt;
- once all the permits for construction of the connection have been obtained, the applicant requests the grid operator to issue the final connection proposal (STMD), which contains the document enabling actual connection of the plant to the grid;
- The applicant is allowed to carry out all the connection works (ie, including those falling within the competence of the network operator). In this case, the applicant is required to obtain the approval (so-called *vidimazione*) of the network operator of the project for the construction of the interconnection facilities, and communicate to the network operator the completion of the interconnection works in order to allow the network operator to test the interconnection facilities and accept them for subsequent take-over.
- The grid connection will be activated after the construction works for the power plant have been completed and the applicant has sent to the network operator all the documentation specified by the TICA. The applicant enters into a connection agreement (*regolamento di esercizio*) providing for the terms and conditions of the connection throughout the operational life of the plant.

4. Legislative/regulatory background and support regimes

Article 7 of Legislative Decree 387/2003, implementing EC Directive 2001/77,[3]

3 The directive established aggressive national targets for each EU member state for the production of electricity from renewable sources in view of the European Union's objective of having 22.1% of total Community electricity consumption produced from renewable energy sources by 2010. Italy's target was to have 25% of total electricity consumption produced from renewable energy sources by 2010.

provided for a general framework on tariffs for PV plants and tasked the Minister of Productive Activities and the Minister of Environment, Land and Sea with designing, through one or more ministerial decrees, the support regime. This was to ensure, among other things, that tariffs were specific (ie, modelled on the technology utilised), decreasing in time and lasting sufficiently long so as to allow a fair remuneration of investments and operating costs.

Italy's Ministry of Productive Activities enacted the first three Conto Energia decrees as instructed in that act and to implement that mandate. Each of those decrees referred to the need to promote PV investment by offering specific tariff rates to ensure fair remuneration and that, once granted to a specific facility, would be fixed for 20 years:

- Conto Energia I – Enacted with Ministerial Decree 28 July 2005 and 6 February 2006 and applying to facilities up to 1 MW realised and entering into operation from 30 September 2005. Tariff rates varied between 0.445 €/kWh for plants between 1 kW and 20 kW to 0.460 €/kWh for plants between 20 kW and 50 kW and 0.490 €/kWh for plants between 50 kW and 1 MW. Tariffs were subject to an automatic 5% reduction for each year subsequent to 2006 for new facilities to enter into operation.
- Conto Energia II – Enacted with Ministerial Decree 19 February 2007 and applying to facilities realised and entering into operation after 13 April 2007 and (after the original deadline was extended) by 30 June 2011. The tariffs ranged between 0.36 and 0.49 €/kWh and were subject to an automatic 2% reduction for each year subsequent to 2008, limited to new applications submitted. The regime effectiveness was conditional upon reaching an overall aggregate incentivised capacity of 1200 MW which occurred on 7 July 2010.
- Conto Energia III – Enacted with Ministerial Decree 6 August 2010, applying to facilities realised and entering into operation after 31 December 2010 and by 31 May 2011. The overall tariffs ranged between 0.251 and 0.402 €/kWh, and were subject to an automatic 6% reduction from 2012 onwards, limited to new applications submitted. The regime effectiveness was conditional upon reaching an overall aggregate incentivised capacity of 3000 MW, which was never reached due to the restriction made by the legislator on the timeframe of application (original end date: 31 December 2013).

Legislative Decree 28/2011 (the Romani Decree) implemented the 2009 EC Directive and operated in a similar fashion and for a similar purpose as Legislative Decree 387/2003 in that it continued the goal of incentivising additional investment in renewable energy production and mandated the MISE and MATTM (Ministry of Environment) to enact further decrees regulating the support regime for PV plant entering into operation after 31 May 2011, which resulted in Conto IV and Conto V decrees being enacted respectively on 5 May 2011 and 5 July 2012.

- Conto Energia IV – Enacted under Ministerial Decree 5 May 2011, concerning plants entering into operation after 31 May 2011. In order to obtain the incentives, plants with a capacity higher than 1 MW had to be enrolled in a specific registry and secure a viable placement in the GSE ranking (a list of eligible candidates published by GSE). Tariffs ranged between 0.172 and 0.387 €/kWh in 2011, while between 0.133 and 0.274 €/kWh in 2012. Starting from 2013 the tariffs were replaced by an all-inclusive feed-in tariff based on the amount of electricity fed into the grid.

- Conto Energia V – implemented under Ministerial Decree 5 July 2012, applying to facilities entering into operation after 26 August 2012 and effective until the overall Conto Energia support costs reached €6.7 billion which happened on 6 July 2013. For plants exceeding 1 MW, the same registry system under Conto IV applied and the incentive tariff was equal to the difference (if positive) between the all-inclusive tariff applicable to plants below 1 MW and the market price (*prezzo zonale orario*) of electricity plus the revenues deriving from the sale of the energy to the market.

As noted above, the Conto Energias mechanism is no longer in force for new facilities, due to the aggregate costs threshold having been reached in July 2013.

Subsequent to Law Decree 91/2014 (*Splamaincentivi Decree*), tariffs already granted to plant above 200 kW have been lowered according to one of the three options illustrated in section 1 above, which the operators had to choose between.[4]

Focus on: Spalmaincentivi Decree, a closed case?

Subsequent to the approval of Law Decree 91/2014, several operators challenged the constitutional legitimacy of Article 26 thereof, providing for the re-setting of incentive tariffs, before the Italian Constitutional Court for violation of the principle of 'legitimate expectation', the measure being a *de facto* retroactive change to terms and conditions.

The Court, however, rejected the claim holding that:

- There was no violation of 'legitimate expectation' given the intervention of the legislator was reasonable, non-arbitrary and foreseeable. The argument is based on the following:
  - while the confidence of a private citizen in legal certainty is an essential and necessary element of a state governed by the rule of law, the protection of legitimate expectations does not imply that the legislator is forbidden from enacting provisions detrimentally modifying a long-term relationship framework even in case of consolidated rights, except for the constitutional limit concerning criminal law, as long as the measure is not irrational and discretionary;
  - the analysis of the reasons underlying the challenged measure and the content thereof does not show such measure has affected irrationally, arbitrarily and unexpectedly the long-term relationships of energy producers as:
    - the legislator acted during a general economic crisis when the

4 A cut in the FIT and extension in tariff period; reduced FITs over 2015–2019 recovered during the subsequent remaining contracted period; or outright cut in FIT by 6% to 8 %. The third option was automatically applied to the operators which had not make any choice by 30 November 2014.

increasing burden of RES costs on end-users (specifically SMEs) had become a matter of significance;
   - there were legislative provisions denying the existence of a 'crystallised' right to incentive stability overtime (stability should be regarded as being in the context of the overall regime rather than to a single tariff), of which a diligent and expert operator should be aware.
- The measure would not have radically affected the investments made as generally argued by claimants without material evidence, but has been modelled in order to take into consideration their viability. Additionally, a number of effective balancing measures has been provided and thus, overall, a fair return on investments would be ensured in the long term,

Notwithstanding the court ruling, operators in Italy (and in other countries affected by similar incentive tariff cuts), have resorted to several arbitration proceedings under the Energy Charter Treaty, a binding multilateral energy investment protection agreement signed in 1994 between European countries and former USSR countries, later extended to extra-EU states. The claims have been for violation of Article 10(1) providing for each contracting party to:

- encourage and create stable, equitable, favourable and transparent conditions for investors of other contracting parties
- fair and equitable treatment;
- not to impair in any way by unreasonable or discriminatory measures the management, maintenance, use, enjoyment or disposal of the relevant investment; and
- not accord investments treatment less favourable than that required by international law, including treaty obligations.

Proceedings are currently ongoing and to date decisions have been varied, with the Spanish cases (the only one for which awards have been rendered to date) resulting in two SCC decisions in favour of Spain (*Charanne* and *Isolux Netherlands, BV*) and one ICSID decision in favour of the investors (*Eiser*).

Therefore, to date, PV plants that qualified under the relevant Conto decree continue operating under reduced tariffs for the outstanding support period. New facilities cannot benefit from the Conto Energia incentives and there are no incentive tariffs in place for them.

Other RES support regimes opened to PV plants include:

- Simplified energy sale and purchase regime (*ritiro dedicato*). Under agreements with the GSE, producers sell the electricity generated and to be injected into the grid to GSE, instead of selling it through bilateral contracts or directly on the power exchange market. GSE purchases and resells the electricity to be fed into the grid at the zonal price or at a minimum guaranteed price (for plants below 1 MW only, 100 kW for photovoltaic plants). On behalf of the producer, it transfers the fees for using the grid (dispatch and transmission fees) to distributors and to the Transmission Service Operator.

- Net metering service (*scambio sul posto*). Under this service, the electricity generated by a consumer/producer in an eligible on-site plant and injected into the grid can be used to offset the electricity withdrawn from the grid. The GSE pays a contribution to the customer based on injections and withdrawals of electricity in a given calendar year and on their respective market values. Net metering is not compatible with either Conto Energia, the simplified energy purchase and resale regime or the all-inclusive feed-in tariff. The service applies to:
  - owners of renewable energy source electricity generation plants with a capacity of up to 20 kW;
  - renewable energy source electricity generation plants with a capacity of up to 200 kW (commissioned after 31 December 2007);
  - high-efficiency combined heat and power (CHP) plants with a capacity of up to 200 kW.

5. Bankability of projects

Bankability of projects in the Italian market context depends on the following elements:

- Stable flows of revenues. In order to obtain access to credit lines, it is necessary to ensure project profitability which depends on whether the project can rely on a stable and certain source of revenues, whether in the form of incentives or energy market prices. In this respect, the existence of feed-in tariffs under the Conto decrees ensured what was perceived to be a reliable source of income; similarly the simplified sale and purchase regime (*ritiro dedicato*) or PPAs solutions are considered to support project bankability;
- Authorisation procedures. The longer the authorisation procedure for building and operating a plant the less attractive the project for investors but also for financial institutions to commit funds. In this respect, Italy has made significant progress in streamlining and simplifying authorisation procedures (since Legislative Decree 28/2011), thus to date while procedures might still be somewhat complex, the overall timetable has now become more acceptable (eg, 90 days under the Single Authorisation procedure);
- Real estate issues. Once the site of a project has been identified, the relevant developer must be able to secure the relevant legal rights for the whole duration of plant operation, or at least for the duration of any incentive granted. In this respect, the developer may secure surface, lease or ownership rights. Usually, for financing purposes, surface and/or ownership rights are preferred as they result in the plant built on the land remaining the property of the developer, all the more important given that banks will likely require security over the land and the assets thereon.
- Access to the market. In order for a project to be bankable, the energy generated must be able to be sold to the market. In this respect, Legislative Decree 79/99 provided for RES dispatching priority, which was of utmost importance since it ensured connected RES generators would be able to

transmit and sell their output at all times when they are operationally able to despatch power.

6. Environmental, real estate, planning and zoning issues

Plant construction requires a preliminary evaluation of the impact on the local environment, which is made under two different procedures:

- The environmental impact assessment (*valutazione di impatto ambientale* (VIA)), an administrative and quite complex procedure carried out by the relevant authority (usually the Province, Region or the Ministry of the Environment) involving several parties beyond the applicant and the authority in charge (ie, any third party having an interest in relation to the project or potentially affected thereby);
- A two-step assessment procedure available to certain categories of plants, which includes a pre-screening assessment (*verifica di assoggettabilità*), the outcome of which is used to determine whether a VIA will be necessary. The pre-screening assessment is a preliminary, streamlined assessment of the impact that a project may have on the environment. Where the pre-screening assessment has a negative outcome, the project is required to undergo a VIA procedure, which is a further and more detailed assessment.

The national legislation exempts projects for the construction and operation of electric power generation plants from the requirement to undergo pre-screening. These exempted plants include RES-E generation plants with a capacity not higher than a specific threshold, which has been changed over the years. Pursuant to the Environmental Code currently in force, as lastly amended by Law 99 of 23 July 2009, plants with a capacity of 1 MW or less are exempt from pre-screening.

Projects to be built within certain areas of high landscape or environmental value because of the presence of specific flora or fauna must undergo an additional assessment. Such additional assessment is the 'environmental incidence assessment' (*valutazione di incidenza ambientale*), which evaluates the impact the project may have on the flora and the fauna living in the area.

The environmental incidence procedure is governed by Presidential Decree 357 of 8 September 1997 (Decree 357), which was implemented in Italy EC Directive 92/43/EEC for the conservation of natural habitats. Pursuant to Decree 357, a project for a plant to be constructed within either an area of significance (*sito di importanza comunitaria* (SIC)), a special preservation zone (*zone speciali di conservazione* (ZPS)), or a zone where certain animal or vegetable species exist (jointly with the SICs and the ZPSs (Areas of Special Environmental Value)), must undergo the environmental incidence procedure.

7. Power offtake arrangements

Italian legislation gives producers of renewable electricity the option to sell the electricity produced pursuant to the so-called 'offtake regime' (the *ritiro dedicato,* (RID)) instead of selling the energy to the market.

Specifically, under agreements with the GSE (AEEG's Decision 280/07), producers sell electricity generated and to be injected into the grid to GSE, instead of selling it through bilateral contracts or directly on IPEX (Italian Power Exchange). The offtake regime agreement has a duration of one year as of the signing date and is automatically renewed through a silent-consent process (ie, unless the owner of the plant decides to choose a different mechanism for the sale of electricity).

Under the agreement, producers, inter alia:

- cannot assign to third parties the receivables arising from the contract;
- shall pay the transmission service fee and imbalance charges in accordance with the relevant law;
- may withdraw from the agreement anytime upon prior written notice to be sent at least 60 days in advance; and
- shall inform GSE of any issue affecting availability and/or functioning and/or productivity of the plant.

In case of non-compliance with the obligations provided, modifications of the authorisations necessary for the operation of the plant, judicial challenges to the authorisation title or measures adopted by the competent authorities affecting the availability and/or functioning and/or productivity of the plant, the GSE retains the right to suspend the effectiveness of the agreement, as well as terminate the contract.

The price applied to the electricity purchased by GSE and injected into the grid is the 'average zonal price' (ie, the average monthly price per hourly band, which is set on IPEX for the market area to which the plant is connected).

PV plants with a capacity lower than 100 kW may benefit from minimum guaranteed prices, updated annually by AEEG. At the end of each year, GSE makes adjustments for plants in respect of which the revenue associated with the hourly zonal prices proves to be higher than the one resulting from the application of the minimum guaranteed prices by paying the producer the difference.

As an alternative to the offtake regime, producers may either sell the energy directly on the market through IPEX, or enter into bilateral over-the-counter contracts with a qualified energy trader (power purchase agreements (PPA)), which must be registered with the *Piattaforma dei Conti Energia a Termine* (PCE) managed by the GME (*Gestore del Mercato Elettrico*), a public entity tasked with the management of power, environmental and gas markets.

8. The future for the solar sector in the region

Despite the end of the Conto Energia support regime, the solar sector will offer interesting opportunities:

- The GSE finally published the new framework on maintenance and renovation of the PV plants benefiting from the Conto Energia support regime (the New DTR) overcoming the uncertainties on repowering's compatibility with the support scheme regulation of the Contos Energia and providing economic support for repowering (the extra power generated by repowering plant may benefit from the energy sale regime), thereby giving the market reasonable economic prospects.
- Withdrawal of government support created an active secondary market –

with a record level of M&A transactions – and a consolidation trend in this very fragmented industry, where fewer owners hold portfolios of increasing sizes (a trend that will likely continue in the near future) with a leading role for financial investors.

- While current market prices would hardly support utility scale plants, PV and wind technology costs are expected to lower further, which would make such projects potentially feasible in Italy, thereby boosting photovoltaic development in light of the EU 2030 targets.
- Corporate PPA – this could be a promising business model, allowing both parties to meet their objectives (on the one hand finding a stable source of supply ensuring demand satisfaction, and on the other finding a stable long-term source of income consuming all the energy generated by the relevant facility).

Public buildings, high energy consumption industries, (pools of) private buildings, large industrial or commercial plants could be the main participants for this business model, as long as certain barriers under the existing legislation are removed. These barriers include, for instance, the 'one user one contract rule', preventing one plant from serving multiple clients under SEU scheme, or the impossibility of extending existing or creating new RIUs (*Reti Interne di Utenza*) (ie, private networks serving industrial units only within a limited territory and managed by one entity). A positive sign for a future substantial development of this kind of contract can be found in the draft of the upcoming EU Directive on Renewable Energy, enabling state members to remove all administrative barriers preventing long-term PPAs. The National Energy Strategy 2017 identifies PPAs as a potential solution to match the need of long-term stability for fixed capital investments that can ensure reliable returns (which the current market structure does not ensure given price volatility).

Japan

Maya Ito
Sadayuki Matsudaira
Edward Rivera
Nathan Schmidt
Nishimura & Asahi

1. Introduction

Japan has consistently been a member of important international treaties on the reduction of carbon emissions including the United Nations Framework Convention on Climate Change, the Kyoto Protocol, the Bali Road Map, the Copenhagen Accords and the Paris Climate Change Agreement. Japan's Ministry of Foreign Affairs (MOFA) is focusing on three key pillars to bring about a global reduction in carbon emissions. The first is increased collaboration with developed nations to bring about more energy-efficient technologies, the second is collaboration with developing nations so they have access to this technology and the third is providing support for particularly vulnerable countries.

Japan has simultaneously been proactive in its low carbon emission policies back home maintaining an impressive and varied approach to reducing carbon emissions over the years. Beginning with its first Promotion of Global Warming Countermeasures Act passed in 1998, Japan engaged in a policy of promoting energy efficiency and alternative local energy supplies. Japan has also taken more innovative approaches to the reduction of emissions via energy conservation. In the summer and winter of 2005 Japan launched the first 'cool-biz' and 'warm-biz' campaigns to encourage companies to engage in energy conservation and allow workers to dress more comfortably. This innovative approach has also been praised for generally raising societal awareness about the issue.

Research and development in solar energy has been active in Japan since 1974 when the first solar R&D in Japan was funded by the government. In the mid-1990s, Japan notably became the first country to implement incentives for small residential solar PV. Despite this early history of engagement, Japan's solar energy industry had remained largely export-oriented up until the introduction of the feed-in-tariff (FIT) regime in 2012. Japan's Diet had introduced legislation to incentivise solar projects in 2002 in the form of the Renewable Portfolio Standard (RPS), which specified a minimum quantity of energy that had to come from renewables. Under this regime, growth was considered sluggish, and projects remained export oriented. Meanwhile, European FIT regimes were being introduced throughout the mid-2000s. During this time frame, the combined renewable energy sources of the country accounted for only 1% of total energy production.

This chapter comprises the following: section 2 provides a summary of the legislative background of the Japanese FIT regime; section 3 discusses the requirements to qualify for FIT rights; in section 4 we review power purchase agreements (PPA) under the FIT regime and curtailment; in section 5 we describe land-use and permitting issues that arise with solar projects in Japan, focusing on the most commonly encountered issues; and in section 6 we provide an overview of the market participants involved in Japan's solar projects. Finally, in section 7 we conclude with a brief look at the future of solar energy projects in Japan.

2. Legislative background of the Japanese FIT regime

The existence of the FIT regime in Japan is often associated with the preceding Fukushima Nuclear disaster that occurred in 2011. In fact, the bill which would become the Act on Special Measures Concerning Procurement of Renewable Electricity by Electric Utilities from Renewable Energy Sources, also known as the Japanese Renewable Energy Act (the FIT Act),[1] had been adopted by the Prime Minister's Cabinet mere hours before the actual earthquake occurred. This disaster may ultimately have played a vital role in motivating the Diet to pass the legislation on 26 August 2011, but it had been proposed as part of a general initiative to increase renewable energy investment prior to the occurrence of the disaster. This can be at least partially attributed to a recognition of the shortcomings of the RPS regime. With the move away from nuclear power, which at the time made up 70% of Japan's energy supply, there was a real impetus to develop alternative energies. In particular, the modifications made to the bill to specify generous price points on tariffs seem to have been driven by the growing popular discontent with nuclear energy.

The FIT Act came into effect on 1 July 2012. The renewable energies targeted by the FIT regime include solar, geothermal, wind, hydroelectric, biomass and other forms of renewable energy. Each type of renewable energy has its own corresponding FIT rate, and these rates have changed over the years. The FIT regime required utilities to purchase energy generated by producers which were qualified under the rules of the FIT regime. By the end of 2016, solar energy came to account for 4.3% of electricity production in Japan, with estimates that by 2050 solar energy would make up 10% of electricity production. Of all the targeted renewable energy sources, solar energy projects account for over 60% of applications under the FIT regime, overshadowing all the other renewables targeted. This is largely because the FIT regime's generous pricing links well with long-term PPAs for solar energy, making for a particularly attractive investment opportunity. Japan's generally low interest rate environment has made debt financing attractive for solar investors, and in turn lenders have been eager to participate in the solar industry.

The original FIT Act was subsequently amended on 25 May 2016 by the Partial Revision of the Act on Special Measures Concerning Procurement of Electricity from Renewable Energy Sources by Electricity Utilities, also known as the 2016 Amendment to the Japanese Renewable Energy Act (or the FIT Act Amendment).[2] The FIT Act Amendment came into effect on 1 April 2017. The FIT Act Amendment implemented substantial changes to the FIT regime. The certification system became stricter as each project must now submit a detailed business plan. Additionally, a

1 Act 108 of 2011.
2 Act 59 of 2016.

reverse auction/bidding process was newly established for projects with a capacity of 2 MW or more, with each bidder required to submit their business plan prior to their bid. Currently, enthusiasm for solar project development continues and there is growing excitement about the future prospects for solar power in Japan.

The primary regulator of Japan's energy markets and utilities is the Ministry of Economy, Trade and Industry (METI) together with its affiliated agency, the Agency for Natural Resources and Energy (ANRE). Accordingly, these agencies are responsible for the development of the renewable energy sector in Japan and the regulation of solar energy projects. They administer the FIT programme (as discussed above) and issue to projects certifications which are required in order to qualify under the FIT programme. The Green Investment Promotion Organisation (GIO) is the government-designated entity that is responsible for administering the reverse auctions that were introduced by the FIT Act Amendment to award FIT rights to projects of 2 MW and larger.

3. The Japanese FIT regime

3.1 FIT prices

Under the FIT regime, power generators for certain renewable energies (ie, solar, wind, biomass, geothermal, and small and mid-size hydro), which qualify under the FIT Act Amendment, are granted the right to sell the electricity they generate at a fixed price for a fixed period set out under the act. Solar projects have a fixed period of 20 years. In March each year METI announces the applicable fixed prices that will apply to projects that qualify during that year (running from April to March). The fixed prices for solar projects with less than 10 kW of capacity are somewhat higher than the fixed prices for solar projects of 10 kW or more. FIT prices for new projects have gradually been reduced each year, as can be seen in the table below. For example, in the 2012 business year, JPY 40 was the FIT price for large solar projects. In the 2017 business year, the FIT price for large solar projects was JPY 21. Since its inception, the Japanese FIT programme was designed to scale back FIT prices over time. However, FIT prices for solar projects have notably been reduced by a greater proportion than other types of generation methods under the FIT programme. This is largely due to the successful growth of the Japanese solar industry. METI and other policymakers determined that solar projects would continue to be developed with lower FIT rates, allowing them to shift FIT funds to support wind and other renewable generation methods.

Business Year	2012	2013	2014	2015	2016	2017	2018	2019
10 kW or more (JPY/kWh)	40	36	32	29 27*	24	21	to be determined	to be determined
less than 10 kW (JPY/kWh)	42	38	37	33 35**	31 33**	28 30**	28 30**	24 26**

* On 1 July or later.

** Those with facilities to respond to curtailment control.

3.2 Requirements under original FIT rules

Under the original FIT Act, a prospective photovoltaic (PV) plant would qualify for rights to sell at FIT prices once it obtained a certification from METI of the modules and other equipment that were specified to be installed at the PV plant, and submitted a formal grid connection application to the applicable utility. METI certification was issued provided that a complete application was properly submitted and all requirements were met. Notably, METI did not restrict or otherwise limit the capacity of prospective PV projects that obtained METI certification. The process to submit a formal grid connection application was administered by each utility and was focused on technical requirements. Under the original FIT Act, the utilities were obligated to accept interconnection applications for prospective PV plants with METI certification with limited exceptions. However, applicants are responsible for the costs of constructing interconnection facilities and related costs to upgrade grid capacity, and utilities may reject applications (effectively cancelling a project's FIT pricing) if payment for such costs is not made when due.

The requirements to qualify for FIT rights were gradually increased by regulatory reforms. These reforms affected both METI certifications and the grid connection documentation. When reviewing whether a solar power project has secured rights to sell at FIT prices prior to the FIT Act Amendment, it is important to review when the rights were secured. Some projects may be grandfathered to qualify under earlier regulations.

For example, in April 2015 a regulatory reform adjusted the date upon which a solar project would secure its FIT price. Projects that had not secured their FIT price as of 1 April 2015 would secure their FIT price as of the latest of:

- the date that its METI equipment certificate was issued;
- the date upon which its grid connection agreement was executed; and
- the 270th day after the formal grid connection application to the applicable utility (if a grid connection agreement could not be executed due to a reason attributable to the utility).

Permitted amendments to a project's METI certification is another area that has encountered several rounds of regulatory reforms. The date of such amendments should be checked carefully to confirm whether they were permitted at the time without triggering a reset of the project's applicable FIT price to the date of the amendment. In particular, amendments to a project's METI certification reflecting changes to its production capacity, its equipment (eg, manufacturer, model, conversion efficiency), and changes to the size of the project site should be checked carefully to confirm that they did not affect the applicable FIT price.

In general, the rules and requirements of the FIT Act Amendment apply to PV plants that qualified for FIT pricing under the original FIT Act prior to 1 April 2017. However, in order to continue to hold their rights to such FIT prices, they must have entered into an interconnection agreement by 31 March 2017 and submitted a business plan to METI for certification by 30 September 2017.

3.3 Requirements following the FIT Act Amendment

The FIT Act Amendment introduced a variety of requirements and changes to the FIT regime. The earlier framework was effectively replaced as of 1 April 2017. A new requirement is that a prospective project must submit a detailed business plan to METI for certification in order to qualify for FIT pricing. These certified business plans replaced the previous concept of METI certification of a PV plant's specified equipment. The elements and requirements for the business plans are set out in the FIT Act Amendment and in regulations that have been adopted. The requirements are too detailed to be summarised here, but they include structure and wiring diagrams, maintenance plans, and plans (and financial contingencies) for demolition of the PV plant upon decommission. In addition, a business plan must confirm that the contemplated PV plant will comply with all applicable regulations, including local government land-use requirements and permitting regulations. Also, part of the process of obtaining METI certification of a PV plant's business plan is entering into an interconnection agreement with an applicable grid company (see the description of utilities and grid connection companies below). The intention of the enhanced requirements to obtain METI certification of business plans is an effort to improve the proportion of certified PV projects that ultimately are actually built and start commercial operations.

As mentioned above, the FIT Act Amendment introduced a reverse auction bidding process for PV projects that are 2 MW or more. Auctions are contemplated to take place twice a year. The process is administered by the GIO. Each respective bidder must submit its business plan to the GIO. Prior to each auction, a maximum bid price and the total volume output that will be available in the auction will be announced. Each bid must indicate the amount of output that it proposes to sell and its bid price. Each bidder must also pay a deposit along with its bid. After winning bids are announced, deposits are refunded to losing bidders. Deposits for winning bids are not subject to reimbursement. Winning bidders must start operations by their contemplated starting date.

Also of note, the FIT Act Amendment added new restrictions against changes to the DC output from PV plants. Previously, restrictions applied only to AC output. If a PV plant changes its output capacity in either AC or DC after obtaining METI certification of its business plan, then an amendment to the METI certification is required, and the applicable FIT price becomes subject to change.

4. Utilities, power purchase agreements and curtailment

4.1 Utilities

Pursuant to amendments to the Electricity Business Act,[3] Japan is in the process of liberalising the generation, transmission, distribution and sale of power within the

3 The reform of the Japanese electricity market is being implemented pursuant to three amendments to the Electricity Business Act. The frist step, pursuant to Act 74 of 2013, which became effective in April 2015, amended the Electricity Business Act to establish the Organisation for Cross-regional Coordination of Transmission Operators (OCCTO) which is a self-regulatory organisation formed to coordinate the development of nationwide transmission and distribution networks. The second step, pursuant to Act 72 of 2014, which became effective in April 2016, amended the Electricity Business Act to liberalise the retail market for electricity. The third step, pursuant to Act 47 of 2015, which will become effective in April 2020, will require the electricity utilities to reorganise themselves into three separate entities: (i) a transmission utility, (ii) an electricity generation company, and (iii) an electricity retail company.

country. Prior to this process, the following 10 general electricity companies controlled the generation, transmission, distribution and sale of power: Hokkaido Electric Power, Tohoku Electric Power, Chubu Electric Power, Kansai Electric Power, Hokuriku Electric Power Company, Chugoku Electric Power, Shikoku Electric Power, Kyushu Electric Power, Okinawa Electric Power and Tokyo Electric Power Company (TEPCO). The deregulation of Japan's electricity market and reforms to the industry contemplate that, by 2020, each of these companies must reorganise into separate entities with their power generation and retail businesses in a company and their transmission and distribution businesses in another company. However, currently, only TEPCO has been reorganised with a holding company, TEPCO Holdings, which has three operating subsidiaries: TEPCO Fuel & Power, TEPCO Power Grid and TEPCO Energy Partner.

Before the Amendment to the FIT regime the purchasers of FIT electricity were the electricity retail utilities. An important change under the FIT Act Amendment is to change this to general transmission utilities in line with Japan's reorganisation described above.

PPAs that are in place with electricity retail utilities will remain in effect until the end of their existing terms. If either of the parties to such a PPA initiates an amendment of a material term of the PPA, the retail company will not be allowed to continue the purchase of FIT electricity, and instead the FIT PV power producer will enter into a new PPA with the applicable energy transmission utility.

4.2 FIT Power Purchase Agreements

The terms of PPAs under the Japanese FIT regime are specified under the FIT Act and are regulated by METI. As a result, the terms of PPAs for PV plants that qualify for FIT rights generally follow a uniform standard based on a form that was developed by METI, with little or no negotiations between developers and utilities. Prior to the FIT Act Amendment, execution of the PPA for a prospective PV plant was often deferred until late in the development process (and sometimes until construction was nearly complete). Following the FIT Act Amendment, PPAs are now often signed together with grid connection agreements.

Under the terms of FIT PPAs, the power generator is not required to generate power and there are no penalties for any failure to generate or otherwise provide power to the utility. The price is fixed, based on the FIT price that the project is eligible for. Neither party is required to post collateral or any other form of deposit or credit enhancement. FIT PPAs can only be assigned if the corresponding FIT rights have been transferred to the assignee. However, they can be assigned to lenders as collateral for financing. Under certain conditions (discussed below), the utility is permitted to curtail the generation by FIT PV plants without compensation.

4.3 Curtailment

The conditions of curtailment are one of the most material terms of a PPA. Curtailment terms are specifically regulated by METI under applicable FIT regulations. Upon the start of the FIT regime in 2012, utilities were only permitted to curtail FIT PV plants without compensation for 30 days a year. Any curtailment during a day would apply (eg, 30 minutes of curtailment was counted as one day of curtailment). This rule was changed in 2015 to allow utilities to curtail output for up to 360 hours a year without paying compensation. However, earlier PV projects were grandfathered under the 30-day rule if they had solar projects that submitted formal grid connection applications prior to 26 January 2015. In some areas, where FIT projects are designated to exceed available grid capacity, FIT PV projects of 10 kW or more are now subject to unlimited curtailment without compensation. These areas include Hokkaido, Tohoku, Hokuriku, Chugoku, Shikoku, Kyushu and Okinawa. Additional areas may also become designated in future.

However, in general, utilities are required to curtail thermal power and similar generation facilities to the extent necessary to avoid, as much as possible, curtailment of output from solar power generation facilities (and other types of FIT eligible facilities). Also, compensation is not required for curtailment in excess of 30 days/360 hours if the curtailment is due to force majeure, safety issues, maintenance and inspections or unforeseen circumstances that are not attributable to the utility.

5. Land use, permitting and other noteworthy regulations

5.1 Local requirements

One of the challenges to the development of PV projects in Japan is the varying legal requirements that surround different land-use permits depending on the project's locality. These tend to vary significantly depending upon how national law is interpreted by the most local form of government. This could cause a particular regulation to vary substantially in its actual requirements depending on where the PV project is carried out. This section provides an overview of the most commonly encountered permits and what they generally require.

The general practice in Japan to help ensure that the development process of a PV project is as smooth as possible is to take detailed project plans to the local municipal and prefectural offices to discuss them with officials before submitting any permit applications. This informal review process allows officials the opportunity to point out potential problems with permitting and land-use that may arise. Ultimately, land and environmental issues are largely manageable when conducting solar power projects in Japan and, with the guidance of local advisers, they are generally navigable.

5.2 The Agricultural Land Act (農地法)

Japan's Agricultural Land Act[4] has been in place since 1952 with the stated aim of protecting farm land and helping to stabilise Japan's agricultural production. In general, it restricts changes to, and any development activities on, agricultural land. Accordingly, if the site of a contemplated PV plant is located on land that is actually used as agricultural land or land that is designated to be agricultural land in the land registry, the project cannot proceed without obtaining prior permission from local officials (eg, a prefectural governor or a mayor of a Designated Municipality

4 Act 229 of 1952.

(designated by the Minister of Agriculture, Forestry, and Fisheries)) to change the designation of the site. To obtain such permission the application is likely to need to show that the development is unlikely to interfere with agricultural promotion, to cause discharge of sediment, or to cause damage to irrigation and agricultural facilities.

5.3 The Forest Act (森林法)

The Forest Act[5] has the simple aim of protecting Japanese forests, and the requirements for permits to carry out development work in forests stem from this act. Development activities in forests tend to present one of the most common challenges to prospective PV projects. Felling trees in areas that are designated as forests is carefully regulated, and proceeding to fell trees without obtaining the proper approvals can result in material disruption to the development of a PV project. An application to carry out development activities on forest land has to show that such development is unlikely to cause environmental deterioration, interfere with water security, cause flooding or cause sediment discharge. The requirements to meet these guidelines are largely left to the local governments to determine. This can result in a variety of interpretations and differing requirements in different locations.

5.4 The Contamination Countermeasures Act (土壌汚染対策法)

The Contamination Countermeasures Act[6] places restrictions on changing soil composition in an area. This Act can require restrictions on certain civil construction activities involving evacuation to various depths (eg, 50 cm) depending on local requirements.

5.5 The Road and River Acts

The River Act[7] (河川法) and the Road Act[8] (道路法) are regularly encountered when using roads and rivers for access routes from a PV project site to the grid substation. These Acts set out the requirements for such use as well as for permitting regarding transportation of heavy equipment. These are sometimes secured during the construction phase of a project.

5.6 Environmental assessments

Environmental assessment requirements will depend heavily on the prefecture in question and, depending on the land and locality, there may be requirements for different procedures. There are two types of environmental assessments that can be required. Certain prefectures require that a natural environment investigation be conducted in order to enter into an environment conservation agreement. These ordinance requirements can take anywhere from several months to a year to fulfil. For the other type, some prefectures require a comprehensive environmental impact

---

5 Act 249 of 1951.
6 Act 53 of 2002.
7 Act 167 of 1964.
8 Act 180 of 1952.

assessment be put in place for any PV plants that are to be constructed if the project site exceeds a certain size. This type of requirement can be managed by reducing the size of the project site to be less than the applicable threshold. Once again, if an environmental impact assessment is required, they can take from several months to over a year to complete.

5.7 Other permits and notices

Some other frequently encountered land type permits are those for conservation areas, natural park areas and residential land. For natural parks and conservation areas there are two types of designations. Those that have been designated on a national level require the permit to come from the Minister of Agriculture. Those that have been designated such on a local level require the permission of the relevant prefectural governor. For Natural Park Areas there are particular prohibitions in place, which could render an area within a Natural Park ineligible for development projects. Residential Land that is subject to prefectural ordinances is particularly subject to the requirements of the locality as set out by the Act on Regulation of Residential Land Development,[9] which states that compliance with regulations depends on the prefectural rules in place.

In addition to permitting requirements, developers must submit a variety of notices depending on the development activities conducted. Notification of changes to be caused by development activities conducted usually must be given prior to carrying out the corresponding development activity and is often coupled with a permit requirement.

It should be noted that this summary is just an overview of some of the permits that developers are likely to encounter, and that in some cases the permits needed may be as specific as those needed for steep slope failure prevention areas, landslide areas and buried cultural property areas. Determining what permits may be necessary to proceed with a project is an important step in determining its viability so that it does not become subject to delays.

6. Overview of market participants

Various market participants are involved in Japanese solar projects, including developers, equity investors, lenders, landowners, EPC contractors and O&M contractors. This section looks at each of these in turn.

6.1 Developers

There is a large and experienced community of solar project developers in Japan. Some are specialised in early-stage and greenfield projects. The principal issues they deal with are obtaining site control and initial permitting at ideal locations. Other developers focus on acquiring initial stage projects and developing them until they are at the ready-to-build stage or until they are ready to provide a notice-to-proceed to the EPC contractor. The development community includes domestic companies as well as international developers. Although there are no official figures which indicate

9 Act 191 of 1961.

the level of in-bound investment into Japanese solar projects, a substantial volume of investment has originated from abroad including large-scale developers from North America, Europe, the Middle East, as well as East Asia, South Asia and South-East Asia. Japan is attractive to in-bound investors since its solar industry is growing from a low base in a jurisdiction that is generally seen as a stable, developed market with no restrictions on foreign direct investment and no capital controls or other restrictions against repatriation of funds by overseas investors.

6.2 Equity investors

Often developers provide the initial equity funding for their projects to develop PV plants. These investments by developers sometimes are intended to be long-term investments, but are more commonly made until another equity investor is brought in at a later stage in the PV plant's development. A fairly broad range of domestic Japanese investors are actively acquiring (or otherwise taking equity positions in) operational projects (or projects otherwise sufficiently advanced to be deemed to have been 'de-risked'). Prices for equity positions in operational projects are often relatively high, as domestic investors find them attractive even at relatively low return-on-investment (ROI) rates. Notwithstanding the relatively high acquisition prices (and correspondingly low ROI rates), there is also a large number of in-bound equity investments by international investors, which results in a competitive market for solar projects. Significantly, there are no restrictions on foreign investment into Japanese solar projects with no requirements for domestic Japanese equity stakes in foreign-owned projects.

Japan is a relatively high-tax jurisdiction, so consultation with a tax adviser is typically an important early step when an international investor is considering a Japanese solar project. One structure that is frequently adopted by in-bound investors for tax purposes is a tokumei kumiai, or Japanese silent partnership. Some Japanese domestic equity investors are interested in the depreciation of PV plants for tax purposes.

6.3 Lenders

Non-recourse project finance loan facilities are available for utility-scale solar projects in Japan. Japanese lenders regularly provide project financing for larger solar projects. International lenders are also actively involved in the Japanese market. The resulting competition results in relatively attractive commercial terms for bankable projects. There are licence requirements in order to originate loans to Japanese entities, so international lenders typically operate through their Japanese affiliates or branch offices. Both domestic and international lenders generally require that security interests are provided in all material project assets, as collateral for facilities. Security interests in collateral is provided in a suite of agreements by asset class (eg, project sites are mortgaged, equipment is pledged as movables, contractual rights (including those under PPAs and EPC agreements) are provisionally assigned, shares in SPV project companies are pledged). There are no restrictions or substantial limitations against providing such security interests, except for third party consent as applicable such as an EPC contractor's consent in the case of the provisional assignment of an EPC agreement. In general, lenders' bankability standards have evolved over time. Notably, Japanese lenders generally accept the creditworthiness of utility offtakers and are comfortable with the terms of FIT PPAs. Lenders look to the general viability of a project with some emphasis on the reputation of the EPC contractors, the financial viability of equipment manufacturers that provide warranties and the expertise of the sponsor. It is not uncommon for sponsors to provide non-recourse carve out letters in which they assume liability for fraud, acts in bad faith and occasionally to address specific issues that are identified in the lender's due diligence.

6.4 Landowners

Almost all of the utility-scale solar projects in Japan are ground-mounted systems located outside of urban areas and industrial areas where other types of real estate development in Japan are generally concentrated. Accordingly, solar developers are often seeking to lease or acquire project sites from rural land owners who insist upon relatively simple documentation, drafted in Japanese. This can often present issues when landowners are hesitant or unwilling to incorporate contract terms which are intended for bankability (eg, non-petition clauses, lender step-in rights, collateral assignment clauses). If a project site is owned by an individual or family, inter-generational inheritance issues are common. If a project site is located on a former golf club owned by an association, then it can be problematic to extinguish the rights held by the members of the association. Obtaining easements and other rights-of-way for access routes from a PV project site to the grid substation must often be obtained from private landowners, and these negotiations present similar issues.

6.5 EPC contractors

Generally, most solar power projects engage Japanese EPC contractors, however, international competitors have also entered the market. While the engineering and procurement components of an EPC contract do not require that the contractor hold a Japanese licence, the Construction Business Act[10] requires a Japanese construction licence in order for a contractor to engage in construction work or to supervise a sub-contractor doing construction work. Providing a guarantee/warranty of a licensed contractor's work can imply that the guarantor has the authority to supervise the sub-contractor which, in turn, implies a licence requirement to provide such a warranty or guarantee.

6.6 O&M contractors

Japanese and international O&M contractors compete to provide O&M services to Japanese solar projects. Technically, an O&M contractor is not required to hold a licence. However, under the FIT programme, solar projects must engage a chief electrical engineer (which is a licensed individual) who is generally provided by the O&M contractor.

---

10 Act 100 of 1949.

7. A look to the future

The solar power industry in Japan has matured significantly since the original FIT Act was introduced in 2012 and is likely to continue to grow and prosper. According to METI, as of the end of March 2017, the installed capacity of non-residential solar plants was 28,750,000 kW in total since the implementation of the FIT Law, whereas the aggregate capacity for PV projects for which METI has issued certifications was 79,050,000 kW. This suggests that for the next few years the major focus of the solar industry in Japan will be on constructing and bringing this unbuilt capacity into commercial operations. These figures include projects that have not yet been developed and may ultimately prove unviable.

Aside from completing development of previously METI certified PV projects, the Japanese solar industry is also ramping up to participate in regular reverse auctions under the FIT Act Amendment. The first auction under the FIT Act Amendment was successfully conducted in October 2017. The results and winning bidders were announced on 21 November 2017 by the GIO. There were nine winning bidders for projects with a total output capacity of 141,366 kW. The lowest bidding price was 17.20 JPY/kWh (for a 7,260kw project). However, there were two winning bids at the maximum bidding price of 21 JPY/kWh. Competition is likely to be more vigorous in future auctions (currently scheduled for June 2018 and November 2018) as many developers took a 'wait and see' approach during the initial auction in order to study how the new rules worked in practice.

Middle East

Keith Bullen
Mona Hammadi
DLA Piper

---

1. Introduction

The potential for development of solar power projects in the Middle East is vast. However, most countries in the Middle East have only recently considered solar power economically viable. The reasons for the Middle East's relatively slow adoption of solar power include:

- state-owned oil and gas reserves;
- energy subsidies for oil, gas and electricity; and
- the need for concurrent power and water desalinated water production from cogeneration plants.

In recent years a combination of demographic pressure, low oil prices and technological readiness primed Middle Eastern states to introduce solar power into their energy mix.

In this chapter, we will consider the case for the recent growth of solar power in the following countries:

- the Hashemite Kingdom of Jordan (Jordan);
- the United Arab Emirates (UAE); and
- the Kingdom of Saudi Arabia (Saudi Arabia).

The drivers for solar power generation in these countries are different. For Jordan, solar power represents an opportunity to reduce its energy dependency. For the UAE and Saudi Arabia, solar power is a way to develop new industries, employment opportunities, and to reallocate hydrocarbons away from power generation towards additional value-producing industries. Jordan's economic imperative has led to greater solar power progress and energy market reform. However, changing economics are creating traction for solar power. This is especially the case in Saudi Arabia where solar power development is happening at the same time as energy market reform.

In this chapter we will also briefly consider the outlook for solar power in Bahrain, Oman and Qatar. These three markets have publicly expressed an interest in developing solar power projects. However, they have not taken the steps we have seen in Jordan, the UAE and Saudi Arabia.

Finally, we will also address rooftop and captive solar PV systems, including the measures adopted to encourage growth and the progress we have seen to date.

2. History of conventional power generation and the growth of independent power projects

Historically power generation was the responsibility of Middle Eastern government entities. However, the oil price crash in the 1990s resulted in the use of Independent Power Project (IPP) procurement structures. IPPs reduced government capital expenditure and encouraged foreign direct investment. The Al Manah project in Oman was the first such IPP. Since then, IPP models have been used by the majority of the Middle East's power and water authorities, including:

- Abu Dhabi Water and Electricity Authority (ADWEA);
- Oman Power and Water Procurement Company (OPWP);
- Qatar General Electricity and Water Corporation (Kahramaa);
- various procurers in Saudi Arabia; and
- Dubai Electricity and Water Authority (DEWA).

IPP models have been successful for a range of reasons, including:

- The economics of IPP models are sound. Investors have achieved reasonable rates of return that have encouraged a large field of developers to invest in IPPs;
- Electricity demand has grown over the last 20 years, tracking the economic development of the Middle East;
- Government support of IPP models is easy to structure and easy to bank. The support has taken various forms, including, in most cases, through explicit government guarantees of offtaker liabilities. Certain procurers, including OPWP and Saudi Electricity Company (SEC) have implicit government support leading to bankable investment grade credit ratings.
- Termination compensation is a typical feature of IPP models in the Middle East. It ensures the offtaker buys the plant in most termination scenarios. Such termination payments typically ensure:
  - Repayment of senior debt; and
  - Depending on the circumstances that caused the termination, equity returns, demobilisation expenses and financing break costs.

Termination payments are addressed in more detail in section 4.

- Well-structured project documentation with bankable risk allocations have been a key to success. Markups of project documentation on IPP bids have become lighter. In more established models mark-ups are confined to the technical and financial aspects of bidder proposals; changes in risk allocation are unacceptable.
- Finally, despite legal and regulatory restrictions, procurers have been able to implement transparent tendering processes. Transparency, including publication of pricing offered by developers, has undoubtedly helped foster market confidence.

The history of IPP models in the Middle East is important. Solar IPPs have been procured on adaptions of conventional IPP models. Some believe this evolutionary approach to solar power procurement has slowed development. Conventional IPP models are complex, time consuming and resource intensive for procurers. This is justifiable for a 2 GW combined cycle power plant, but perhaps less so for relatively small and simple solar PV plants. In practice, regulatory regimes, procurer resources and the demand for new power plants do not allow for anything but the re-tooling of the conventional IPP models developed for conventional plants. In fact, one could argue that trying to implement solar power on a new procurement model would have further slowed progress.

3. Middle Eastern solar power markets

3.1 Jordan

(a) *Context*

Jordan does not produce oil or gas and relies on imports for more than 90% of its energy needs. Supplies are often disrupted. Developing domestic solar power could help ensure Jordan's long-term energy security. To that end, Jordan has committed to reducing its reliance on energy imports to 70% by 2020.

Like most other Middle East markets, Jordan is experiencing significant growth in load (circa 7% year on year). Summer peaks typically require power generation facilities to run at full load. Power surges and network failures have caused load shedding and blackouts.

(b) *Jordan's energy market*

In the 1970s, Jordan established its Electricity Authority (JEA). The JEA was a government-owned monopoly responsible for all aspects of the electricity value chain. In 1996, JEA was converted into a public shareholding company, the National Electric Power Company (NEPCO). NEPCO was subsequently unbundled in 1998. The unbundling resulted in two new government companies:

- Central Electricity Generating Company (CEGCO) – responsible for generation assets; and
- Electricity Distribution Company (EDCO) – responsible for distribution.

NEPCO retained responsibility for IPP procurement, wholesale power contracts and transmission.

In 2002, Jordan established the Electricity Sector Regulatory Commission (ERC) to regulate and license the electricity value chain and to set electricity tariffs. At the same time, a framework was introduced to allow IPP procurements. In 2007, Jordan privatised EDCO and CEGCO. Ownership of these companies has changed since their privatisation, allowing experienced international operators in Jordan's electricity markets.

Jordan's electricity market structure has allowed for the introduction of IPPs in an efficient and integrated manner. IPPs typically sell power to NEPCO (as sole offtaker). NEPCO then sells power to the three distribution companies and certain bulk consumers.

(c) *Solar power project models*

Jordan introduced the Renewable Energy and Energy Efficiency Law (REEEL) in 2012. Since this landmark legislation, solar power projects have followed two distinct paths.

First, REEEL allowed the Ministry of Energy and Mineral Resources (MEMR) to procure solar power projects in 'rounds'. Round 1 procured 450 MW of renewable capacity (solar and wind) based on feed-in tariff power purchase agreements (PPA). Round 2 provided a further 200 MW of solar power through a reverse auction process that allowed developers to bid for PPAs allocated to four preselected sites; 24 prequalified companies bid for the four PPAs. The reverse auction process resulted in a drop in tariff of approximately 60%. Jordan commenced the 200 MW Round 3 in 2014 but due to grid stability concerns it was postponed to 2016. In Round 3, developers submit bids to develop projects of up to 50 MW within the Ma'an Development Area.

Build, own, operate is the most common model for solar projects in Jordan and is consistent with its conventional predecessors. The project company is 100% owned by the developer. It enters into a PPA and standardised grid connection agreement with NEPCO. A government guarantee secures NEPCO's obligations. The project company and the land owner sign a land lease agreement.

Secondly, REEEL empowers NEPCO and distribution companies to negotiate PPAs with developers. There are a number of examples of direct proposal projects with sizes ranging from 10 MW to 200 MW. PPAs and grid connection agreements in these transactions are generally consistent with the MEMR projects.

3.2 UAE

(a) *Context*

The emirates of Abu Dhabi and Dubai are home to the majority of the UAE's population. Although they have few solar power projects in number, those projects are among the Middle East's largest renewable energy developments. Reasons for renewable energy developments in the UAE include:

- to partially remove gas from the electricity supply chain. Although the UAE has the seventh largest reserves of natural gas in the world, production of dry gas (the cheapest gas to convert to power) is comparatively low;
- to match supply and demand. The daily solar generation curve closely matches the daily demand curve in the UAE;
- to meet growing demand. The annual economic growth rate has been around 5–7% in both Abu Dhabi and Dubai.

(b) *The UAE's energy market structure*

Federal: The Federal Ministry of Energy, the UAE's primary energy regulator, was formed in 2004 by merging the Ministry of Petroleum and Mineral Resources with the Ministry of Electricity and Water. In 2008, the Ministry of Energy was restructured making it responsible for water and electricity sector policy.

In 2014, the federal government further restructured the Ministry of Energy, including the introduction of three new departments:

- the Clean Energy and Climate Change Department;
- the Rationalisation and Energy Usage Efficiency Department; and
- the Regulation and Control Department.

Abu Dhabi: Abu Dhabi's electricity market is regulated under the Abu Dhabi Electricity Law of 1998. The Abu Dhabi Regulatory Standards Board (ADRSB) is the body responsible for implementing the legal framework. ADRSB's authority includes the power to:

- issue licences for the conduct of regulated activities;
- monitor and ensure compliance with licences; and
- introduce such regulations as it sees fit to achieve its legislative obligations.

ADWEA controls the entities responsible for the generation, transmission and distribution of electricity in Abu Dhabi. Abu Dhabi Water and Electricity Company (ADWEC), a wholly owned subsidiary of ADWEA, is the sole offtaker of power and water in the emirate.

Dubai: DEWA was established in 1992 as a vertically integrated monopoly. It is the sole offtaker of power and water in Dubai. Subsequently Dubai has enacted laws to encourage private investment, including laws that established the Dubai Supreme Counsel for Energy (DSCE) and the Dubai Regulatory Standards Board (DRSB).

DSCE regulates the exploration, production, storage, transmission and distribution of petroleum products and electricity. Its remit includes privatisation and pricing of electricity.

DRSB is primarily responsible for regulating, licensing and supervising electricity generating service providers. It also determines and establishes standards and controls for electricity generation in the emirate and proposes legislation governing Dubai's electricity market.

Generation, transmission and distribution of electricity in the UAE are regulated and require specific licences from the relevant regulatory authorities. In Abu Dhabi regulated activities include electricity generation, transmission, distribution and supply to premises. In Dubai regulated activities are more widely defined, including "any activity related to generating electricity ... for the purpose of supplying the Transmission System with electricity".

(c) Solar power project models

In Abu Dhabi and Dubai, build, own, operate IPP concessions are awarded through competitive bidding processes managed by ADWEA and DEWA respectively. The primary evaluation measure is the levelised cost of electricity (LCOE) to ADWEC or DEWA. It is an evaluation measure only, not the actual price of electricity. LCOE is calculated by dividing the net present value of the offtaker's payments over the PPA term by the volume of electricity generated. When making the calculation, assumptions are made about availability and production.

ADWEA and DEWA's competitive bidding processes entail qualification of developers, issue of request for proposals and selection of the successful developer. The successful developers can be a single company or a consortium of companies.

The ADWEA and DEWA project document structures are drafted differently but are broadly consistent with regards to risk allocation. The request for proposals for an IPP in Abu Dhabi and Dubai typically include a PPA, land lease or usufruct agreement and a grid connection agreement.

In Abu Dhabi, project companies are established by the successful developer (40%) and a government-owned shareholder (60%). In Dubai, project companies are established by the successful developer and DEWA on a 49:51% basis. Accordingly, the request for proposals also contains a draft of the shareholders' agreement to be entered into with the successful developer. The project company has a board of directors composed of members appointed by the shareholders. In Abu Dhabi the government holds a majority on the board whereas in Dubai the successful developer holds the majority. Operational management of the IPP is granted to the successful developer.

The land for IPPs is acquired by ADWEA and DEWA. ADWEA and DEWA grant land rights to the project company through a land lease or usufruct. Other matters addressed in this agreement include:

- easement rights for access to the land typically form part of the same agreement;
- environmental responsibilities;
- termination and surrender provisions; and
- reinstatement.

Grid connections in Abu Dhabi are provided under a standard form grid connection agreement. Typically the substation and cabling form part of the scope of work of the project company's EPC contractor, however, the substation and cabling are transferred to the transmission company at no cost once they have been commissioned. The project company then recoups the cost through the energy charge that is paid by ADWEC. The situation is slightly different in Dubai. Dubai's solar power projects have been located on the Mohammed bin Rashid Al Maktoum Solar Park (the Solar Park). The Solar Park has a 380 kV substation that all of Dubai's solar projects will connect into. Accordingly, in Dubai the grid connection scope of work for the project company's EPC contractor is limited to laying cables along pre-defined corridors and connecting into the substation.

3.3 Saudi Arabia

(a) Context

Saudi Arabia has the world's largest oil reserves and is home to the fourth largest concentration of natural gas resource in the world. The country produces approximately 12 million barrels of crude oil per day, of which 3 million barrels are consumed domestically.

The key drivers for developing solar power projects in Saudi Arabia are similar to those of other Gulf Cooperation Council countries and include:

- to reallocate hydrocarbons to other parts of the Saudi Arabian economy where they will produce additional value for Saudi Arabia. Saudi Arabia's domestic consumption of hydrocarbons is rising rapidly. Export revenues are decreasing. Recent price depression has amplified the effect of domestic hydrocarbon consumption;
- to match supply and demand. Saudi Arabia's power sector is the biggest in the Middle East, with installed generating capacity of over 60 GW. Saudi Arabia is planning to double its capacity to 120 GW by 2032. Peak power demand has grown at 7–10% per year in the last decade;
- to grow and diversify the economy. The current rate of unemployment in Saudi Arabia is more than 6%. Renewable energy can contribute to reducing this, engaging and developing talented young employees. A key focus of Saudi Arabia's renewable energy plan is to develop local supply chains that will create employment opportunities for its young and growing population.

In 2014, Saudi Aramco's then CEO (and now Saudi Arabia's Energy Minister), noted that Saudi Arabia's domestic energy consumption could rise to 8 million barrels of oil a day by 2030. At that time, Saudi Arabia's oil production averaged around 9 million barrels a day. Citigroup concurred with that assessment, suggesting that Saudi Arabia could cease to be an oil exporter by 2030.

Significant reforms are underway. Reducing oil and gas subsidies and raising the price of energy are considered to be the most effective ways to restrain domestic consumption. However, this is a very sensitive topic in Saudi Arabia. Implementation of energy price reform is expected to be gradual even though it is an important element of its National Transformation Plan and Vision 2030 plan.

The King Abdullah City for Atomic and Renewable Energy (K.A. Care) was established in 2010 to oversee the realisation of the country's renewable and nuclear energy ambitions. In 2012 it published a white paper, proposing a roadmap towards 41 GW of solar power capacity. Despite K.A. Care's white paper, the renewable energy industry waited until 2016 for the announcement of a renewable energy programme. SEC issued an IPP tender for 100 MW of solar power at Rafha and Al Jouf. The tender was subsequently cancelled with the announcement of a new procurer for renewable energy projects, the Project Development Office of the Ministry of Energy, Industry and Mineral Resources (REPDO).

REPDO is tasked with delivering the 9.5 GW of renewable energy projects set out in the Vision 2030 plan. Vision 2030 did not specifically reference a date for the 9.5 GW, however, it was subsequently clarified the target date is 2023. Development of 9.5 GW of renewable energy is ambitious, but it appears to be more meaningful and realistic than K.A.Care's roadmap.

(b) Energy market structure

From 2001 to 2016, the Ministry of Water and Electricity (MWE) was responsible for Saudi Arabia's water and power. However, in 2016 the water and electricity sectors were reorganised and MWE was abolished. The Ministry of Petroleum and Mineral

Resources was renamed the Ministry of Energy, Industry and Mineral Resources and took responsibility for the power market.

SEC was incorporated in 2000 by consolidation of Saudi Arabia's regional electricity companies. SEC has the power to:

- generate, transmit and distribute electrical power in Saudi Arabia directly or through companies that are wholly or partially owned by SEC;
- buy, sell and provide electrical power services in Saudi Arabia directly or through companies that are wholly or partially owned by SEC; and
- participate and invest in projects for electrical power generation, transmission and distribution inside or outside Saudi Arabia.

In order to achieve its objectives, SEC may:

- establish or own affiliate companies operating in the fields of power generation, transmission and distribution;
- establish other companies or participate with Saudi or foreign companies in carrying out business activities similar to its activities;
- own shares or stock in such companies; and
- undertake all industrial or commercial activities associated with its objectives including sales, procurement, leasing and renting of movable and immovable assets.

SEC was granted wide-ranging powers to consolidate the power sector and usher in a new era including an unbundled electricity market and privately owned power projects.

Until recently, SEC had been the procurer and offtaker of power from IPPs. However, as efforts to unbundle the electricity market intensify, SEC has incorporated the Saudi Power Procurement Company (SPPC) to undertake this role. SPPC will also act as offtaker for power projects procured by other entities, including REPDO.

The Electricity and Cogeneration Regulatory Authority (ECRA) was established in 2002. ECRA's charter and implementing regulations set out its duties as regulator. It also identifies ECRA's objectives and responsibilities and grants ECRA the powers to achieve these requirements. As part of its duties as regulator, ECRA issues licences to power and water sector participants. ECRA also monitors performance of the licensees, sets the regulations that govern relationships between licensees, and between licensees and consumers.

(c) Solar power project models

Since the 2005 Electricity Law, the Water and Electricity Company, Marafiq, Saudi Aramco, SEC and now REPDO have gone to market to procure grid connected IPPs in Saudi Arabia. Their models are broadly consistent, adopting a common risk allocation that ring fences projects in a bankable way. When SEC went to market with the Rafha and Al Jouf solar power projects in 2016, the project documentation was an adaption of its conventional model. When REPDO went to market with the Sakaka solar IPP, the form of project documentation was similar to SEC's conventional documentation.

Similar to the models in the UAE, Saudi Arabia's solar IPPs are awarded on the basis of an LCOE competition managed by REPDO. The successful developer will incorporate a project company in Saudi Arabia. The project company will enter into a PPA, land lease and a grid connection agreement. Historically these agreements have been with SEC, although in REPDO's case these project documents are with SPPC (PPA and land lease) and SEC (grid connection agreement).

Ownership of conventional IPPs in Saudi Arabia has been shared between the public and private sectors in a similar way to the UAE. However, solar IPPs have brought about 100% developer-owned projects. To ensure transparency into the ownership and operation of 100% developer-owned solar power projects, a founders' agreement has been added to the suite of project documents. The founders' agreement includes covenants to implement the IPP in accordance with the successful developer's proposal and share transfer restrictions.

Similar to Abu Dhabi, in Saudi Arabia substations may only be operated by the licensed transmission service operators. Accordingly, in most IPPs the project company has been responsible for constructing the substation and has transferred it to SEC at no cost. In conventional power projects the cost of the substation has been recovered through capacity payments and in solar projects the cost has been recovered through the energy charge.

4. Common PPA protections

PPAs in the Middle East share many common risk mitigation strategies. At their core is the goal of balancing risk in a way that insulates project companies from economic and political concerns so that projects can be banked by the full spectrum of lenders and LCOEs included only a limited 'risk premium'. Risk mitigation manifests in the following PPA provisions.

4.1 *Force majeure* relief

Force majeure is subdivided into natural *force majeure* and political *force majeure* (also referred to as 'offtaker risk'). Natural *force majeure* typically includes weather, natural disasters, epidemics and explosions at adjacent unrelated facilities. Political *force majeure* includes acts of war, sanctions, changes in law, discovery of hydrocarbons at the site and the actions or inactions of government departments and regulatory bodies. Provided that the project company has clean hands, has taken all reasonable precautions to mitigate loss and has followed all contractual notice requirements, the project company is typically entitled to relief from the performance of its obligations under the PPA.

In the case of political *force majeure*, construction timetables are amended, however, the duration of the PPA is not extended. Financial compensation is paid to the project company by the offtaker in the form of deemed energy payments. The quantum of the deemed energy payments for political *force majeure* keeps the project company whole (ie, the full amount that the project company would have received for each day that the *force majeure* exists). Increased cost payments are also made should the project company suffer from such.

In the case of natural *force majeure*, project companies may be entitled to changes to construction timetables and an extension to the duration of the PPA. Deemed energy payments may be whole (where no extension to the PPA is given) or in part (where the PPA is extended). Partial payments are sufficient to allow the project company to meet debt service obligations during the period of the natural force majeure. During the period of the extension, the project company will recover its fixed and variable operating expenses, thereby keeping it whole. No increased cost payments are made for natural *force majeure*.

4.2 Termination payments

Termination payments are of utmost importance in a wide range of PPP projects in many jurisdictions. This is certainly the case in the Middle East power markets where difficulties in taking effective security and enforcing that security lead to even greater reliance on termination payments as the arbiter of 'bankability'. The termination payment is made in instalments and result in the plant reverting to offtaker ownership once senior creditor claims are satisfied. The typical quantum of termination payments is set out below.

(a) Termination for offtaker events of default and political force majeure

The compensation payable in such scenarios includes: senior creditor claims (senior debt, accrued but unpaid interest and senior debt termination costs including interest rate hedge breakage); shareholder equity contributions (and equity return if the project had achieved commercial operation); and certain termination costs such as demobilisation.

(b) Termination for project company default after commercial operation date

The compensation payable for project company event of default is limited to the senior creditor claims as described above.

(c) Termination for prolonged natural force majeure

The compensation payable for prolonged natural *force majeure* includes senior creditor claims and termination costs (each as described above).

(d) Termination for project company default prior to commercial operation

Offtakers generally have the right but not the obligation to make termination payments if the PPA is terminated for project company default prior to commercial operation. Project companies typically ensure that a right of rejection is included in their EPC contracts such that they can 'put' the plant back to the EPC contractor and obtain sufficient reimbursement from their EPC contractor to repay their lenders. If rejection-reimbursement is not obtained by the project company, the lenders would then seek to enforce reimbursement for rejection against the EPC contractor through an EPC direct agreement. Should the lenders be unsuccessful in such a claim, their primary backstop would be enforcement of security over plant assets.

4.3 Currency

Although energy charges are paid in local currencies, project companies do not take substantial currency risk. The Jordanian dinar, UAE dirham and Saudi riyal are all pegged to the US dollar. Should a currency become unpegged or the pegged exchange rate change, the US dollar component of the energy charge will be automatically adjusted using a formulae contained in the energy charge calculation.

In Jordan, the UAE and Saudi Arabia, termination payments are paid in US dollars to ensure that lenders will not suffer a partial recovery 'haircut' if a termination payment becomes payable following a currency de-pegging or revaluation.

5. Future IPP markets

At the time of writing, Oman, Bahrain and Qatar are starting to procure their first utility scale solar projects through IPP models. Like Jordan, the UAE and Saudi Arabia, they have all established IPP programmes and are expected to adapt their conventional programmes to reflect the requirements of solar power projects.

5.1 Oman

Oman was the first Arab country to turn exclusively to the private sector to build, own and operate a major power project. In 2004 the Authority for Electricity Regulation (AER) was established to regulate the electricity sector. Additionally, electricity generation was opened to competition, requiring all new electricity projects constructed, owned and operated by the private sector.

OPWP is the single buyer of power for all IPPs. Land for IPPs is typically provided by the Ministry of Housing and a grid connection is provided to the IPP by the Oman Electricity Transmission Company. IPPs are wholly owned by the private sector. As a condition of being awarded an IPP, the initial shareholders must undertake to perform an IPO of project company shares within a certain period of time.

Oman's peak demand is expected to grow at about 9% per year, from 5 GW in 2014 to 9.5 GW in 2021. Personal income growth and continuing government investment in infrastructure projects are major contributors to continued growth in electricity demand. In 2015 AER announced that it would look to renewables to provide approximately 10% of Oman's power within the next decade.

Oman has substantially decoupled its power and water generation projects. IWPP procurement has given way to standalone IPPs and IWPs. This shift has arguably enhanced the potential for solar power projects. It gives OPWP flexibility to dispatch power projects without affecting water production.

5.2 Bahrain

The Electricity and Water Authority (EWA) oversees power production in the Kingdom of Bahrain. In 2003, Bahrain began liberalising its electrical production. IPPs have been procured with a minority government ownership. At the time of writing, EWA is procuring a conventional IWPP at Al Dur. This new IWPP may influence timing for the future development of solar power projects.

In May 2012, EWA signed an agreement with an American company, Petra Solar, to install a 5 MW pilot project at the Bahrain Petroleum Company's facilities. Bahrain has set a goal of producing 10% of domestic energy from renewable sources by 2030.

5.3 Qatar

Kahramaa was established in 2000 to act as a generator and procurer of independent power and water projects. A number of IWP and IWPP projects have been procured by Kahramaa. The 3 GW Facility D was the latest such project. A further 2.8 GW, Facility E, is currently being procured. The Kahramaa model is similar to other IPP models in the Middle East, with Kahramaa itself acting as offtaker. Land is provided to the project company by way of a usufruct. Qatar Petroleum provides fuel for conventional plants. The project company is owned by the successful developer and Qatar government entities, typically in a 40:60 ratio. A government guarantee supports Kahramaa's ongoing payment obligations as well as the termination payment following event of default.

Between 2006 and 2016, power consumption recorded an average growth rate of 10% per year according to data released by Kahramaa.

Research suggests that Qatar will need to attract US$9 billion in its power sector between 2016 and 2020 to keep pace with electricity demands. In 2017 it was announced that Siraj Power, a joint venture between Kahramaa and Qatar Petroleum, will develop a 200 MW solar plant in Qatar under an IPP model based on Kahramaa's typical conventional IPP structure. Plans are already in place to expand the plant to 500 MW.

Qatar has announced plans to set up 1,800 MW of solar power capacity by 2020, which is expected to contribute up to 16% of total power generation. Qatar has expressed its intention to hold auctions for solar power projects on multiple occasions, however, these intentions have not yielded substantial results.

6. Rooftop and captive solar power generation

6.1 Regulations

Regulatory support for rooftop and captive solar power generation has grown significantly in recent years. Jordan was at the forefront of rooftop and captive solar power when it issued the REEEL. The REEEL allows consumers to install a rooftop or captive solar PV system for their own use, and export power to the grid when their consumption is lower than their generation. Distribution companies pay for power exported to the grid under a feed-in tariff structure.

Dubai followed Jordan's lead in 2015 by introducing the 'Shams' net metering scheme for consumers. However, unlike the Jordanian model, DEWA will not pay a premium for power exported to the grid. Instead, export electricity is netted off against future consumption. Electricity netting does not need to occur in the month that the power was exported to the grid. It may roll over into subsequent billing periods. The Shams scheme does not allow for wheeling power. Instead, netting is done on a meter by meter basis. Accordingly, if a consumer has two meters it may simultaneously be in both credit and debit to DEWA.

Abu Dhabi issued regulations for connecting solar PV systems to the grid in 2016. However, they do not provide compensation for surplus electricity transmitted to the grid. Proposed regulations allow surplus electricity from rooftop solar PV systems to be exported to the grid. Similar to Dubai's 'Shams' scheme, exported power will be netted against power consumed in periods when a consumer's consumption is greater than its own generation. Current understanding is that if a solar PV system is larger than 5 MW, it would not qualify for net metering.

In 2017 Saudi Arabia introduced the Small Photovoltaic Solar System Regulations (the SPSS). The SPSS set out the framework for the connection of solar PV systems to the grid. Effective from 2018, the SPSS will apply to solar PV system installations smaller than 1 MW, connected in parallel to the distribution system. Similar to Dubai, SPSS does not permit wheeling of power. However, the rollover period for netting is limited to three years. If any exported power has not been netted off after three years, it will be paid for by the distribution service provider at a tariff approved by ECRA. In the absence of specific legislation, Solar PV systems larger than 1 MW are seemingly treated in the same way as captive conventional projects. Accordingly, if the solar PV system is not connected to the grid and only provides power to its owner, then it is exempt from ECRA's licensing process. Small generation licences are required for plants up to 30 MW that do not meet the above requirements. If a solar PV system is not exempt and does not meet the criteria for a small generation licence, a full ECRA generation licence is required.

Oman's Sahim scheme incentivises the installation of solar PV panels in homes, as well as public and private facilities. Similar to the Jordanian rooftop solar model, the Sahim scheme allows for the export of power to the grid with payments made to the generator on a feed in tariff structure.

6.2 Rooftop and captive solar models

Although many of the Middle East's markets are only just beginning to embrace rooftop solar PV systems, business models are emerging that allow third parties to act as solar PV system providers. Structures vary, however, we typically see operating lease models whereby the consumer will lease its roof space to a contractor/operator for a period (often 15 years). The contractor will install, operate and maintain the solar PV system. Payment structures vary within the operating lease model. Commonly payments are based on the output of the solar PV system in the previous month, multiplied by an agreed, indexed, tariff. For larger systems we have also seen build, own, operate transfer models. In these models the contractor undertakes to install, operate and maintain the solar PV system until capital costs and accrued operation and maintenance expenses are paid in full.

Energy service companies (ESCOs) are becoming increasingly common. Although relatively expensive compared to many energy efficiency measures, rooftop solar PV systems are an effective way of allowing ESCOs to meet the energy efficiency guarantees they have given to their customers.

7. Closing thoughts

Jordan's solar power market has taken significant strides in recent years. Reducing its reliance on neighbours for energy has forced it to adopt feed-in tariffs and large-scale procurement of solar power projects. Roll back of subsidised power for consumers has also played a large part by making the 'business case' for solar power much more clear cut.

Although the UAE and Saudi Arabia have made significant progress in solar power, the undoubted solar energy potential in these countries has not been tapped in any meaningful way. There are a number of reasons for this.

UAE and Saudi Arabian consumers benefit from some of the world's lowest electricity prices. This is due to government subsidies for conventional power producers through the provision of cheap oil and gas. Subsidies are also applied at the consumer level.

Cheap conventional power has, until recently, proven to be a barrier to solar power. However, the UAE and Saudi Arabia have started to make electricity price reforms. As the removal of subsidies shifts market dynamics, it is inevitable that solar power projects will become increasingly attractive. Ultimately, however, utility subsides are a very sensitive subject in the Middle East, and governments tread carefully when it comes to reform.

A key factor in the growth of solar power globally has been government support in the form of feed-in tariffs or other government-backed incentives for renewable energy investment. Various commentators on this subject have championed the need for support in the Middle East similar to the European models. Apart from a few notable and quite recent exceptions referred to in this chapter, governments in the Middle East have chosen not to embrace feed-in tariffs.

It would be wrong, however, to suggest that solar power is not supported by governments simply because there are no feed-in tariffs. There are four primary mechanisms through which governments subsidise their solar IPP programmes.

First, solar IPPs are normally located on land provided for free by the government.

Secondly, grid infrastructure is provided by the government at no charge to the IPP. Jordan has a policy of not charging solar project developers for grid connections. The Maktoum Solar Park in Dubai has an offtaker funded 380 kV substation in to which the various IPPs will connect. Similarly, the Sweihan IPP in Abu Dhabi has a preconstructed substation. REPDO's programme has a different approach, however, the effect is the same. The IPP builds the substation and transfers ownership to SEC. SPPC pays for the substation over the term of the PPA.

Thirdly, IPPs benefit from sovereign or quasi-sovereign credit support through longterm PPAs. In the case of the UAE, a government guarantee is provided to the project company and lenders. In Saudi Arabia, government support is provided if the electricity market is restructured in a way that alters the current (and previously banked) offtake arrangements. The government guarantee in the UAE and the government support in Saudi Arabia covers the termination payment should an event of default occur. The termination payment covers cost of debt in all scenarios. Should the PPA be terminated due to an offtaker event of default, demobilisation and equity return compensation is also provided.

Finally, in the case of the UAE, governments take equity positions in the project companies with preagreed minimum equity IRRs. The equity positions subsidise the IPPs by reducing developer equity investment burden. The government shareholders also take below-market return on their equity.

Market liberalisation is an essential ingredient for the development of solar industries. Away from rooftop solar power systems, only Jordan has a mechanism for the private sector to instigate the development of solar power projects. Vision 2030 and the unbundling of SEC may yet result in the adoption of direct proposal models. However, given the market structures that exist in Dubai and Abu Dhabi, direct proposals are unlikely in the UAE.

Feed-in tariffs are closely allied to direct proposal models as they allow developers flexibility in their business plans. The era of feed-in tariffs is coming to an end in many established solar power markets. However, for the new markets of the Middle East, feed-in tariffs for utility scale solar power projects would be a significant boost to the growth of solar power. We have already seen this in Jordan where we can say without doubt that its feed-in tariffs for Round 1 and rooftop solar have been critical.

Equally, if not more important, without direct proposals, it seems unlikely that the tremendous potential for utility scale solar power projects will be fully realised. Solar IPPs would remain large-scale procurements exclusive to all but the largest developers. A direct proposal model allowing small and mid-size developers will require substantial energy market reform in both the UAE and Saudi Arabia. Such reform is slowly occurring in Saudi Arabia, however, it is highly unlikely in the UAE. Rooftop solar programmes will therefore shoulder a large burden if solar power is to be a significant part of the energy mix.

Morocco

Ghalia Mokhtari
Mokhtari Avocats

1. Introduction

With solar irradiance levels among the highest of any country in the world – over 2,300 kWh/m^2/y, or 30% higher than the best sites in Europe[1] – and a national political commitment that reflects this potential, Morocco is set to claim its position as a solar powerhouse. Coupled with its political stability and attractive business climate, Morocco's future production and foreign investment prospects are unquestionably bright.

For nearly 25 years, the government has demonstrated its willingness to open the power and renewables sectors to such investment. The national electric utility's monopoly on large-scale power generation ended in 1994, laying the foundation for independent power producers (IPPs) by granting private developers the right to build and operate plants up to a capacity of 10 MW. This self-production threshold for private development capacity was later raised to 50 MW in 2008,[2] two years after promulgation of the first official framework for the delegated private management of public utilities. Considering Morocco's dependency on expensive foreign energy imports (upwards of 90% of total consumption[3]) and the projected quadrupling of energy demand by 2030[4] due to steady increases in GDP and the continued electrification of the rural countryside, reforms of this nature are certainly in the best interest of the state.

By the time the Kingdom announced a revamped national energy strategy in 2008, Morocco was thus no stranger to ambitious regulatory overhauls. The new strategy's focus on the Moroccan renewables sector follows suit – in addition to a total capacity target of over 50%[5] from renewable sources by 2030, as well as energy efficiency goals and a reduced dependency target of 85% by 2025, the strategy outlines an estimated US$9 billion investment to install 2,000 MW of solar capacity by 2020, adding in terms of power generation a capacity equal to the current electricity consumption of Casablanca.[6] Referred to as the Moroccan Solar Plan (the MSP), this component of the new national energy strategy is currently under

1 Linklaters, "Lighting Up the Kingdom of Morocco", 29 March 2013.
2 This threshold was raised to 300 MW (under certain conditions) by Law 54-14 of August 2015 (see Renewables framework, below).
3 United Nations Economic and Social Commission for Western Asia, Renewable Energy Policies Case Study for Morocco, Baku Azerbaijan, 19–20 October 2016.
4 *Ibid.*
5 42% by 2020, 52% by 2030
6 Solar GCC Alliance Morocco Solar Profile.

development and includes the use of both photovoltaic (PV) and concentrated solar power (CSP) technologies. To realise the MSP and the national energy strategy's larger vision for the Moroccan renewables sector, the Kingdom has created a number of new competent authorities and substantially reformed its relevant legal framework.

2. Energy entities

Since 2010, the Kingdom has created energy institutions to meet the demands of the national energy strategy and oversee its reforms relative to long-term challenges like energy efficiency and sustainable development.

2.1 Moroccan Agency of Sustainable Energy (MASEN)

MASEN is a public-private agency first created as the Moroccan Agency of Solar Energy in 2010 that since 2016 has been the designated lead for renewable power projects and the institution responsible for overseeing clean production targets under the national energy strategy. The promulgation of Law 37-16 in August 2016, amending and supplementing Law 57-09 of June 2010 establishing MASEN, expanded the scope of the Agency's duties and responsibilities from solar projects to all renewables.[7] This modification was published alongside two other laws modifying the charters of ONEE and ADEREE (see below) to reflect the new balance of responsibilities. MASEN now acts as the designated offtaker of electricity generated from renewable projects[8] and also reserves the right to act as a minority equity partner for these projects. MASEN obtains financing for its projects via a unique lending structure that involves borrowing from international finance institutions (IFIs) and direct on-lending to project companies (see description of MASEN's contractual structure below). Relative to solar developments, MASEN's primary objectives include the implementation of the MSP and the promotion of solar energy projects in Morocco.

As part of the expansion of MASEN's duties and responsibilities, a transition period is now in effect during which ONEE must hand over its renewables projects to MASEN. For now, MASEN works in tandem with ONEE to facilitate the acquisition and distribution of electricity generated from the projects it oversees. However, ONEE still manages projects developed under the framework provided by Law 13-09, as well as its own group of NOOR projects (see Ongoing projects, below).

2.2 National Office of Electricity and Potable Water (ONEE)

ONEE is the public electric and water utility created in 2011 by a merger of the national water utility ONEP (*Office National de l'Eau Potable*) and the national electric utility ONE (*Office National de l'Electricité*). ONEE's primary responsibilities include the transport and distribution of electricity, as well as the management of generation projects from non-renewable sources. As referenced above, ONEE's responsibilities were modified by Law 38-16 in August 2016 to reflect the redistribution of

7 Excluding projects developed under Law 13-09 (see ONEE description and Renewables framework, below).

8 Excluding projects developed under law n°13-09 (see ONEE description and Renewables framework, below).

responsibilities between ONEE, MASEN and ADEREE. Before this modification, ONEE was the main entity in charge of thermal and renewable IPPs.

ONEE is also in charge of the management of the electric grid and grid connections.

2.3 Moroccan Agency of Energy Efficiency (AMEE)

Created as the Moroccan Agency for the Development of Renewable Energy and Energy Efficiency (ADEREE) in 2010, the Agency's name and mission were updated by Law 39-16 of August 2016 to reflect its primary focus on national and regional plans to develop, implement, monitor and coordinate the energy efficiency of projects in Morocco. The Agency's responsibilities include conducting national energy audits and mobilising the financial resources necessary to realise its mission. Under the national energy strategy referenced above, the Agency is tasked with achieving energy savings of 15% by 2030.

2.4 Société d'Investissements Energétiques (SIE)

SIE is a Moroccan investment fund created in 2010 from Dh 1 billion of capital from the Energy Development Fund (FDE) to promote the development of the Moroccan energy sector through both investment and project development in renewables and energy efficiency. SIE is a quasi-national joint stock company controlled by the state and the Hassan II Fund for Economic and Social Development. According to the fund, "SIE develops and finances projects contributing to public policy: it intervenes in response to either a priority need of public authorities, or to address the lack of initiative of private (emerging sectors)".[9]

2.5 National Authority for Electricity Regulation (ANRE)

ANRE was created by Law 48-15 of May 2016 to act in a regulatory capacity, but to date the entity has not taken physical form.

3. Renewables framework

Further to its goal of ensuring the success of the national energy strategy, the government has introduced a number of updated laws and frameworks since the liberalisation of the energy sector.

3.1 Revised Foreign Investment Charter (1995)

Law 18-95 of November 1995 established an incentivised investment charter aimed at attracting foreign investment to the Kingdom.

3.2 Law 16-08 modifying the charter of ONE (2008)

Promulgated in October 2008, this law modified the charter of what was then ONE (now ONEE), as set out in Dahir 1-63-226 of August 1963. It raised the self-production threshold for private development projects from 10 MW to 50 MW. In addition, it granted ONE the right to enter into agreements with private

9 SIE Investment Principles, https://www.siem.ma/en/investment-principles.

companies for the development of solar projects with a total installed capacity greater than 50 MW.

This threshold was later raised to 300 MW by Law 54-14 of August 2015 modifying Dahir 1-63-226 referenced above and amending Law 40-09 creating ONEE (see below).

3.3 Law 57-09 creating MASEN (2010, revised in 2016)

This law (the MASEN Law), which officially created MASEN and outlined its obligations relative to the MSP, also granted the Agency the authority to enter into agreements with private companies for the development of solar projects with a total installed capacity greater than 2 MW. The MASEN Law was amended by Law 37-16 in August 2016 to reflect the Agency's expanded scope of work and new responsibilities (see MASEN profile above).

3.4 Law 13-09 on Renewable Energy (2010, revised in 2016)

The development of private renewable energy projects in Morocco is governed by Law 13-09 of February 2010 on renewable energy, as amended by Law 58-15 of January 2016 (the Law). In addition to prioritising development from renewable sources, the Law liberalised the renewables sector by providing a legal framework for the production and export of renewable energy. As of the 2016 update to the Law, this includes granting private investors the right to sell electricity generated from renewable projects to clients connected to the low voltage, medium voltage, high voltage and extra high voltage grid infrastructures. In addition, the Law also provided a framework for selling excess electricity (up to 20% of annual power production) produced from renewable sources to ONEE.

The Law and its implementation decrees (see below) established two different authorisation regimes for new and upgraded projects based on generation capacity, whereby submission of a preliminary declaration is required for project capacities between 20 kW and 2 MW and receipt of prior authorisation from the Ministry of Energy is required for project capacities greater than 2 MW. Authorisations granted by the Ministry are valid for 25 years.

Projects developed under the Law use the following structure:

- The IPP enters into a grid connection agreement with ONEE that dictates the grid connection and commercial conditions for the transmission of power to the consumption facility;
- The IPP enters into a PPA with private consumers that provides the conditions of supply and the commitment of the consumer to use the power to satisfy their own needs. The private consumers thus amend the initial tariff options stipulated in the electricity supply agreement with ONEE;
- The IPP enters into a PPA with ONEE to handle the reselling of excess power supply.

Two implementation decrees for the law have been published, namely Decree 2-10-578 of April 2011 relative to the application of the Law and Decree 2-15-772 of October 2015 relative to the MV grid (the MV decree). Among the provisions outlined in the former, developers are required to build solar projects within high potential areas specified by the Ministry of Energy based on opinions from AMEE, ONEE and local collectives.[10]

The MV decree focuses on the conditions of access to the medium voltage (MV) grid for renewables projects, and it addresses four key issues:

- A description of the declaration/authorisation process described above;
- The opening of the MV grid by defining a 10-year roadmap (*trajectoire*) consisting of annual allocations (*enveloppes anuelles*) of the volume of green energy to be absorbed by the grid;
- The idea of capacity reservation (*reservation de capacité*) by the developers; and
- The technical conditions and rules of operational management (*gestion opérationnelle*).

According to the MV decree, every grid operator (*gestionnaire de réseau de distribution*) has 12 months starting from the decree's publication date to publish such a 10-year roadmap[11] in which it defines the annual volume of green energy (allocation) that can be integrated into its distribution system. These allocations must cumulatively consist of between 5% and 10% of the energy distributed by the grid operator, and the grid operator may choose to distribute these allocations between renewable sectors (eg, wind, solar). Subsequently, the Ministry of Energy and the Ministry of the Interior will fix a global roadmap for the next 10 years with the option to raise the volumes every two years, and the option for operators to reduce their allocations, but only in exceptional circumstances where such a reduction would affect installations already benefiting from a capacity reservation.

As with the technical conditions and rules of operational management set out in the MV decree, the grid operator has the authority to interrupt the injection of energy from renewable sources for any reason related to the security/integrity of the grid without the obligation to pay compensation to the projects.

In addition, as authorised by Law 58-15 amending the Law, the decree grants the grid operator the option of procuring energy from renewable projects under the following terms:

- to satisfy the operator's 'ancillary services' (*services auxiliaires*), namely internal consumption needs, and grid losses (*pertes réseaux*);
- procurement of up to 7% of energy distributed; and
- procurement of up to 20% of the energy produced by an individual project.

Note, however, that ONEE has yet to complete any solar projects under the framework described above. As a result, there is a much larger body of information available on the contractual structure of MASEN's NOOR projects (developed under MASEN's specific contractual structure, not Law 13-09) than ONEE's.

---

10 Article 14 of Dahir 2-10-578.

11 To date, no such roadmap has been published.

3.5 Law 16-09 creating ADEREE (2010, updated 2016)

Originally promulgated in 2010, this law creating ADEREE was updated by Law 39-16 of August 2016 alongside similar revisions to the charters of MASEN and ONEE. Among the many changes introduced to the structure of the Agency, it officially changed the name to AMEE to reflect the Agency's focus on energy efficiency.

3.6 Law 40-09 creating ONEE (2011, updated 2016)

This law created ONEE via a merger of the national water utility ONEP and the national electric utility ONE.

This law was updated by Law 38-16 in August 2016 alongside similar modifications to the charters of MASEN and ADEREE. While Law 38-16 technically modified the charter of ONE as set out in Dahir 1-63-226 of August 1963, it effectively altered the responsibilities of ONEE.

3.7 Law 86-12 on PPPs (2014)

Promulgated in December 2014, this law established the legal and contractual framework for public-private partnerships (PPPs) in Morocco.

4. Recent successes and ongoing solar projects

The regulatory framework in the Kingdom has enabled both MASEN and ONEE to launch their respective solar projects as part of the MSP. Highlights of each entity's recent successes and ongoing projects in this context are as follows.

4.1 MASEN's co-generation Noor projects

The evolution of the now-defunct DESERTEC initiative that collapsed in 2013 after losing the support of most of its shareholders, these five co-generation facilities in Ouarzazate, Laayoune, Boujdour, Midelt and Tata are being developed under a consortium led by Saudi Arabia's ACWA Power and are expected to utilise both CSP and PV technology. The total investment for the project is estimated to be US$2.2 billion, with funding provided by the German development bank KfW (largest contribution), the *Agence Française de Développement*, the Clean Technology Fund, the African Development Bank, the European Investment Bank and the World Bank.

The Noor Ouarzazate Project was originally selected as the first project site for the DESERTEC initiative before its collapse. It is the first of the five sites to reach partial commercial operation and has been online since the completion of Phase 1 (Noor I) in February 2016, which included the installation of 160 MW of CSP capacity and three hours of energy storage capability. The site's three remaining development phases are all underway, with a target commissioning date of February 2018 for the final stage and a total production capacity of 580 MW (70 from PV). As of now, construction on the other four sites has yet to begin, and the current state of each project is as follows:

- Noor Laayoune and Noor Boujdour – ACWA Power has been awarded a 20-year build-own-operate-transfer (BOOT) agreement to develop 80 MW of PV in Laayoune and an additional 20 MW of PV in Boujdour;

- Noor Midelt – MASEN is currently finalising the list of companies that will be invited to participate in the first phase of the project, including 150–190 MW of CSP technology and a minimum of five hours of energy storage technology. The project's total capacity is slated for 800 MW. Five pre-qualified bidders have been announced: ACWA Power, EDF Energies Nouvelles, ENGIE, Innogy SE and JGC.
- Noor Tata – MASEN has announced the project's target capacity of 800 MW, though further details have yet to be revealed.

4.2 ONEE's solar projects

ONEE has also launched its own separate project programme calling for the development of 500 MW of PV capacity by 2020. The project will be completed over three phases: Tafilalt, Atlas and Argana. All project tenders are required to be completed within a five-year period, after which operation will be transferred to MASEN. The state of each planned installation is as follows:

- Noor Tafilalt – The first phase of ONEE's solar project envisions the development of 120 MW of PV capacity over three separate installations of 40 MW each in Arfoud, Missour and Zagora. This phase of the project has an estimated cost of US$150 million and will be financed by the World Bank through the International Bank for Reconstruction and Development and the Clean Technology Fund. In November 2016, ONEE announced the names of 11 pre-qualified bidders.[12] Commissioning of the project is scheduled for 2018.
- Noor Atlas – With a total planned capacity of 200 MW of PV over eight installations of 10 to 30 MW each, the second phase of ONEE's solar project has an estimated investment cost of €300 million and will be financed by KfW, the European Investment Bank (EIB), and the European Commission.
- Noor Argana – The third phase of ONEE's solar projects includes a planned capacity of 200 to 225 MW. The original estimated investment cost for the project was €250 million excluding the cost of preliminary studies, though this figure is likely to rise given the decision to increase the size of the project from its original capacity of 125 MW. Pre-qualification was expected to commence at the start of 2017, though a list of bidders has yet to be announced.

5. Contractual structure of solar projects

In Morocco, there is a significant difference between the contractual structure used for most IPPs and the structure used by MASEN for its solar projects. These two distinct structures are summarised below.

5.1 Classic project finance structure

Most energy projects in Morocco follow the classic financing structure for IPPs, whereby the project company is charged with developing, constructing, operating

12 Article 14 of Dahir 2-10-578.

and maintaining the generation facility, and its shareholders are private sponsors. In this structure, the project company enters into an Engineering, Procurement and Construction Contract (EPC) with the contractor; and an Operation and Maintenance Agreement (O&M) with the operator. The following graphic summarises the classic project finance structure:

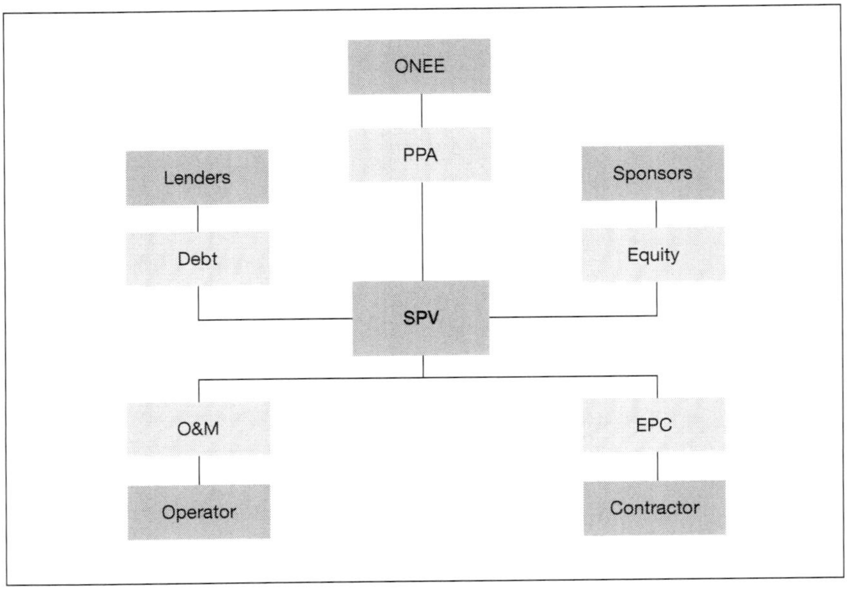

Financing for energy projects is provided by both Moroccan and international banks, as well as by concessional funds and IFIs. Moroccan law provides a number of possibilities for investment security, including the assignment of receivables, pledges over bank accounts, share pledges, pledges over the business of the project company, mortgages over real property, pledges over going business, and the transfer of insurance policy proceeds, among others.

Securities over real estate assets have traditionally been difficult to secure in Morocco, given the predominant use of a build, transfer, operate (BTO) scheme, whereby the project company is only granted usage rights to the project site that remains in the public domain of the state. The lack of definitive ownership over the land has historically prevented project companies from issuing securities in the form of mortgages or related real estate assets.

Recent projects, however, have adopted a more bankable build, own, operate, transfer (BOOT) scheme, whereby the state reclassifies land from the public domain of the state to its private domain and transfers full ownership rights over the land (*droit de superficie*) to the project company via a reclassification procedure. This allows the project company to acquire a form of ownership over the land and thus to provide stronger guarantees over the project site to financing banks (ie, mortgages).

In addition, for the purposes of each project, the project company may enter into an investment convention (*Convention d'Investissement*) with the Kingdom of Morocco, which would provide it with certain tax advantages. This principle of entry into an investment convention is set out under the Investment Charter.

Based on recent IPPs in Morocco, as a consequence of entering into an investment convention, the project company may be granted certain tax incentives, in particular exemption from some onshore taxes and certain import duties for a limited period of time.

In Morocco, note that foreign investment financed in hard currencies does not require the prior authorisation of the Moroccan authority in charge of foreign exchange (*Office des Changes*). Such foreign investment can take several forms, including in particular foreign loans, shareholder loans, incorporation of companies and subscription in share capital increases.

Also note that any profits, dividends, directors' fees dividends distributed by the project company and any proceeds resulting from the sale of the shares of project company may be freely transferable without any specific authorisation from the *Office des Changes*.[13]

5.2 MASEN contractual structure

For solar projects in which MASEN acts as primary offtaker, the contractual structure is noticeably different – in these projects the IPP typically enters into a PPA with MASEN, which in turn enters into a separate PPA with ONEE to sell the power it purchases from the IPP. A schematic overview of MASEN's IPP structure is summarised in the diagram below:

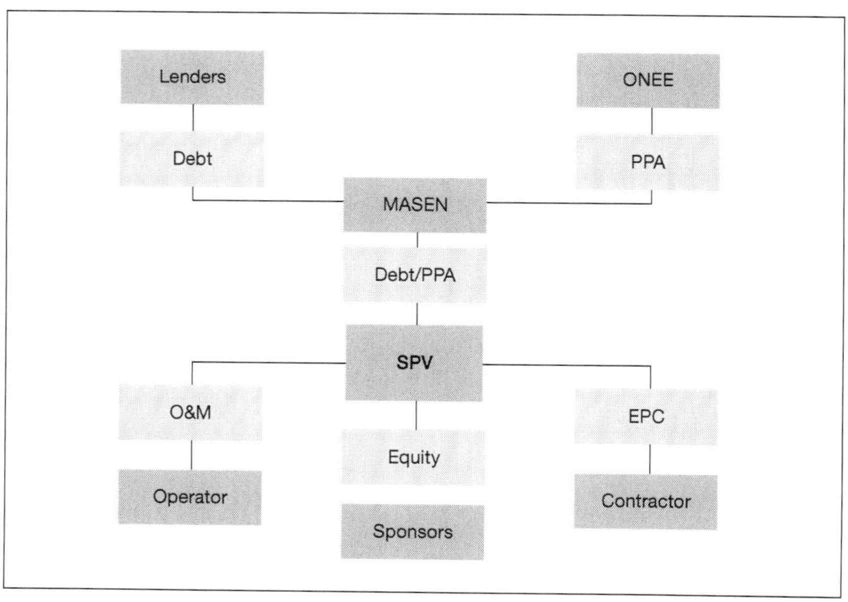

13 In practice, the project company requests a formal authorisation from the *Office des Changes* even if the law does not require it.

Taking MASEN's approach to the Noor I Ouarzazate project (see below) as a model, the PPP for MASEN's solar projects is based on the creation of a special purpose vehicle (SPV) that consists of a consortium between the private developer and MASEN. This consortium is responsible for the design and development of the generation facility, as well as its operation for the duration of the PPA (25 years for the Noor I project).

As referenced in the description of the Agency above, financing for MASEN's projects is obtained through a unique multilateral borrowing mechanism, whereby MASEN borrows funds from a collection of IFIs and acts as the sole on-lender to the project company. The additional risk inherent to this structure is addressed on a project-by-project basis in the financing documents. Under this arrangement, MASEN thus acts as borrower, lender, offtaker, and shareholder in the context of its projects.

6. Power Purchase Agreements (PPAs) for Moroccan solar projects

For solar projects, MASEN and ONEE each submit their respective RFPs and launch competitive bidding processes for tenders. Note that MASEN and ONEE have each run their own procurement programme that reflects and falls within the original national energy strategy implemented by the Kingdom.

Both MASEN and ONEE have developed their own PPA precedent, each of which has been accepted as a bankable and market practice document. The key terms of Moroccan PPAs in the context of solar projects can be summarised as follows.

6.1 Financing and contract term

Recent energy projects in Morocco have adopted a BOOT scheme made possible by the state's willingness to reclassify sites in the public domain to the state's private domain, after which the state transfers full ownership to the developer. Project sponsors are required to maintain the facility for a period of 20–30 years depending on the renewable energy source. Most PPAs for solar projects involve an operation period of 20 years.

During the bidding process, MASEN does not require developers to secure their own debt financing. As referenced in the description of the Agency given above, MASEN obtains financing through multilateral borrowing by the Moroccan government and on-lends it to the project company on a sole lender basis. In this case, the debt financing is passed through the PPA tariff.

6.2 Equity

RFPs issued by MASEN and ONEE have been known to reserve the right to equity participation by MASEN and ONEE in the project company. In addition, share transfers typically require the offtaker's approval.

6.3 Credit and government support

Payments under Moroccan PPAs are backed by letters of credit. Additionally, the government typically agrees to guarantee the offtaker's termination payments. The terms and conditions under which the government agrees to pay termination payments to the project company are negotiated on a case-by-case basis.

6.4 Project site

As referenced above, the government has begun to transfer ownership rights to the project company via a reclassification process. ONEE and MASEN both tend to attempt to pass on pre-existing sub-soil and environmental risks to the project company, a condition which is subject to negotiation and potentially more balanced risk allocation (eg, temporary assumption of this risk by the offtaker) during the bidding process. After the expiration of the PPA, site ownership is automatically transferred to the offtaker.

6.5 Industrial integration

There is precedent for both MASEN and ONEE to require bidders to include a minimum industrial integration rate for the project and relevant measures to realise this integration, though no such obligations have yet to be practically implemented.

6.6 Liquidated damages

The project company may have to pay delay liquidated damages and/or performance shortfall liquidated damages. Delay liquidated damages are applicable for the delay in achieving the initial commercial operation date. Performance shortfall liquidated damages are applicable if the performance of the plant as certified during the final tests is lower than the agreed contractual performance.

6.7 *Force majeure* and other risks

Risk of *force majeure* is borne by the offtaker, who commits to pay deemed availability payments if the commercial operation date is delayed due to *force majeure* or if the facility's operation is impaired during the operational phase, and to reimburse the project company for additional EPC contract costs incurred by reasons of *force majeure*. MASEN and ONEE tend to cap these deemed availability payments at the debt service amount due by the project company under the terms of the financing documents. For projects where MASEN acts as sole lender, the project company's repayment obligations are not triggered until after commercial operation.

However, neither MASEN nor ONEE offers payment protection with respect to *force majeure* events – such events only entitle the project company to additional time with respect to commercial operation. Additionally, the term of the PPA is typically not extended to allow for recovery of revenue.

Risk of financial close is assumed by the sponsors, including loss of the development bond (typically 10% of the sponsors' equity investment) if financial close is not reached by the prescribed deadline, and/or the sponsors' direct liability to the offtaker in such a case.

For solar projects, radiation level risks rest with the sponsor.

6.8 Termination

In case the project company defaults before commercial operation, Moroccan PPAs tend to follow regional precedents by permitting the offtaker to draw the performance bond and obliging the project company to return the project site to its original condition. Offtakers are traditionally exempt from paying indemnities in this case.

Should such a default occur after commercial operation, Moroccan PPAs tend to favour the rights of the lender more than their regional counterparts. This is influenced by the fact that some termination payment to the project company covering outstanding debt is typically compulsory, regardless of whether such termination occurs due to default of the project company.

If the PPA is terminated due to the offtaker's default, the project company is usually entitled to compensation covering the outstanding debt plus the non-amortised equity and other termination costs. However, this compensation does not take into account return on equity.

7. Conclusion

Given the government's commitment to the national energy strategy and the success of solar projects constructed to date, Morocco has emerged as a regional hotspot for solar IPPs. The Kingdom's renewables progress does not stop there – with ONEE buying power from wind projects like the 300 MW Tarfaya Wind Farm commissioned in 2014 and the 850 MW wind farm currently under construction by the Nareva-Enel-Siemens consortium, Morocco continues to demonstrate its commitment to a clean energy transformation. The renewables market should continue to expand towards 2020 and 2030 capacity targets, with the backing of the state and an increasingly robust institutional framework paving the way.

The author would like to acknowledge the kind assistance of Thomas George in the preparation of this chapter.

South Africa

Rajen Ranchhoojee
Orrick, Herrington & Sutcliffe LLP

---

1. History

The South African electricity market has historically been dependent on its monopoly, vertically integrated state-owned utility of Eskom Holdings (SOC) Limited. Eskom's responsibilities for generation, transmission and distribution were further expanded to that of offtaker with the introduction of the South African Renewable Energy Independent Power Producer Procurement Programme (REIPPP) in 2011.

The introduction of the REIPPP indicated a changing tide from the monopoly electricity generation functions of Eskom, by opening up the electricity generation market not only to Independent Power Producers (IPPs), but also to new generation technologies, such as solar PV. This was a welcome shift from the status prior to the REIPPP, a period through which over 90% of South Africa's 44 000 MW was generated by coal-fired power stations.

The birth of the REIPPP was informed by South Africa's Integrated Resource Plan 2010–30 (IRP 2010), an ever-evolving planning document administered by the Department of Energy, that seeks to map electricity sector development from 2010 to 2030. The IRP 2010 initially provided for approximately 18 GW of renewable energy generation by 2013, of which solar PV accounted for 8,400 MW.

In determining an appropriate procurement mechanism for electricity through the REIPPP, the Department of Energy gave consideration to the implementation of a Renewable Energy Feed-In-Tariff versus a competitive pricing process. In the end, the competitive pricing process was preferred, and it was on this auction basis that the REIPPP was designed. The REIPPP follows a phased approach of implementation, in which a certain number of megawatts are allocated, per technology, at predetermined bid intervals termed 'bid windows', and through which IPPs tender their projects. The REIPPP envisaged five bid windows, and as at the end of 2017, 45 solar PV projects have been awarded through the four bid windows concluded, with a generation capacity of 2,191.83 MW. It is worth noting, however, that the 12 projects awarded in bid window 4, with a generation capacity of 762.9 MW, were the subject of a tariff impasse with Eskom and various other challenges, the end result of which has jeopordised the ultimate signature of PPAs for these projects.

Below is a table of successful solar PV projects grouped by bid window in the REIPPP.

Bid window	Submission date	Number of projects	PPA signature date	Capacity in MW
1	November 2011	18	November 2012	762.9
2	March 2012	9	May 2013	417.12
3	August 2013	6	December 2014	385
4	August 2014	12	Unsigned	762.9

With the introduction of the REIPPP, ancillary solar PV opportunities have been unearthed, including corporate and industrial roof-top and ground-mounted installations, corporate PPAs and even contemplation of regulating small-scale grid-tied rooftop installations. While corporate and industrial as well as private utility-scale installations remain a popular concept, the major challenges in implementation arise from the long-term commitment to the take-or-pay principle (this is particularly prevalent in the mining industry), as well as the past uncertainty in obtaining generation licences. The generation licence and small-scale grid-tied generation issues are discussed in section 3 below.

2. Overview of market participants

The advent of the REIPPP brought with it keen interest and participation of the world's major sponsors, developers, contractors and equipment manufacturers to the South African renewable energy market.

There is widespread sponsor representation from China, Canada, France, Germany, Italy, Norway, Spain, United Kingdom and United States, among others. These include the most well-known participants in global solar energy generation. The common position would be for these global players to take a majority stake in project-specific special purpose vehicles, with minority shareholding going to South African entities with which the global participants have partnered. Local equity ownership in projects is both a requirement under the REIPPP as well as a consideration in the application for an electricity generation licence. Participation within the South African solar market includes public and privately owned solar companies, as well as various state utilities, and includes the likes of First Solar, Solar Reserve, X-Elio, Canadian Solar, Enel, SunPower, Gransolar, Juwi, Mainstream Power, Yingli Solar and Scatec Solar as examples.

The contractors in the South African solar market tend to follow the trend of the sponsors in large part, with international entities partnering with local contractors to construct and service successful projects. It is safe to say that contractor participation is extremely competitive and diversified across global market participants.

A number of companies have sought to take advantage of the growth of the solar sector in South Africa, as well as the regulations requiring local equipment to be used in REIPPP projects and, as such, have set up panel manufacturing and assembly operations in South Africa. The success of these businesses is wholly reliant on the continued roll-out of the REIPPP programme, in accordance with the IRP 2010, and the continued growth of the solar power industry across Africa.

The brewing, mining, industrial and even large-scale retail industries have all shown tremendous interest in solar energy as a viable compliment to the conventional energy currently being consumed.

3. Legislative and regulatory background

3.1 Electricity Regulation Act

The Electricity Regulation Act No 4 of 2006 (ERA), is foundational legislation regulating electricity generation in South Africa. The ERA governs generation by IPPs and state-owned companies alike, and expands across all technologies. The ERA addresses licensing requirements and, importantly, any licensing exemptions.

Generation licences are granted by the National Energy Regulator of South Africa (NERSA), on an individualised application basis. The process for application includes providing complete information on the facility and its required permitting, as well as holding public participation processes. Inclusion of the PPA that will govern the project is also required. The Minister of Energy allocates the number of megawatts to be awarded for solar energy in any particular bid window or grouping of bid windows, by issuing determinations against the scheduled roll-out according to the IRP 2010, and against which each megawatt of solar energy is recorded. Successful bidders in the REIPPP find the process of a generation licence application much easier than those applying for a 'private' generation licence, based on the simple fact that the REIPPP is a government-sponsored programme designed to roll-out the government's renewable energy strategy. Applicants do still have to defend against public challenge to any particular project, though these are few and far between insofar as solar projects are concerned. Private PPAs for commercial and industrial or utility-scale facilities, however, tend to be held up in the process, given that each new renewable energy facility, including solar installations, must be recorded against the targets established in terms of the IRP 2010 (as revised from time to time) and, as such, must not be seen to clash with the government roll-out via the REIPPP.

NERSA's award of generation licences must be done in consultation with the Minister of Energy, and with the keeping of a 'scoreboard' tally of solar generation against the targeted total generation of the IRP 2010. Decisions on the impact of such private PPAs, no matter how large or small, are scrutinised and decided upon not only on the circumstances as at the time of application, but having due consideration on the impact on future implementation to be determined by the Minister of Energy.

Numerous private projects and installations have sought to take advantage of the exemptions provided for in the Electricity Regulation Act, pertaining to the generation licence. Historically these applied to any generation plant constructed and operated for demonstration purposes, constructed and operated for own use, or non-grid connections except for commercial use. Though not formally recorded, there was also a generally accepted exemption for installations of less than 1 MW. The 'own-use' exemption was the one most commonly exploited in which corporate structures were created through which the facility was owned by the offtaker, however the mechanism for payment was linked to energy generation and took the form of an instalment payment, lease or some other form of performance fee. The veracity of these structures were tenuous at best, with the risk of ERA non-compliance a distinct possibility.

3.2 Small-scale renewable embedded generation

In 2015, NERSA issued a discussion document for the potential implementation of small-scale renewable embedded generation (SSREG). The discussion document was directed at rooftop solar generation for residential and business grid-tied users, and limited to installations of up to 1 MW, through which such users would be able to take advantage of net-metering to the point of becoming net-neutral users of the grid. Generation credits would be given up to a maximum actual usage, resulting in the SSREG being a net-neutral participant on the grid.

Through the public participation process, distributors, mainly in the form of municipalities, raised the issue of substantial revenue losses stemming from inevitable electricity sales losses should the SSREG programme be implemented. Further to this, technical and administrative challenges in implementing such SSREG regulations have been raised. The process had not yielded a suitable compromise between SSREGs and distributors and as such it would appear that the net-metering element of SSREG has been removed.

Following on from the ever-increasing interest in commercial and industrial rooftop facilities, and having due regard to the uncertainty of the generation licence exemptions in the ERA, an overhaul of the relevant exemptions schedule[1] was published on 10 November 2017. In terms of the 'new' exemptions schedule, a generation licence will not be required if the facility has a generation capacity of less than 1 MW, is supplied to a single customer (with or without wheeling), if grid tied has the approval of the relevant licensed distributor, and provided that the Minster of Energy has not published a notice stating that the amount of MW allocated for embedded generation in terms of the IRP 2010 has been reached. Other licence exemptions include those for facilities built solely for demonstration purposes or for backup generation.

In terms of the November 2017 'new' exemption publication, all exempt facilities must be registered with NERSA. As at February 2018, NERSA had yet to publish its rules, requirements and forms for registration. The process of preparation for the applicable registration rules began in December 2017, with NERSA putting out a notice that a public consultation process would ensue, and that it would not consider any registration applications until such time as the registration rules had been finalised and published.

---

1 Schedule 2 to the ERA.

3.3 REIPPP

In understanding the holistic purpose of the REIPPP, consideration must be given not only to the energy needs of the country, but also government's mandate to address the socio-economic inequalities created by South Africa's apartheid past. It is with an understanding of this complete mandate that informed the REIPPP process. As mentioned above, the REIPPP is a phased approach to renewable energy procurement, managed through determinations made by the Minister of Energy in accordance with the IRP 2010, and through which Eskom as the state-owned electricity utility is the offtaker. In terms of the REIPPP, solar PV projects must be have a generation capacity of at least 1 MW and are restricted to an effective maximum generation capacity of 75 MW per project.

Projects in the REIPPP undergo a two-stage evaluation approach, the first being compliance qualification followed by actual evaluation based on a points system, as discussed below. The REIPPP has six substantive qualification criteria that must be met, before a bidder is progressed through to the evaluation phase. These six qualification criteria broadly grouped are:

- legal;
- land acquisition;
- environmental;
- financial;
- technical; and
- economic development.

A further two criteria of structure and value for money are also included in the qualification criteria. For each of the aforementioned criteria, bidders are required to submit comprehensive details and information as required per evaluation criterion.

The REIPPP encapsulates the various procurement legislation governing public procurement which includes the South African Constitution 108 of 1996 (the Constitution), the Preferential Procurement Policy Framework Act 5 of 2000 (PPPFA) and the Public Financial Management Act 1 of 1999 (PFMA), read in conjunction with the referenced legislation therein including the Broad-Based Black Economic Empowerment Act 53 of 2003 (BBBEE Act). It is generally irregular for the government to engage in any unsolicited procurement, unless the strict rules applicable to such processes are applied. The solar power industry does not lend itself to such exception to the unsolicited procurement rules.

Section 217 of the Constitution of South Africa provides that:

when an organ of state in the national, provincial or local sphere of government, or any other institution identified in national legislation, contracts for goods or services, it must do so in accordance with a system which is fair, equitable, transparent, competitive and cost-effective.

In this construct, the state must always be able to show value for money insofar as its procurement is concerned.

The PPPFA, through its regulations, provides for a points system to be implemented for procurement by any organ of state. The points system attributes a certain quota of evaluation points to price and the remainder to broad-based black economic empowerment (BBBEE) as envisaged in the BBBEE Act and its implementation codes. The applicable points system to be used is dependent on the value of procurement and leads with the price weighting followed by the BBBEE weighting, as follows:

- procurement value R30,000 to R50 million[2] – 80/20;
- procurement of R50 million and above – 90/10.

In terms of Section 3 of the PPPFA, the Department of Energy successfully applied for an exemption to the points system weighting in respect of the REIPPP which has resulted in a 70/30 weighting being applied to the adjudication of projects. Given the significant increase and importance of the BBBEE component for adjudication, it is worth understanding the elements making up this component.

The BBBEE component in the REIPPP has been expanded to address broader economic development imperatives, and is made up of the following seven elements, each carrying its own weighting within the 30-point BBBEE component of adjudication. Certain qualification thresholds are required per element, which thresholds must be achieved in order for the project to be compliant in terms of the REIPPP. The economic development elements are as follows.

(a) ***Job creation [7.5/30]***[3]

Job creation is measured to a baseline of person months according to a formula provided by the REIPPP, and is split between the construction and operations phases of the project. It incorporates sub-elements with scoring based on the skill level and education, demographic, citizenship, local community participation and other such elements. All commitments require quarterly reporting throughout the life of the project.

(b) ***Local content [7.5/30]***

Local content is the measure through which localisation of goods and services are attributed to the project. This supports the government's drive for increased job creation. Local content is the amount of expenditure on South African goods and services as a percentage of the project construction cost (excluding certain costs such as funding costs, land and others). In future programmes it is likely that the element of local content will be split between key components and equipment (modules, inverters, metal structures and transformers) and the remainder of the plant. In this way there would be a specific focus on increasing manufacturing and job creation in the key component industries. It is worth noting that for purposes of the local content calculation, raw steel is deemed to be locally produced, regardless of where it is sourced from.

(c) ***Ownership [4.5/30]***

The ownership element is divided into various sub-elements, each with differing thresholds. There is a minimum threshold of 8% black[4] ownership for the EPC and

---

2 The threshold amount was drastically increased in April 2017 from R1 million to R50 million.
3 Point allocation per element.
4 As defined in the BBBEE Act 53 of 2003.

O&M contractors respectively and 12% for the project company itself. Though these thresholds represent the minimum levels of ownership required for compliance, the levels of black ownership have proven to be substantially higher. The verification of ownership pierces the corporate veil, and relies on the actual ultimate beneficiary being black. To this end, proof of identification through the submission of identification documents of the ultimate beneficiaries is required in order for a bidder to claim compliance with this element.

The REIPPP also requires at least 2.5% of the equity in any project to be held by a local community (defined as a community within a 50 km radius of the project). This requirement, however, has not achieved the levels of community upliftment as initially envisaged, and it is likely that in the future, equity equivalence programmes will be initiated in order to ensure real-time benefits to local communities. The equity held by communities is usually held through community trusts specifically created for this purpose.

There is also a threshold requirement of 40% South African ownership in any given project. The elements of black ownership and local community ownership are incorporated into the requirement of South African ownership.

The Implementation Agreement in the REIPPP also restricts change in control of ownership in the facility. In this regard, no change in control (very widely defined) in the project may occur without the approval of the Department of Energy, and no dilution of ownership of any kind is allowed within the first three years of operation. These restrictions extend to 'conduit' shareholders as well. Further restrictions are placed on equity held by BEE shareholders, which not only require the approval of the Department of Energy, but as a general rule, in selling, must not dilute the BEE shareholding in the facility.

(d) *Management control [1.5/30]*

Management control is calculated as a percentage of the top management of the project company, and reflects the percentage of black person, Women, and black Women, in the top management of the project company.

(e) *Preferential procurement [3/30]*

The preferential procurement element evaluates the vendors that service the project. This element is made up of three sub-elements: black ownership, women ownership, and qualified small enterprises or exempt micro enterprises. The level of black and women ownership as well as the BBBEE status of each such vendor is considered in the scoring process.

(f) *Enterprise development [1.5/30]*

In order to meet the enterprise development requirements of the REIPPP, each project must commit a percentage of its annual revenues towards enterprise development. Scoring for this element is not uniform, in that greater recognition points are given for enterprise development of qualified small enterprises or exempt micro enterprises, women-owned enterprises and black-owned enterprises. The target for enterprise development is 0.6% of revenue.

(g) *Socio-economic development [4.5/30]*

The socio-economic development element is met by committing a percentage of revenues to address the needs of communities surrounding the project. The threshold for the socio-economic development element is 1% of revenue.

Earlier bid windows of the REIPPP saw bidders achieving the maximum number of points for a particular element if they achieved the target for that particular element, however this was later adjusted to a system whereby the bidder with the highest commitment level for any particular sub-element would set the 'new' target for that sub-element, which if achieved, would be awarded maximum points. All commitments below this 'new' target would be proportionately scored against the 'new' target.

Compliance with the economic development commitments by successful IPPs is monitored on a quarterly basis and enforced through an Implementation Agreement which the IPP executes with the government. Should an IPP fail to completely meet its economic development commitments, termination points may be issued against the PPA. On reaching a threshold number of termination points, the PPA may be cancelled.

3.4 Small Projects Renewable Energy Procurement Programme

Following on from the successes of the REIPPP, the Department of Energy launched the Small Projects Renewable Energy Procurement Programme (Small Projects Programme) in August 2013. The Small Projects Programme was designed for projects with a generation capacity of no less than 1 MW and no greater than 5 MW. The initial allocation of 100 MW for small projects participating in the Small Projects Programme was later increased to 400 MW. Unfortunately, since inception, the Small Projects Programme has stalled, and as at the end of 2017 no bids have been finally evaluated and no PPAs signed through this programme.

The Small Projects Programme has sought to lessen the formal procurement burden on smaller IPPs, by bifurcating the submission process. The first stage of the process involves a fairly basic submission of the project information based on the qualification criteria generally grouped as legal, land acquisition and land use, environmental, technical and economic development. Fulfilment of these qualification criteria through the first stage of the bid process are much less onerous than the requirements in terms of the REIPPP.

On successfully meeting the qualification criteria of Stage 1, bidders proceed to Stage 2 of the programme, which very much follows the line of the REIPPP in that extensive and comprehensive submissions on the qualification criteria submitted through Stage 1 are required, with the additional requirement of meeting financial criteria.

4. Environmental, real estate and planning issues

Before being awarded a PPA or being granted a generation licence, IPPs must have secured a project site and obtained an environmental authorisation in respect of such site, in accordance with the National Environmental Management Act 107 of 1998 (NEMA).

4.1 Environmental authorisations

The appropriate environmental authorisation process to be followed is informed by the studies required to be performed and which make up the largest part of the application for awarding of a record of decision for environmental authorisation. These studies may be fully-fledged Scoping and Environmental Impact Assessment studies for larger projects as envisaged in Listing 1 of the listed activities in accordance with NEMA (20 MW or greater), or a Basic Assessment process for smaller projects in accordance with Listing 2 of NEMA (10–20 MW or less than 10 MW with a footprint greater than 1 hectare). It is worth noting that solar PV installations in urban areas are exempt from the environmental process.

Barring complications, obtaining an environmental authorisation following the Environmental Impact Assessment process should take 14–18 months, while the Basic Assessment process usually takes around nine months.

The South African Department of Environmental Affairs announced in April of 2017 that it had implemented a process of streamlining the application for environmental authorisations in the eight Renewable Energy Development Zones (REDZ) identified by the government in February 2016. The REDZ are located in the following regions: Overberg and Kornsberg (Western Cape), Cookhouse and Stromberg (Eastern Cape), Kimberley (Free State and Northern Cape), Vryburg (North West), and Upington and Springbok (Northern Cape). The Department of Environmental Affairs has carried out its own preliminary studies on each of the REDZ, which have resulted in it being able to accept the Basic Assessment process for obtaining an environmental authorisation in respect of large utility-scale projects within the REDZ. This dispensation also covers transmission lines and substations. Accordingly, the period of time within which the Department of Environmental Affairs must accept or reject an application falling within the REDZ is reduced from 300 days to 57 days.

4.2 Real estate

With vast open spaces available in prime solar resource areas, the securing of land for project development purposes has proven to be unproblematic. Land owners have shown a willingness to work with developers in meeting project needs, many agreeing to enter into option to lease or buy arrangements, which extend over a few years, allowing the developer the right to proceed with the required environmental processes at low cost, without committing to any long-term obligations through the development phase. Lease arrangements are secured through notarial deeds which are registered against the property, and protect the real rights of the project over the land.

In the event that only a portion of land (agriculturally zoned) will be secured for the facility, an application must be made for the subdivision of such agricultural land in terms of the Subdivision of Agricultural Land Act 70 of 1970. It is also necessary to obtain approval from the Minister of Mineral Resources in terms of the Minerals and Petroleum Resource Development Act 28 of 2002, as the construction and operation of the project may sterilise a mineral resource or impede future mining operations. These approvals are generally readily obtainable.

Given the colonial history of South Africa, one of the government's objectives is to drive fair and lawful restitution to indigenous South Africans for actions of the past, including those pertaining to land ownership. As such, it is worth ensuring that no land claims have been instituted on the proposed project sites. This may be done through confirmation from the office of the Land Claims Commissioner for the region in which the project is intended.

5. Power offtake arrangements and bankability

5.1 Currency

PPAs in South Africa have been exclusively Rand-based, both through government programmes and in the private sector. The volatility of the Rand has thus inherently increased the risk profile of projects in South Africa.

Sponsors have in part managed the currency risk for foreign denominated equipment and services by hedging through the construction phase of the project, however the Rand-based revenue risk through the life of the project remains unmitigated.

5.2 Repatriation of funds

South Africa's established exchange control rules, as implemented by the South African Reserve Bank, allows for ready and efficient repatriation of funds whether generated through dividends of an operating project company or for service fees pertaining to projects. Adhering to the initial registration process for investments where equity is concerned and/or registering relationship agreements for sales and/or services, is essential for securing the trouble-free repatriation of funds.

5.3 Capitalisation of investments

Shareholders can capitalise their investment in projects either through shareholder debt (shareholder loans) or acquisition of equity (subscription of shares) in project entities. Dividends are subject to a withholding dividends tax (subject to applicable tax treaties) currently set at 15%, while the nature of interest deductions in the hands of the project entity, pertaining to shareholder loans, is subject to the general transfer pricing rules, which broadly require any interest charged to be marked-to-market, based on the currency in which the capital has been borrowed and is to be repaid.

5.4 Import duties and other tax considerations

While import duties do apply to equipment shipped into South Africa for generation purposes, there are various tax incentives that are available to tax payers, ranging from CER (certified emissions reduction) exemptions to renewable energy production allowances. This latter incentive allows for the acceleration of capital allowances for machinery, plant and implements used for the first time by the taxpayer, for the purpose of its trade.

5.5 Offtake arrangement

PPAs generally endure for a period of 20 years. This applies to utility-scale projects as well as those awarded under the REIPPP. Shorter PPA terms of 10–15 years have been considered for smaller corporate PPA projects.

The take-or-pay regime has been implemented in government PPAs, with Eskom as offtaker, and has carried over to the corporate PPA market. Deemed energy payments are enforced should the offtaker not consume all of the electricity made available to it in accordance with the PPA.

Deemed energy payments are also accepted in circumstances where there is a 'system event'. In such instances, guaranteed grid availability is agreed, and any unavailability outside of the agreed period will result in deemed energy payments. Grid connection delays and grid outages are the most common examples of a 'system event'.

'Compensation events' being breaches of the offtaker's obligations under the PPA in accordance with the REIPPP (excluding circumstances of payment default), will result in an extension of the PPA for the period through which such breach continues, though in corporate PPA relationships these have also resulted in deemed energy payments.

There is generally no penalty for failure to produce electricity under the take-or-pay regime. The generator suffers loss through the inability to generate revenues, while the clock on the PPA continues to run. There is accordingly no credit support expected of the generator.

Delay in reaching commercial operation does however run the sanction of penalty, specifically in contracts where Eskom is the offtaker. In such instances, the commencement date for generation (scheduled commercial operation date) remains, while the PPA is shortened on the back end for every day of delay. Should the commercial operation date be delayed by one month, not only does the generator lose out on one month's revenue between the scheduled commercial operation date and the actual commercial operation date, but the PPA term is reduced by one month.

5.6 Price escalation

Price escalations in the REIPPP and Small Projects Programme are indexed to the South African Consumer Price Index. Bidders have an opportunity to fully index or partially index their tariffs at bid submission in order to allow for the most competitive pricing proposal.

In the private sector, there has been a drive to link PPA tariff escalation to Eskom's megaflex tariff, to the extent that IPPs could always offer an Eskom megaflex less X% tariff. This approach, however, has not been well received by the market. The rationale for rejecting such a proposal is that the annual Eskom megaflex tariff increases are so high, that even though the escalation would result in less than grid tariffs, such extensive increases cannot be justified by solar IPPs, which have no significant cost escalations throughout the life of the solar facility.

5.7 Government support

There is currently a government support regime in place for successful projects under the REIPPP. This support takes the form of a PPA Payment Guarantee, and is enforceable if any payment is overdue by 25 business days; the amount is undisputed; and a claim has been brought against the offtaker and has not been paid within three months of the due date. In such circumstances, the government will make payment within 40 business days of the demand. The reality of the government support is that payment is unlikely to take less than 3–5 months from the initial breach.

5.8 Grid access

The South African transmission system is run by Eskom. In 2008, through the 'electricity crisis', in which South Africa experienced rolling black-outs, the government implemented a strategy to 'keep the lights on at all cost'. The repercussions of this policy was not fully apparent at the time, and now a decade later, the impact is stark. Through the period of 'keeping the lights on at all cost' priority was given to generation, and very little to transmission and distribution. Maintenance was allowed to slip as funds were channelled to new generation projects, and upgrades to the grid did not keep pace with the new generation. These factors, coupled with the fact that most solar generation plants are situated outside of the 'traditional' electricity generation areas of the country, has left the grid severely strained.

The strained transmission system has meant that grid access has become disproportionately critical in determining the viability of projects. The first step to determining grid access availability is for developers to request a Cost Estimate Letter from Eskom, which process allows the system operator to consider the proposed project site in relation to the existing transmission infrastructure, and to provide a non-binding estimated cost and timeline for connection. Once a project is awarded preferred bidder status in terms of the REIPPP, or determines that it otherwise has a secure offtake arrangement, the project would apply for a more detailed budget quote, through which process Eskom is able to consider its standing work-orders as regards grid connection, and provide an accurate budget quote and timeline for completion. The trend has been for Eskom to increase capacity infrastructure at any given connection point, with the downside being that project developers have been burdened with the additional cost.

Most projects opt for a 'self-build' arrangement with Eskom, whereby the project contracts out the connection works, from the project site up to the grid, to a third party independent contractor, thus allowing Eskom to focus on the work required on the transmission system alone. The independent contractor would in turn work with Eskom for the ultimate connection of the project to the grid. Once complete, the developer transfers ownership of the connection works to Eskom, and in this way avoids necessary maintenance obligations on the connection works through the life of the project. This self-build option has been preferred for reasons of cost control and mitigation against delays in bringing the project to commercial operation.

5.9 Wheeling

At various stages in the South African renewable energy industry development, wheeling arrangements for corporate PPAs have been considered. These have failed for commercial reasons. The fee levied by Eskom as the transmission system owner and operator ultimately results in the offtaker paying more for the electricity than simply relying on the grid. There is an appetite for wheeling arrangements in South Africa, however without a commercially reasonable wheeling fee, such arrangements are unlikely to develop.

6. Challenges and the future for the solar sector

Although tremendous strides have been made in the solar energy sector in South Africa since 2011, particularly through the REIPPP, progress seems to have halted on government programmes and utility-scale corporate PPAs. Extensive delays have occurred through almost every bid window of the REIPPP, culminating in a refusal by Eskom to sign PPAs in the fourth bid window and after more than a one-year delay in closing PPAs with IPPs, the Department of Energy reaffirmed its commitment to the renewable energy industry through the end of 2017, having overcome numerous hurdles, including a unilaterally proposed 77c (Rand)/kWh cap on all tariffs, despite the successful competitive tariffs on which the projects were awarded at bid submission. This unilateral amendment would have automatically rendered all solar PV projects ineligible as they would fall outside of the unilateral tariff cap. PPA execution for the 2.2 GW of outstanding projects, including the 12 solar PV projects representing 762.9 MW in bid window 4 of the REIPPP, are expected to be closed in the first quarter of 2018.

Notwithstanding the remarkable progress made to date in the renewable energy industry through the REIPPP, in August 2017 the Minister of Energy announced that future bid windows would be stayed until the outcomes of the IRP 2010 revision had been finalised, thus rendering the future of the REIPPP programme, and its ancillary industries uncertain.

Grid connection continues to be a major issue, with Eskom being unable to keep pace with the generation industry in connecting facilities to the grid. This has resulted in the delayed roll-out of facilities as well as, in most instances, significant increases in connection costs. This has been exacerbated by Eskom's cash flow challenges.

On the commercial and industrial side, the outright purchase of panels and facilities has been preferred over more viable PPA-type structures given the various factors mentioned herein. Grid connectivity and perceived oversupply has also been a point of contention in this regard.

Despite these challenges, solar technology has been proven to be a viable and affordable technology in South Africa, and with the right political will to put renewable energy back on track, the successes of the REIPPP from 2011–2016 will stand the country and IPPs in good stead to gain momentum where progress has slowed.

Spain

Elena Aldave
Nasif Hamed
Fernando Quicios
Belén Wert
Pérez-Llorca

1. Brief history of the solar sector in Spain

Renewable energy has been, and still is, one of the most controversial aspects of the energy sector in Spain. In particular, the regulatory approach adopted in the field of PV solar energy. The Spanish solar market represents about 5% of the global solar photovoltaic (PV) market. Favourable meteorological conditions and public subsidies have contributed to the development of the PV sector in Spain. The country has optimal renewable energy resources, with high levels of solar irradiation that make Spain one of the most attractive European countries for PV investment.

The first Spanish PV solar installation connected to the electricity grid was built in San Agustín de Guadalix (near Madrid) in 1984 by the former Hidroeléctrica Española, SL (now Iberdrola, SA[1]). Since then, pilot projects have gradually been installed in Spain in order to develop this type of renewable energy.

Following the approval of Law 54/1997, 27 November, on the electricity sector (Law 54/1997), incentives were introduced for renewable energy facilities, including solar PV: the feed-in-tariffs (FiTs) and the feed-in-premium (FiP). This led to a proliferation of PV facilities throughout Spain.

From 2000 onwards, the development of PV solar energy grew significantly. Both Royal Decree 436/2004, 12 March, on the methodology to actualise and systematise the economic and legal regime of electricity production activity under the special regime (RD 436/2004) and Royal Decree 661/2007, 25 May, on the regulation of the electricity production under the special regime (RD 661/2007), paved the way for Spain to become one of the countries with the highest installed PV power capacity in the world by 2008. This period in Spanish history became known as the 'photovoltaic boom'.

However, the electricity sector reform in Spain in 2012 modified the entire system of incentives (FiT and FiP). The economic crisis and the need to reduce the tariff deficit led the government to develop a new system. In order to do so, it was necessary to pass a law to ensure effective competition and financial sustainability in the electricity sector. For that reason, the Spanish Parliament passed Law 24/2013, 26 December, on the electricity sector (Law 24/2013).

This new economic regime removed the incentives and premiums mentioned above and introduced a new remuneration system linked to investment and operation. As a result, investment in the solar PV sector was significantly reduced.

Ongoing regulatory changes have resulted in the slowdown in the development

1 Iberdrola, SA is one of the five biggest electricity companies in Spain.

of solar PV plants. Currently, and despite the high levels of irradiation, Spain has less installed PV power than other countries with lower levels of irradiation, such as Germany.

2. Overview of market participants

Market participants in the Spanish solar sector can be broadly divided in two areas: public participants and private participants.

2.1 Public participants in the Spanish Solar Sector

The Ministry of Energy, Tourism and Digital Agenda (the Ministry) has the role of proposing and implementing energy policies. The State Secretariat for Energy, which is under the direction of the Minister of Energy, Tourism and Digital Agenda, has the following main functions: to develop energy policy; to develop and implement measures to ensure the adequate supply of electricity; and to analyse and adjudicate on electricity sector auctions. The General Directorate for Energy Policy and Mines, which also has a wide jurisdiction on energy regulation in Spain, reports to the State Secretariat for Energy.[2]

Article 3 of Law 24/2013 gives the Ministry the capacity to grant administrative authorisations for electricity generation facilities, including its connection infrastructure, with an installed capacity of more than 50 MW, as well as those electricity generation facilities that, together with its connection infrastructure, exceed the territorial scope of an autonomous community.[3] According to Article 3 of Law 24/2013, the Ministry is also competent to guarantee the economic and financial sustainability of the electricity system, and to establish the general and the regulated remuneration regime.

The Spanish electricity system market operator and system operator are also important participants in the Spanish solar sector.

- The Spanish system operator is Red Eléctrica de España, SA (Red Eléctrica), which also acts as the transmission system operator (TSO). Its main function is to guarantee the continuity and security of electricity supply and to coordinate the production and transmission of electricity.[4]
- The market operator is OMI-Polo Español SA (OMIE), a company which manages the Iberian electricity market (MIBEL) in Spain and Portugal. It assumes the management of the system of electric power purchase and sale bids.[5]

Red Eléctrica is responsible for the management of grid access and connection procedures for the agents that intend to connect a new installation to the transmission network (or to the distribution network if it affects the transmission

2 Article 3 of Royal Decree 903/2017, 13 October, implementing the basic organic structure of the Ministry of Energy, Tourism and Digital Agenda.

3 Hence, the competent authorities of the Spanish autonomous communities grant administrative authorisations for electricity generation facilities, including its connection infrastructure, with an installed capacity of less than 50 MW and which do not exceed the territorial scope of the autonomous community.

4 Article 30 of Law 24/2013.

5 Article 29 of Law 24/2013.

network or the security or quality of supply) or for the agents that intend to modify the conditions of an existing installation.[6]

When the grid access and connection procedure is completed and the corresponding authorisations are granted, a Technical Access Agreement (*Contrato Técnico de Acceso* or CTA) is signed between the agent and the distribution or transmission operator. Such CTA shall be communicated to Red Eléctrica, as operator of the system and manager of the transmission network.

When several generation installations, such as solar PV plants, are connected to the same point of connection, the CTA is signed between the agent, the grid node single contact (*interlocutor único del nudo* or IUN) and the owner of the transmission network point of connection. In such cases, there is the possibility that all installations owned by the same generator have just one CTA.

Finally, the National Commission for Markets and Competition (CNMC) is another significant market participant. The CNMC is a public body with an independent legal status which promotes and defends the proper functioning of all markets in the interests of domestic consumers and commercial participants. Its main functions relating to the electricity system are:

- issuance of circulars and reports on energy regulation;
- supervision of the functioning of the wholesale electricity market and the retail electricity market, as well as the operation of the system;
- management of the electricity settlement system and the specific remuneration regime; and
- resolution of disputes and conduct of disciplinary proceedings.

The joint and coordinated action of these Spanish public authorities ensures that the electricity sector is properly regulated and provides guarantees for both domestic consumers and commercial participants operating in the electricity market.

2.2 Private participants in the Spanish solar sector

The most important private sector participants in the Spanish solar market are the power generators, those individuals or legal entities which have the function of building, operating and maintaining power generation facilities.

Private power generators in the solar PV sector must comply with some specific obligations in addition to the general obligations for power producers. Having a metering point certificate (*certificado de puntos de medida*), a certificate of submission of real-time measurements (*certificado de envío de telemedidas en tiempo real*) or to be duly registered with the RRRE (administrative registry for electricity generation facilities from renewable sources) are some of these obligations.[7]

6 The procedures for grid access and connection to the electricity transmission network are established in Law 24/2013. In particular, grid access and connection for electricity generation from renewable sources, cogeneration and waste are established in further detail in Royal Decree 413/2014, 6 June, on electricity generation by means of renewable, cogeneration and waste facilities (RD 413/2014). The technical details, including the commissioning stage, are developed in operating procedures called 'PO 12.1' and 'PO 12.2'.

7 All obligations for private power generators from renewable energies are established in Article 7 of RD 413/2014.

The CNMC publishes annually the list of the main private operators in the field of generation and supply of electricity. The 2016 main operators were: Endesa, SA, Iberdrola, SA, Gas Natural SDG, SA, Hidroeléctrica del Cantábrico, SA and Viesgo Infraestructuras Energéticas, SL.[8]

It is also important to note the significant role of private equity funds as significant market participants. Many PV facilities that have been constructed and are operating in Spain were financed by private equity funds, both domestic and foreign.

In recent years the solar sector has seen renewed interest from foreign investors. Such foreign investors (mainly distressed funds, private equity and infrastructure funds) have taken advantage of regulatory instability to target company and asset acquisitions at depressed prices. By way of example, Cerberus acquired Renovalia for €1.1 billion in 2016, Oaktree Capital Management acquired Eolia and KKR acquired Gestamp Asetym Solar.

Moreover, in recent years there has been a phenomenon of refinancing in Spain due to the economic difficulties of small and medium-sized operators due to the reduction in regulatory tariffs. These regulatory changes have been seen by some as an opportunity for investment by these private funds.

In relation to energy storage, it is important to note that currently there is no specific legal framework in Spain. The lack of regulation has resulted in some difficulties for the development of technological solutions linked to energy storage. However, Red Eléctrica has opted for investment in energy storage systems, with the ultimate goal of maximising the integration of renewables and providing greater efficiency and security to the entire electricity system.

Red Eléctrica has developed the *Almacena* Project, consisting of electrochemical energy storage connected to the grid, as well as the installation of a flywheel prototype in the Canary Islands. The *Almacena* Project has been developed in collaboration with NEC Energy Solutions and co-financed by the European Regional Development Fund (ERDF).

3. Legislative/regulatory background and support regime

3.1 Regulatory background of the solar sector

The liberalisation of the electricity sector, the establishment of third-party access to the grid, the formation of an organised energy trading regime and the reduction of state intervention in the market were all brought about as a result of the introduction of Law 54/1997.

The law also established a system to support the production of solar PV and thermoelectric energy.[9] A new framework was established, differentiating between an Ordinary Regime and a Special Regime. These regimes involved two different

8 Resolution of the CNMC, of 31 October 2017, establishing and making public, for the purposes of the provisions of Article 34 of Royal Decree-Law 6/2000 of 23 June, the list of main operators in the energy sector. The 2017 list of main operators in the energy sector will be published in 2018.

9 It should be noted that the Spanish regulation for solar PV facilities and thermoelectric facilities is almost the same. Law 54/1997, Law 24/2013 and their implementing regulation consider solar PV and thermoelectric facilities within the same group: electricity generated from solar power.

economic frameworks. The Ordinary Regime related to electricity production from nuclear and fossil sources, while the Special Regime related to electricity production from facilities with installed capacity not exceeding 50 MW using renewable energy, waste or cogeneration sources.

The Special Regime involved the payment of tariffs or premiums per kWh and represented an incentive for the production of electricity from renewable resources. The Special Regime tariffs and premiums were updated annually and reviewed every four years. In this context, the 1999 Plan for the Promotion of Renewable Energies (PFER) was passed, whose primary objective was to ensure that renewable energy sources net at least 12% of primary energy demand in Spain by 2010.

RD 436/2004 amended this incentive regime, updating and unifying the methodology for the calculation of incentives. Under the Special Regime owners of facilities such as PV solar and thermoelectric plants could choose between two alternative means of payment for the electricity generated:

- Selling electricity to the electricity distributor in exchange for a fee in the form of a regulated tariff: the FiT.
- Selling electricity freely in the daily market, in the forward market or through a bilateral contract, in exchange for a market price and incentive payment for participation in the market, plus a premium: the FiP.

The PFER was replaced by the Renewable Energy Plan (PER) 2005–2010. This Plan maintained the government's commitment to meet at least 12% of total primary energy demand from renewable sources in Spain by 2010. It also introduced the objective of producing 29.4% of electricity generation from renewable sources by 2010, in accordance with European targets. The PER 2005–2010 objective for the implementation of solar PV in Spain by 2010 was 500 MW.

RD661/2007, which introduced an incentive regime for the production of electricity from renewable sources, repealed RD 436/2004 but replicated its structure, as well as the system of two options for payment for output.

The significant changes when compared to the previous regulation were:

- The remuneration framework would be reviewed in 2010 and every four years thereafter. Remuneration would also be reviewed if a specific volume of installed capacity was reached in Spain (371 MW for PV installed capacity and 500 MW for thermoelectric installed capacity).
- Mandatory guarantees construction to prevent speculation.
- The approval of the Renewable Energy Plan for 2011–2020.
- A cap-and-floor price system for installations participating in the market.

Options for payment for output were modified to allow the owner of the plant to choose between:

- Selling electricity to the electricity distributor in exchange for a fee in the form of a regulated tariff: again, the FiT.
- Selling electricity freely in the daily market, in the forward market or through a bilateral contract, in exchange for a market price. The incentive payment to participate in the market and the premium element were removed.[10]

On November 2011, the Renewable Energy Plan (PER) 2011–2020 was passed, replacing the PER 2005–2010. The PER 2011–2020 established that renewable energy should meet 20.8% of total primary energy demand in Spain by 2020.

This system of incentives for renewable energies, and specifically for solar PV energy and thermoelectric plants, represented a considerable expense for the state and precipitated tariff deficit issues.[11]

The tariff deficit was regarded by the government as a debt owed by consumers to the electricity sector, which increased considerably over time due to the level of incentives afforded to renewable energy generators under the Special Regime. The deficit issue was not addressed and became a critical issue from 2006 onwards. During the Spanish economic crisis the Spanish government opted for deferring the payment of the tariff deficit by consumers. However, the situation became unsustainable and the government had to implement a substantial reform of the Spanish electricity sector as a result.

RD 661/2007 was first modified by Royal Decree 1614/2010, 7 December, which regulates and modifies certain aspects related to the activity of electricity production from thermoelectric and wind technologies, which limited the number of working hours of thermoelectric plants that were being remunerated through FiT.

Afterwards, the first cuts to the incentive system of solar PV plants were introduced by Royal Decree 14/2010, 23 December, on urgent measures to correct the tariff deficit in the electricity sector, which aimed to modify the economic regime of RD 661/2007. In particular:

- The number of annual working hours of solar PV installations that were being remunerated through FiT was limited.
- From 1 January 2011, electricity generators had to pay a fee of 0.5€/MWh for access to the electricity grid.

3.2 Current regulatory framework for solar PV facilities

From 2010 to 2012 economic incentives under the previous Special Regime were generally reduced and some were suspended. Royal Decree Law 9/2013, 12 July, on the adoption of urgent measures to guarantee the financial stability of the electricity system (RDL 9/2013) established a new regulatory framework and economic regime for electricity generated from renewable sources, cogeneration and waste. The incentive scheme established in Law 54/1997 was abandoned and replaced by the so-called Specific Remuneration Regime, which was regulated in RDL 9/2013, Law 24/2013 and RD 413/2014.[12]

During this period changes in the electricity sector were introduced which were necessary to establish a new regulatory framework in order to achieve the complete inclusion of renewable energy in the framework and to generate more

10 The value of the premium was set at 0c€/kWh, which means that there was no premium.

11 The tariff deficit is an annual imbalance in the difference between revenues and expenses of the regulated activities of the electricity sector.

12 The new specific remuneration regime arising from the electricity sector reform applies retroactively. In other words, the new economic regime applies to both new solar PV installations and those already existing before the reform.

competitiveness in the sector as a whole. The new economic regime was based on a market price for the sale of electricity plus a regulated additional payment. This additional payment is no longer linked to production, as under the previous system, but rather is by reference to the investment in, and the operation of, the relevant plant. The ultimate goal is that renewable energy installations such as solar PV facilities can participate in the market under the same conditions as other technologies, and achieve a reasonable return[13] for investors. The Specific Remuneration Regime is based on:

- Incentive for investment: An amount per unit of installed power (€/MW) that allows the facility to achieve a reasonable return as determined by the government.
- Incentive for operation: An amount relating to plant operation (€/MWh) which is intended to cover the difference between the operating costs and the operating revenues from the sale of electricity in the market.[14]

In order to calculate the level of payments, reference is made to the standard of an efficient and well-managed installation. The plants that receive the payments will stop receiving them when the plant exceeds its useful life, thereafter receiving only remuneration obtained from the sale of the energy generated in the market.

The enactment of RDL 9/2013 included other regulations regarding the methodology used to calculate payments. In addition, Law 24/2013 was enacted which repeals almost in its entirety Law 54/1997, leaving only limited provisions in force. Law 24/2013 confirms the status of Specific Remuneration System and states that, in exceptional circumstances, the government may implement compensation schemes to promote the generation of electricity from renewable sources, cogeneration and waste in circumstances where there is an obligation to meet energy targets established under directives or other European standards, provided that they help to reduce the cost of energy and reliance on external energy sources.

The aforementioned measures must also be compatible with the principle of economic and financial sustainability of the electricity system. Currently, due to the European targets set by Directive 2012/27/UE of the European Parliament and of the Council of 25 October 2012 on energy efficiency,[15] the government is implementing a revised regime through the introduction of renewable energy auctions. The regime will be established through the use of competitive bidding procedures whereby additional payments will be made relating to installed capacity and not to the production of energy. Bidders will receive a specific amount for each megawatt that is bid in the auction ultimately constructed, regardless of the level of generation carried out by the installation.

Moreover, such auctions are carried out using a marginal pricing system whereby

---

13 This new system is based on the reasonable return of the investment made. The specific remuneration regime should allow the facilities to cover the costs necessary to compete in the market on equal terms with other technologies and obtain a reasonable return from the whole project.

14 The renewable energies, cogeneration and waste facilities will not receive this incentive for operation if the revenues from the sale of electricity in the system are higher than its operating costs.

15 The Energy Efficiency Directive established an objective of 20% of renewable energy consumption by 2020.

bids are ordered from lowest to highest cost for the electricity system. The highest marginal bid will set the price for all plants to be installed.

To date, there have been three renewable auctions: the first, on January 2016, for 200 MW of biomass technology and 500 MW of wind technology; the second, on May 2017, for 3,000 MW of wind technology; and the third on July 2017, for 3,909 MW of solar PV technology and 1,128 MW of wind technology.

As a result of the renewable auction system, Spain is making some progress towards achieving the 20% renewable energy European target by 2020. In this regard, Spain's penetration ratio on renewable energy is higher than the EU average and most of its surrounding countries, having achieved in 2015 a 17.3% of renewable energy consumption over final energy consumption. Considering the latest 2017 renewable auctions, it is anticipated that Spain will be able to fulfil its renewable energy objectives in advance of the relevant deadlines.[16]

Law 24/2013 improved the existing legal framework and was based on the following principles:

- Separation of transmission and distribution (as regulated activities) from generation and supply (which can be carried out in conditions of free competition).
- Liberalisation of contracting and selection of suppliers by end consumers.
- Freedom of access to transmission and distribution networks through the payment of charges.
- Economic and financial sustainability of the electricity system.

Subsequently, RD 413/2014 was enacted, further developing RDL 9/2013. RD 413/2014 establishes the methodology for calculating the new remuneration scheme (including the rate of return). The methodology incorporates periodic reviews, defining a six-year regulatory period divided into two, three-year regulatory semi-periods. Furthermore, installations must be included in the register of the Specific Remuneration Regime in order to receive the aforementioned payments.

The themes of RD 413/2014 are further developed and complemented by Ministerial Order IET/1045/2014, 16 June, adopting the remuneration parameters for the facility type applicable to certain energy production facilities from renewable energies, cogeneration and waste (Order IET/1045/2014) which regulates the payment parameters to be considered in order to determine payments available for each facility type. Order IET/1045/2014 makes a classification of the facility type according to the technology, the capacity, the number of years of operation of the installation, the electricity system and other criteria. Moreover, it establishes that each type of facility should be assessed against a set of parameters on the basis of an efficient and well-managed company. The most relevant parameters are: cost of investment per unit of power; cost of operation; regulatory useful life; and the amount of working hours.

---

16 See the Spanish government website at www.lamoncloa.gob.es/serviciosdeprensa/notasprensa/minetur/Paginas/2017/260717-renovables.aspx.

3.3 Current regulatory framework for thermoelectric facilities

In Spain the majority of the energy generated from thermoelectric facilities comes from concentrated solar power (CSP), which is a thermoelectric technology that uses mirrors or lenses to concentrate solar radiation in order to power a steam turbine connected to an electricity generator.[17]

The Specific Remuneration Regime established in RDL 9/2013 and in Law 24/2013 is applicable to CSP plants under similar terms as solar PV plants. Hence, CSP plants have the option to opt for receiving additional payments (the incentive for investment and the incentive for operation) through participation in the renewable energy auctions.

Moreover, CSP plants are included in the RD 413/2014 scope and are affected by the periodic reviews, the six-year regulatory period divided into two, three-year regulatory semi-periods, and the mandatory inclusion in the register of the Specific Remuneration Regime in order to receive regulated payments. CSP plants' remuneration is also affected by Order IET/1045/2014 parameters.

In particular, CSP plants in Spain are authorised to use an 'auxiliary fuel' (*combustible de apoyo*), which is usually natural gas in liquid form (liquefied natural gas or LNG) in order to maintain the temperature of the heat transfer fluid used in the energy generation process. Such auxiliary fuel is also used to compensate for lack of solar radiation by generating electricity from LNG, for instance, instead of generating electricity from solar power. However, the percentage of auxiliary fuel used in CSP plants is limited.

RD 661/2007 first regulated a limit of 12–15% of auxiliary fuel that could be used in CSP plants. The limit depended on the alternative chosen: a 12% limit if selling electricity in exchange for payment of the FiT; or a 15% limit if selling electricity at the market price.

Law 15/2012, on tax measures for the energy sustainability (Law 15/2012) established that energy generated using auxiliary fuels, such as LNG, in CSP plants would be excluded from the Specific Remunerations Regime. This means that the percentage of the energy generated in the CSP plant that comes from auxiliary fuel would be remunerated by a market price, while the percentage of energy generated that comes from solar power would continue receiving the regulated payment.

Law 15/2012 did not establish a specific consequence for the infringement of the 12–15% limit of production using auxiliary fuel. The consequence of non-compliance with the annual limit of production using auxiliary fuel used to be that the CSP plant did not receive the regulated revenues for the energy produced above such limit.

Since RD 413/2014 came into force, this infringement is sanctioned with the suspension of the regulated payment during the year in which the excess has occurred. Additionally, if a second breach of the limitation occurs, the plant could be excluded from the Specific Remuneration Regime and a sanctioning proceeding may be initiated.

RD 413/2014 also modified the former limit of 12–15% and established a new limit of 12% of auxiliary fuel that could be used in CSP plants' electricity generation.

17 Spain is the CSP world leader in terms of number of operating plants as well as installed capacity.

3.4 Proceedings before the Spanish Constitutional Court and the Spanish Supreme Court

RDL 9/2013, RD 413/2014 and Order IET/1045/2014 were all appealed before the Spanish Supreme Court and the Spanish Constitutional Court by plant owners, especially in relation to the retroactivity of the new legislation passed after the 2012 reform of the electricity sector. The Spanish Constitutional Court dismissed the constitutionality claims against RDL 9/2013 because it was considered that the principle of legitimate expectations could not be equivalent to a regulatory immutability, especially in the context of an economic crisis and with the electricity sector deficit issue. Thus, the Spanish Constitutional Court validated the regulatory change under the electricity reform and the new regulation applicable to renewable energies.[18]

The Chamber for Administrative Proceedings of the Spanish Supreme Court also validated the reductions to renewables incentives which resulted from the new remuneration regime after the electricity sector reform. The Supreme Court considers that the new regulations are constitutional and in compliance with European law.[19]

Having no opportunity under Spanish national courts, plant owners have also initiated international arbitration proceedings against Spain to claim compensation for breaching the European Energy Charter (ECT). So far there have been two arbitral awards in favour of Spain,[20] concluding that the 2010 regulatory amendments in the renewable sector did not infringe any legitimate expectation under international law.

However, there is one recent arbitral award against Spain (*Eiser Infrastructure Limited and Energía Solar Luxembourg SARL v Spain*[21]) which concluded that Spain has infringed the ECT due to a lack of fair and equitable treatment of Article 10 of the ETC and a lack of complete protection and security to investors (for changing the regulatory and economic regime of the renewable energy sector in Spain between 2012 and 2013). Such arbitral award was the first one in which a tribunal found in favour of investors on this issue and it has raised the hopes for investors who were affected by the 2012–2013 reform of the Spanish electricity sector.

Currently, there are more than 25 claims pending before international arbitration courts under the provisions of the ECT. It seems that there are more chances of success for investors in international investment arbitration than in the national courts. However, it should be noted that an investing company (regardless of the place of its incorporation) that is controlled by national investors of the state where the investment was made cannot submit to arbitration within the scope of the ECT.[22] Spanish investors are not entitled to file an arbitration claim against Spain and they shall only have recourse to the national courts. This criteria has been highly criticised within the sector.

---

18 Resolution of the Constitutional Court of 270/2015 of 17 December.

19 Among others, Resolution of the Supreme Court of 1272/2016 of 1 June (RJ 2016\3477).

20 *Charanne BV and Construction, Investments SARL v Spain* (Case 062/2012).

21 *Eiser Infrastructure Limited and Energía Solar Luxembourg SARL v Spain* (Case ARB/13/36).

22 Nevertheless, if a company evidences that it is controlled by foreign investors, even if the company has the nationality of such a state, that company can file an arbitration claim against the state due to its foreign control.

3.5 Regulatory changes summary

Year	Law	Regulation	Incentives for renewable energies
1997	Law 54/1997	Liberalisation of electricity sector.	Differentiation between Ordinary and Special Regimes.
2004	RD 436/2004	Methodology to calculate Special Regime incentives.	(i) FiT (ii) Market price + incentive+ FiP
2007	RD 661/2007	New methology to calculate Special Regime incentives.	(i) FiT (ii) Market price
2010	RD 14/2010	Haircuts to the Special Regime incentives.	(i) FiT (limited incentivised number of hours for solar PV) (ii) Market price
2013	RDL 9/2013 Law 24/2013	Specific Remuneration Regime. End of the Special Regime.	(i) Incentive to investment (ii) Incentive to the operation
2014	RD 413/2014 Order 2 IET/1045/2014	Methodology to calculate the specific remuneration. Regulation of parameters.	(i) Incentive to investment (ii) Incentive to the operation

3.6 Self-consumption in Spain

Finally, it is necessary to consider the Spanish 'self-consumption' issue in the solar PV sector, which is regulated under Royal Decree 900/2015, 9 October, regulating administrative, technical and economic types of electricity supply and generation with self-consumption. The Royal Decree is applicable to any facility connected to the electricity grid; however, isolated installations are not covered.[23]

The Royal Decree distinguishes between two types of self-consumption facilities:

- Supply with self-consumption: facilities with a capacity of less than 100 kW

23 Isolated installations are those which are not connected to the grid.

and whose electricity exported to the grid is not remunerated. These facilities are only for self-consumption and there is only one party: the consumer.[24]

- Generation with self-consumption: facilities with a capacity of up to or more than 100 kW and whose electricity exported to the grid is remunerated. They are registered with the Register of Installations for the production of electricity and there are two parties: the generator and the consumer.[25]

Remuneration for the excess of electricity generated can only be received if the self-consumption facility meets the criteria of the second example above. Furthermore, facility owners must comply with the procedure established in Royal Decree 900/2015, including the procedure to obtain the licences required to participate in the market.[26]

4. Practicalities from a transactional standpoint

4.1 Acquisition of operating PV facilities

In the past few years the sale and purchase of operating PV projects in Spain has been increasing despite recent cuts to regulatory tariffs. This kind of transaction is usually carried out by one of two means: share deals, where the company operating the PV project is the object of the transaction; and asset deals, where the PV project, the relevant contracts and the administrative licences and authorisations are the objects of the transaction.

In practice, share deals occur more often than asset deals given that they present a significant regulatory advantage. Since the ownership of the asset would not vary (as the company would remain owner of the project and as titleholder of the administrative licences and authorisations), there is usually no need to obtain further authorisations and approvals from the Public Administration or from third parties for the change of ownership that would occur in the shareholding structure of the company. However, in an asset deal, the transfer or replacement of these authorisations and approvals would be necessary, thereby delaying the completion of the transaction and running the risk of not obtaining all the consents or approvals required. Consequently, while in share deals the signing and closing of the transaction is often simultaneous, in asset deals it is often necessary to establish an interim period between signing and completion of the transaction in order for the parties to comply with all the applicable regulatory requirements.

4.2 Real estate

In Spain it is fairly typical for PV facilities to be built on rural land made up of a number of different plots. Such land usually belongs to private individuals, although on rare occasions the company operating the PV facilities is also the landowner.

It is important that companies operating the PV facilities hold valid and

24 It corresponds to the modality of supply with self-consumption of Article 9.1(a) of Law 24/2013.

25 It corresponds to the modalities of self-consumption of Article 9.1(b) and (c) of Law 24/2013.

26 Royal Decree 900/2015 also includes the payment of a fee for the access to the grid, for those facilities connected to the grid, known as 'sun tax'.

sufficient occupancy titles to use all the plots on which the PV facilities are located (not only in relation to the PV plant, but also in terms of any associated infrastructure). In Spain, occupancy titles are usually granted by means of lease agreements and surface rights agreements entered into with the owner of the plots. In PV projects under construction the contractor usually undertakes the actions needed to formalise the lease agreements and/or surface rights agreements with the owner of the plots on behalf of the company, up until their registration with the appropriate Land Registry, prior to the beginning of the PV facility's operations.

Given that the PV facilities are conceived and built to be run on a long-term basis, the occupancy titles should, at a minimum, regulate the following: the term that covers, at least, the regulatory life span of the PV facility; and the possibility of extending the term of the agreement at the discretion of the owner of the facility, once the initial term of the agreement has come to an end.

In practice, surface rights are as common as lease agreements and both titles are perfectly valid in order for a company to occupy plots. However, the following differences regarding the requirements for the validity of the titles and the consequences once the agreement has come to an end should be considered.

(a) *Validity of the occupation title*

Under Spanish law surface rights are only deemed to be validly created if they are formalised in a notarial deed and registered with the Land Registry. Nevertheless, surface rights are often not notarised and registered in order to avoid triggering a stamp duty liability. If the surface rights have not been granted by means of a notarial deed and have not been registered with the Land Registry, any third parties interested in acquiring the plot (and even the owner of the plot) could potentially question if the company making use of the plots has a valid title to occupy the relevant plots subject to the surface right.

In order for lease agreements to be valid, their formalisation in a private agreement is considered sufficient and having a notarial deed or registration with the Land Registry is not legally required. However, registration of lease agreements with the Land Registry is very useful since it protects the company's rights *vis-à-vis* third parties who might seek to challenge the existence of a lease agreement when acquiring the plot. It should be noted that if the landlord is an entrepreneur and the lease agreement is formalised in a notarial deed, such formalisation would also be subject to stamp duty. The taxable amount would be the value of the lease agreement and the applicable tax rate would depend on the region in which the real estate asset is located (generally varying between 0.5% and 2%).

(b) *Ownership of the facilities*

Once a lease agreement for a plot of land reaches its expiration date, the facilities built on the plot will continue to be the property of the lessee. At this point the lessee should negotiate the extension of the agreement or remediate the leased plot.

In the case of surface rights, which, in accordance with Spanish law, cannot have a term of more than 99 years, once the agreement reaches its expiration date, all the facilities built on the plot become the landowner's property. However, it is common practice to negotiate and agree on clauses with the owner in order to avoid the application of this rule. In most cases surface rights agreements establish that all the facilities built on the land will continue to be owned by the lessee after the expiration date.

4.3 Land planning and environment

With regard to solar PV facilities' sales and purchase transactions, from a land planning and environment standpoint, the legal analysis is mainly focused on the review of the corresponding authorisations and licences granted by the relevant local and regional administrations which are mandatory in order to legally operate the projects.

(a) Land planning permits

As mentioned in section 4.2, in Spain solar PV facilities are mostly built on rural land, which entails the prior grant of the relevant favourable resolution by the regional government (*calificación urbanística*) that authorises and settles the conditions of usage of the land.

Furthermore, operation of the solar PV facilities requires the following land planning licences granted by the local government (city council) which are regulated by the specific by-laws of the Spanish municipalities:

- the municipal works licence (*licencia de obras*), which must be granted prior to the initiation of construction works, and guarantees that the project complies with the land planning regulations in force;
- the municipal activity licence (l*icencia de actividad*) which must be granted before setting up the specific installations, and confirms that the planned installations comply with the regulations in force;
- the municipal first occupation/utilisation licence (*licencia de primera ocupación*), which must be granted once the construction of the given installation is terminated, and confirms that the works executed comply with those initially authorised by the works licence; and
- the municipal functioning licence (*licencia de funcionamiento*), which should be granted once the installations are finished, and confirms that the activity to be undertaken corresponds to the one initially authorised by the activity licence.

The works licence and the activity licence are considered the 'initial licences'. The first occupation/utilisation licence and the functioning licence are considered the 'final licences', and evidence that the works and activities undertaken under the initial licences have been executed in compliance with the requirements therein. Therefore, holding the final licences usually involves that the initial licences had been duly granted. Both the initial licences, on the one hand, and the final licences, on the other, are simultaneously granted via a joint procedure.

The lack of any licence could be deemed as an administrative infringement and administrative penalties could be imposed as a result thereof (eg, order to rectify deficiencies, fines, suspension of the activity etc). In general terms, penalties arising out of such infringements depend on their classification as minor, serious or very serious offences. The amounts of the penalties vary from a minimum to maximum amount for each classification and depending on each regional regulation in that regard.

As mentioned in section 4.1 above, in an asset deal it is necessary to action the transfer of licences to the new owner of the solar PV facilities by means of a communication to the relevant city council signed by the former and the new owner. Generally, the mere communication has legal effect and no express approval is required by the city council (unlike the regulatory authorisations). In any event, the procedure depends on the specific local regulations provided by each municipality.

(b) ***Environmental permits and procedures***

In accordance with Spanish law, the implementation of solar PV facilities requires a prior favourable environmental assessment of the projects, which is provided in the relevant Environmental Impact Declaration (EID), granted by the regional government.

Furthermore, it is necessary to obtain an Integrated Environmental Authorisation (IEA) granted by the regional government or an environmental licence (*licencia ambiental*) granted by the city council, depending on the technical parameters of the project (power output, scale, etc) and their environmental impact. In that regard, the regional regulations provide the specific procedures required in each individual Spanish region.

Moreover, it is compulsory to obtain an Environmental Start-up Authorisation (*autorización de inicio de actividad*), which verifies that the activity has been implemented according to the IEA and/or the environmental licence. With regard to projects authorised by the city council by means of an environmental licence, this Environmental Start-up Authorisation could be substituted with the corresponding municipal land planning permit (ie, functioning licence – *licencia de apertura*).

As well as land planning permits, the lack of any environmental authorisation and/or environmental licences or the breach of their conditions and obligations could be deemed as an administrative infringement and administrative penalties could be imposed as a result thereof (eg, order to rectify deficiencies, fines, suspension of the activity, etc).

Likewise, as well as the land planning permits, in an asset deal it is necessary to action the transfer of EIA or environmental licences to the new owner of the solar PV facility by means of a communication to the relevant Administration. Generally, the mere communication has legal effect and no express approval is required by the city council and/or the regional government (unlike regulatory authorisations). In any event, the procedure depends on the specific local and regional regulations provided by each municipality and region.

Finally, with regard to environmental matters, in the sale and purchase of solar PV facilities' transactions, consideration is also given to the potential production of pollutant waste and discharge that may be subject to registration with the Register of Waste Producers, as well as other obligations such as the submission of annual declarations concerning such waste and discharge before the relevant

Administration. In general terms, solar PV facilities do not produce such pollutant waste and discharge which are normally subject to such registrations and periodic declarations, but it depends on the particular project and the appropriate analysis is required in that regard.

5. Power off-take arrangements

Also known as power purchase agreements or PPAs, these are long-term contracts by which the seller agrees to export a volume of energy into the electricity grid and the buyer agrees to consume that volume of energy at a specific price established in the contract.

In Spain, generators can opt between selling their energy in the wholesale market or through private contracts. Specifically, generators of renewable energy such as solar PV have the option, as we have seen before, to obtain a specific payment from the renewable auctions that are organised by the government.

In relation to private contracts, Article 24 of Law 23/2014 provides for 'physical bilateral contracts', which is supplemented by Articles 19 and 20 of Royal Decree 2019/1997, 26 December, which organises and regulates the electric power production market. These physical bilateral contracts correspond to the PPA concept. Through these contracts, generators receive a price agreed with the buyer at a discount to the market price.

Sale in the electricity forward market is an alternative to the sale of energy in the wholesale or spot market. In the forward market, two methods are distinguished: physical bilateral contracts, and financing contracts, for energy price risk hedging.

Bilateral contracts may be made directly between consumers and generators, between generators and marketers (energy trading companies) and/or as between marketers. Bilateral contract parties must be registered with the corresponding administrative registry, comply with prescribed technical requirements, and provide certain guarantees to the electricity system operator.

Solar PV PPAs are quite common in other European countries and in the United States, but they are unusual in Spain. This kind of contract involves a medium or long-term sale and purchase of solar PV energy directly from the generator. The developer or the owner of the solar PV facilities offers the energy that will be produced during the agreed term at a discount to the market price, since there is a guarantee of a stable and fixed income, something that is not guaranteed in renewable auctions.

One of the main problems for the implementation of PPAs in Spain is that the bilateral contracts that are signed do not usually have a term of more than one or two years, whereas the usual approach with PPAs is to use a much longer term, perhaps of 10 or 15 years.

The promotion of PPAs could result in a resurgence of the construction of solar PV facilities. This would help stem their decrease in number following the reductions in incentive payments. Likewise, PPAs improve the prospects for sellers securing financing contracts in order to construct and operate solar PV facilities, in addition to guaranteeing project income and profitability. PPAs should play a bigger role in Spain in the coming years. Companies are entering into PPAs not only for economic reasons but also for sustainability reasons.

6. The future for the solar sector in Spain

The short-term future of the sector lies in renewable auctions, in order to obtain the regulated payments. Generators may opt between building new facilities under the auction regime, or go to the wholesale market and obtain a market price for output which is, therefore, not regulated. It has been suggested that the auctions may also be applied to repowering projects and not only to new facilities, which would give a greater incentive for the production of renewable energy.

Another system to encourage the production and construction of renewable technologies projects, including solar PV facilities, is the promotion of PPAs already mentioned in the previous section. The difficulty that solar PV generators occasionally have in obtaining the specific payments available under the energy auctions increases the likelihood that PPAs will become a viable and interesting alternative for generators. Moreover, it is an interesting alternative for solar PV facilities that have already reached the level of a reasonable return on investment and, therefore, cannot obtain the specific payment anymore.

Self-consumption is another way to promote the construction of solar PV facilities in Spain. Nevertheless, it is crucial to develop the regulatory framework and improve the remuneration system for the excess of electricity generated, and offering improved arrangements for generators in relation to licences and procedural requirements.

Moreover, Spain is interested in promoting renewable energy and solar PV energy due to its commitment to meeting the environmental objectives set by Europe. Specifically, the 2015 Paris Agreement requires reducing CO_2 emissions in the atmosphere in order to ensure that any rise in the planet's temperature does not exceed 2°C by 2050. There is a key commitment: by 2020 at least 20% of the energy consumed in the European Union will need to come from renewable sources. The 2020 Climate and Energy Package sets specific 2020 targets for renewable energies, energy efficiency and CO_2 emission reductions:

- 20% cut in greenhouse gas emissions (from 1990 levels);
- 20% of EU energy from renewables; and
- 20% improvement in energy efficiency.[27]

In addition, the so-called 'Winter package' establishes new objectives by 2030. In particular, at least 27% of the energy consumed in the European Union should come from renewable sources by 2030.

The generation of renewable energy in Spain must be increased to meet the environmental European targets.[28] Furthermore, and as a consequence of these objectives, the population is increasingly aware of the need for renewable energy in the market, which explains the growing popularity of 'green energy'.

The so-called 'green energy' contracts guarantee that the energy supplied has a 100% renewable origin. The CNMC has the authority to issue certificates of guarantee of origin of electricity from renewable energy sources and high-efficiency

27 See the European Commission website at https://ec.europa.eu/clima/policies/strategies/2020_en.
28 In 2015, 16.4% of the energy consumed in Spain came from renewable sources.

cogeneration.[29] Marketers can buy certified energy packages of renewable origin in order to sign green energy contracts with consumers.

These contracts are closely related to PPAs. In fact, major consumers who want to guarantee a 100% renewable source of the energy they consume have the option to sign long-term green energy PPAs.

The future of the renewable energy sector and, in particular, the solar PV sector in Spain is uncertain. The European objectives to promote renewable energies mean that Spain has to promote renewable production either through the further promotion of energy auctions, or by facilitating the conditions for PPAs and the regulation of self-consumption.

Another way to achieve European renewable objectives is to enhance electricity interconnection. International electricity interconnections allow countries to connect to each other, so that energy can flow from the country where it is cheaper to produce it, to where it is more expensive to produce it, promoting the integration of renewable energies and ensuring security of supply and support between countries. The purpose of this economic principle is to lower the energy price in the country where it is more expensive to produce energy by increasing competition between generators.

Given the low level of interconnection between Spain and other European countries (the interconnection ratio of Spain with other European countries is below 5%, much below the 10% European target for 2020 and the 15% target for 2030), it is necessary to explore new mechanisms of interconnection. A complete interconnection between the Spanish electricity system and the North African electricity system would promote the achievement of a 100% renewable electricity system in the future. In this context, generation with solar PV and thermoelectric power would be crucial.

Spain is determined to accomplish the European targets in the coming years. However, the uncertainty regarding the current regulatory framework may discourage investors, an issue that could compromise the future development of solar PV energy.

---

29 Circular 6/2012 of 27 September.

United Kingdom

Elisabeth Blunsdon
Orrick, Herrington & Sutcliffe LLP

1. Introduction

Despite the United Kingdom being a rather damp collection of islands situated in the North Atlantic, known rather more for its abundant rainfall than its sunshine, it has seen nothing short of spectacular growth in the amount of installed solar generation capacity in recent years. In the period between 2012 and 2016 the installed solar photovoltaic (PV) capacity in the United Kingdom increased nearly sevenfold from 1.7 MW to 11.9 MW.[1] With an average load factor of 10.88 over the same period,[2] in 2016 solar PV[3] produced 10,420 GWh of electricity.

As in many other markets, the strong growth and development of the UK solar PV market in the last decade or so is inexorably linked to the rise, and fall, of government subsidy schemes. In 2002 the Renewables Obligation Scheme was introduced which brought into being a comprehensive support system for renewables in the United Kingdom, which was to last for 15 years and have a profound effect on the sector, not just solar PV.

The Renewables Obligation, or RO, imposed a quota obligation on licensed electricity suppliers, as described in more detail in the "Economic drivers" chapter. Initially, Renewables Obligation Certificates or ROCs (the tradeable certificates through which suppliers discharged their obligations) were issued on a straight one-to-one basis: 1 MWh of generated renewable electricity was given one ROC. However, it became apparent that while this level of subsidy was adequate for some of the less risky, lower-cost renewable technologies such as onshore wind and the co-firing of energy crops, new, more capital-intensive technologies such as larger-scale (≥50 kW) solar PV, could not compete. As a result, following consultation in 2007/2008, the concept of banding was introduced, with the aim of providing certain technologies with more support. Solar PV at that time was classed as an 'emerging technology' in recognition of the fact that the United Kingdom's climate means that irradiation levels are low and the existing technologies in the first decade of the century simply did not yield sufficient electricity to be economically viable. As an emerging technology solar PV was eligible for two ROCs per MWh produced, along with such technologies as Advanced Conversion Technology, dedicated biomass with CHP and geothermal.

1 Digest of United Kingdom Energy Statistics (Dukes) 2017, Chapter 6, Table 6.4.
2 Dukes 2017, Chapter 6, Table 6.5.
3 Solar PV is by far the most common solar technology in the United Kingdom.

The coming into force of ROC banding certainly encouraged increased levels of investment, but it also coincided with the start of the sharp fall in the cost of installing solar PV equipment, as described in more detail in the "Economic drivers" chapter. As a result, by 2011 the UK government was becoming concerned that the levels of subsidy being provided to larger-scale solar PV projects was becoming too expensive, so ROC support was reduced to 1 ROC/MWh.

The entire RO scheme was closed to newcomers as of 31 March 2017 (save for certain projects which could demonstrate that they had incurred a level of financial commitment prior to a fixed date and therefore were eligible for a 'grace' period). However, the closure date for solar PV was brought forward to 31 March 2016 for small-scale solar PV (≤50 MW) and 31 March 2015 for large-scale solar PV (>5 MW). These rather abrupt and accelerated changes to the RO scheme caused a massive rush to construct and commission projects by the revised deadline, and put considerable strain on distribution network operators (who had to sign off the grid connection on each project) and on Ofgem as it was deluged with a wave of accreditation applications.

Feed in Tariffs (FiTs) were introduced in 2010 to support the installation of small-scale renewables.[4] The UK scheme is aligned with the principles set out in the "Economic drivers" chapter and provides payments to generators which generate electricity from eligible renewable sources for a period of 20 years, starting from the accreditation of the facility.[5]

FiTs are aimed at smaller projects (the maximum limit on capacity is 5 MW) and the majority of the installations that have been accredited are below 50 kW, including many small-scale domestic rooftop solar installations. The value of the FiT for solar PV has decreased significantly in recent years, following numerous government interventions including the introduction of digression mechanics which reduce generation tariffs over time[6] and deployment caps which limit the amount of capacity that can be accredited in each quarter, which has resulted in a FiT 'queue' of projects waiting to be accredited. While the solar PV FiT may still be attractive to domestic consumers installing rooftop panels, it is increasingly less attractive for ground-mounted and large rooftop schemes.

The way that changes to the FiT scheme have been introduced in recent years has been challenged in the courts. In particular, the changes in respect of small-scale solar were successfully challenged by an environmental pressure group, Friends of the Earth, and two solar companies, Solarcentury and HomeSun, which jointly made an application for judicial review as the proposed changes would have been retrospective. The application was successful, which meant that any eligible solar PV systems installed before 3 March 2012 received the higher rate of FiTs at 43.3p/kWh rather than the proposed rate of 21p/kWh.

For completeness, we should also mention the Renewable Energy Guarantee of Origin or REGO, which was introduced as a result of an EU requirement that all

4 The statutory basis for FiTs is contained in Sections 41–43 of the Energy Act 2008, and further detail is set out in the various statutory instruments promulgated under those sections

5 Accreditation is managed by the energy regulator Ofgem for larger projects under the ROO-FiT scheme.

6 Tariff rates, and associated digression, have been set up to March 2019. Beyond that we currently have no visibility as to what, if any, FiTs will be available for solar PV.

member states must implement a scheme for the issue of certificates that evidence the production of renewable energy.[7] One REGO is issued per MWh of renewable energy generated, and the scheme is administered by Ofgem. The main use for REGOs in the United Kingdom is in Fuel Mix Disclosure (FMD), which is a requirement on licensed suppliers whereby they must be able to prove to their customers how much of the electricity that they supply is from renewable sources. REGOs currently have no monetary value but are usually transferred with the electricity sold under the PPA to the offtaker so that they can be either used for FMD or, if the offtaker is a corporate entity, they may be retired.

As the RO scheme entered its demise, the UK government was implementing its Energy Market Reform or EMR process which entailed significant changes aimed at, among other things, the decarbonisation of electricity in the United Kingdom. As part of EMR, the Contract for Difference with Feed in Tariff (the CfD-FiT) was introduced, as further described in the "Economic drivers" chapter. The CfD-FiT is a contract between a generator and the Low Carbon Contracts Company (LCCC), a private limited company wholly owned by the Secretary of State for Business, Energy and Industrial Strategy (BEIS). As described in the "Economic drivers" chapter, CfD-FiTs are allocated via an auction. Initially it was thought that auctions would be held on a regular basis, however government policy has changed, partly over concerns about costs. As a result there have been only two auctions since the start of the scheme (the first launched in late 2014 and the second in 2016). A further auction will take place in 2019, with specific details to be provided by LCCC in due course. Crucially for solar PV, neither the second nor the third auction was or will be open to solar PV. In the first allocation round, five solar projects were awarded a CfD-FiT, three with a strike price of £50 per MWh for the delivery year 2015/2016 and two with a strike price of £79.23 per MWh delivery year 2016/2017. Only the 2016/2017 delivery year projects have been developed, leading to speculation that the strike price for the earlier delivery year was too low for the projects to be economically viable.

Therefore, as things stand, there is no government subsidy available for solar PV projects and no realistic prospect of any future subsidies. There are no targets for levels of renewable generation or any other policy drivers aimed specifically at renewables or solar PV. As always in such situations, and depending on one's outlook, what to some is the death of the industry to others is the start of greater opportunity. The future of solar PV is discussed in greater detail in section 6 below.

2. Structure and industry regulation

The Electricity Act 1989 (as amended) (the 1989 Act) provides the legislative framework for the electricity sector. The 1989 Act established the current structure of the industry, and established specified activities such as generation, transmission, interconnector operation, distribution and supply, replacing, by privatisation, the previous state-owned, vertically integrated monopoly. The 1989 Act has been amended significantly since it was first enacted, with many such amendments relating to the implementation of rules relevant to the renewables sector.

---

7 The generic term in the wider European Union is 'Guarantee of Origin' or 'GoO'.

Initially, post-privatisation the electricity market was run on a pool model.[8] Following competition concerns on the part of the regulators,[9] the pool was replaced in 2001 by the New Electricity Trading Arrangements (NETA) in England and Wales and subsequently extended in 2005 to Scotland (and therefore the whole of mainland Great Britain) by the British Electricity Trading and Transmission Arrangements (BETTA).[10] NETA introduced the concept of a bilateral market whereby market participants trade directly with each other rather than through a pool. Crucially for renewables, generators and suppliers are required to ensure that their overall positions in each Settlement Period[11] are balanced (ie, overall inputs equal overall outputs). If a participant is not balanced, the system operator takes action to ensure the market overall is physically balanced, and the market participant bears the financial consequences of being out of balance by paying (or being paid) a settlement amount. These settlement amounts are designed to incentivise balanced positions and therefore a long position will result in a deemed sale to the system at a low price and a short position will result in a buy from the system at a high price. The mechanics of the balancing mechanism are outside the scope of this chapter, but the important point for solar PV generators is that their output is intermittent, and relatively unpredictable. In addition, solar PV facilities are, in the general scheme of things, of very small capacity. As a result, solar PV projects will always need to pass imbalance risk to another party that is better placed to manage the risk, which will always represent a cost to the project. The management of imbalance risk is discussed in more detail in the "Securing project revenues" chapter.

The small size of solar PV projects (in relation to the more traditional thermal assets on the system) means that they are almost always connected to the distribution network rather than the transmission network,[12] and are thus characterised as embedded generators. As a result of being embedded, certain benefits relating to use of system charges are available to solar PV projects, either by way of rebates on use of system charges, or payments from the transmission system operator. Historically these payments have been substantial, but the regulator is currently reviewing the situation and the general view is that so-called 'embedded benefits' will decrease over the next few years. Further details on the mechanics of connection is set out in the "Securing project revenues" chapter.

2.1 Market participants

As described above, the 1989 Act establishes the *dramatis personae* in the UK industry,

---

8 The pool model is still seen in markets such as Nordpool and some of the US markets.

9 The nature of the pool model means that generators must bid a price into the pool on a day ahead basis. The market operator then collates those bids against volumes required and publishes the pool price that all generators receive, regardless of their original bid. There was an ongoing, and never proven, concern on the part of the regulator that generators were somehow manipulating the pool price through collusion in their bidding strategies.

10 Northern Ireland has its own electricity trading arrangements, known as the Single Electricity Market (SEM) which combines the electricity markets of Northern Ireland and Eire in a single market. Note, however, that the arrangements for renewables are set by the UK regulator, not under the SEM.

11 In the UK market each Settlement Period is 30 minutes starting on the full or half hour – see the Balancing and Settlement Code for further details

12 Further details on the mechanics of connection is set out in the "Securing project revenues" chapter.

with the same basic categories as other jurisdictions, namely generators that produce power and on-sell it into the market, suppliers that buy power from the market or generators and supply it to end users and entities that operate infrastructure such as the transmission system or interconnectors. Each of the defined activities requires authorisation, which either takes the form of a licence, or an exemption from the requirement to hold a licence. Note, for completeness, that if an entity is licensed as a transmission network, distribution network or interconnector operator it cannot also be licensed as a generator or a supplier. Further, the EU unbundling rules will also apply, meaning that certification as an independent operator is required for the entity concerned.[13] While this would not appear to impact solar PV projects directly, care should be taken where multiple assets across the European Union are held by a common owner such as a structured fund.

The operator of any solar PV project will be a generator under the 1989 Act. However, to date most solar PV projects are exempt from the requirement to hold a generation licence as they have an installed capacity of less than 50 MW.[14,15] As well as relieving such projects from the regulatory burden of holding the licence itself, not being licensed also means that there is no requirement to sign up to the various industry codes that govern generation, the principal codes being the Grid Code, the Connection and Use of System Code and the Distribution Code. As a result there is also no requirement on the solar PV project to provide Mandatory Ancillary Services.[16] Conversely, there is no opportunity to provide ancillary services either. Given that the nature of the system is changing as large thermal plant is decommissioned and smaller, embedded facilities increase in number and the value of ancillary services is increasing, there are currently various initiatives under way looking at how ancillary services are procured and paid for. This process may well prove to be beneficial for solar PV projects as a potential additional revenue stream.

It is worth noting that there is a trend towards larger solar PV farms emerging, with plans reportedly being made for projects of up to 350 MW.[17] The cessation of subsidies for solar PV is certainly driving the move towards larger projects, but the lack of suitably large sites may prove to be a limiting factor in the development of many such behemoths.

In terms of the entities that can buy power in the UK market, or, to put it the other way around, to whom a generator can sell its output, traditionally there were two options: a licensed supplier or an energy trader. A licensed supplier is, clearly, licensed under the 1989 Act, a trader is not, however both will be party to the

13 See Sections 10A et seq of the 1989 Act.

14 The Electricity (Class Exemptions from the Requirement for a Licence) Order 2001 sets out the full set of standard exemptions and includes the exemption whereby a generating facility with declared capacity of less than 50 MW does not require a licence.

15 It is common in UK PPAs to see a warranty that the operator of the solar PV will ensure that it remains exempt from the requirement to hold a licence.

16 Mandatory Ancillary Services enable the system operator to require the provision of certain services which ensure the stability and security of the system. The relevant provisions are in the Connection and Use of System Code, see www.nationalgrid.com/sites/default/files/documents/Complete%20CUSC%20-%20%2015%20November%202017_0.pdf.

17 Hive Energy and Wirsol have announced a potential 350 MW project in Kent, England.

Balancing and Settlement Code or BSC.[18] The BSC governs the operation of the UK market and is administered by Elexon in its role as the BSC Co. The balancing mechanism described briefly above is set out in the BSC, together with the operational rules and requirements that govern the import and export of electricity onto and off the system. The operator of a solar PV project could become a signatory to the BSC and make inputs into the system itself, but the obligations that come with being a BSC signatory are sufficiently onerous that this almost never happens. Therefore the project needs a suitably qualified counterparty to which it can sell its output – and that entity will need to be a signatory to the BSC or have the ability to pass obligations to an entity that is a signatory.[19] The answer is, of course, the signing of a PPA, as further described in the "Securing project revenues" chapter.

2.2 Regulator

The 1989 Act contains the statutory basis of the regulatory regime in the United Kingdom. The Gas and Electricity Markets Authority (GEMA) is the regulator of the gas and electricity industries, and it operates through the Office of Gas and Electricity Markets (Ofgem). GEMA's duty under the 1989 Act is set out in Section 3A as follows:

1 *The principal objective of the Secretary of State and the Gas and Electricity Markets Authority (in this Act referred to as 'the Authority') in carrying out their respective functions under this Part is to protect the interests of existing and future consumers in relation to electricity conveyed by distribution systems or transmission systems ...*

(1A) those interests of existing and future consumers are their interests taken as a whole, including—

(a) their interests in the reduction of electricity-supply emissions of targeted greenhouse gases; ...

(b) their interests in the security of the supply of electricity to them; and

(c) their interests in the fulfilment by the Authority, when carrying out its functions as designated regulatory authority for Great Britain, of the objectives set out in Article 36(a) to (h) of the Electricity Directive.

The principal objective has been amended over the years, but as can be seen, the reduction of greenhouse gases is the first on the list, even above security of supply. Whether it remains there, time will tell.

In practice, there is little contact between the industry and GEMA itself. Ofgem is the principal point of contact for all industry participants. Ofgem not only administers and regulates the energy markets, it is also responsible, through its e-serve division, for implementing the government's environmental and social programmes. As a result, participants in the solar sector have a good deal of contact with e-serve as it is responsible for administering the RO, REGO and FiT schemes, including accreditation, the issue of ROCs and REGOs, management of FiT arrangements and the maintenance of the registers that are associated with the various environmental programmes.

---

18 The BSC and associated procedures and rules can be found at www.elexon.co.uk/bsc-and-codes/balancing-settlement-code/consolidated-bsc/.

19 The latter qualification comes about due to the relatively recent phenomenon of the corporate PPA, described in more detail in the "Securing project revenues" chapter.

Ofgem is constantly active in developing the energy markets and as a result issues numerous consultations on potential changes to the industry. It is, however, worth noting that any changes to government policy are driven by BEIS and any changes to system operation are dealt with by National Grid as system operator. As a result it is necessary to monitor consultations from all these entities, not just Ofgem.

3. Authorisation requirements

There are no requirements or restrictions on the ownership of solar PV assets in the United Kingdom and there are no restrictions specified as to who may invest in the UK solar PV market.[20] If the operator of a solar PV facility is a licensed generator it will need to have an address for service of documents in the United Kingdom.

4. Environmental and planning matters

4.1 Environmental and planning

Planning in the United Kingdom is under the jurisdiction of local authorities unless the proposed development qualifies as a nationally significant infrastructure project, when a development consent order (DCO) is required under the Planning Act 2008. DCOs are awarded by the Secretary of State after public examination and in this context would be required for any onshore generating facility with a capacity of more than 50 MW. Planning consent for projects with an installed capacity of 50 MW or less is granted under the Town and Country Planning Act 1990 by the relevant local authority.

A project will require planning permission if it does not qualify as a permitted development pursuant to the Town and Country Planning (General Permitted Development) (England) Order 2015. Given that most commercial solar plants are standalone ground-mounted developments, planning permission will usually be required.

The process of submitting a planning permission application will involve the project developer conducting numerous studies and considering various factors such as location and design plans, connection details, flood and ecological assessments, and environmental impact assessments to assess how the project fits in with the current use of neighbouring land. The government has published guidelines for solar projects[21] to ensure proposals are appropriately sited, proper environmental considerations such as landscape and visual impact are considered and that local communities to have a say in the process.

Planning decisions may be subject to appeal, and it is only once all appeal proceedings have concluded that a developer will actually start to construct a project.

Planning permission will contain conditions with it that will need to be satisfied at different stages of the project, for example pre-commencement planning conditions will be need to be satisfied and discharged prior to the commencement of

---

20 Note, however, the unbundling rules referred to above which may affect the structure of ownership were there are multiple assets held within a common group.

21 UK Solar PV Strategy Policy Paper, GOV.UK, 4 April 2014, www.gov.uk/government/uploads/system/uploads/attachment_data/file/302049/uk_solar_pv_strategy_part_2.pdf.

construction on site. The project will also be subject to various ongoing conditions throughout the operational life of the development.

5. Power offtake arrangements

The key role played by the power purchase agreement or PPA in any project is described more fully in the "Securing project revenues" chapter. Traditionally, PPAs in the United Kingdom have been entered into between generators (as seller) and licensed suppliers or electricity traders (as buyer) and have provided for the sale of electricity generated by the plant along with the associated ROCs and REGOs. The end of the RO, the removal of other subsidies and a general contraction in the number of market participants following the credit crunch, has meant that securing a PPA has become more difficult in recent years.

There has, however, been an increase in the number of corporate PPAs being entered into. Similar to the structure of a traditional PPA, corporate PPAs are long-term contracts where a generator agrees to sell its electricity to a business, rather than a utility. As the reduction of carbon emissions and sourcing renewable energy has gradually made its way into many corporations' long-term sustainability plans, the corporate PPA not only assists this goal but can also enable corporations to fix the price of their electricity on a long-term basis, providing financial benefits where rising prices may otherwise work to their disadvantage. For a more detailed analysis into offtake arrangements and corporate PPAs, see the "Securing project revenues" chapter.

6. The future

The removal of subsidies from solar PV has undoubtedly had an effect on the market – it would be extraordinary if it had not. Latest figures show a dramatic fall of 56% in investment in solar and wind power in 2017,[22] marking the second year in a row that financing has fallen in the United Kingdom, despite a global upward trend of a 3% increase. Now that the RO scheme has finally closed to new entrants,[23] there will doubtless be a period of consolidation as increasingly large portfolio sales of operational solar assets, often with mixed subsidies, are bought and sold. Indeed, this process is already well under way. The drop in new operational assets coming to the market has shifted the solar M&A market from a buyer's to a seller's market. The scale of assets and portfolio sizes has also changed the pool of potential buyers increasingly to pension funds and infrastructure funds.

That said, rumours of the demise of solar PV in the United Kingdom are definitely greatly exaggerated. Another few years of subsidy would have been helpful, a more gentle progress to a subsidy-free world would have been less stressful, but ultimately there is still a market for solar PV in the United Kingdom, and where there is a market, people will find a way to make it work, whether or not the government is providing support. As has already been mentioned, one solution is to scale up, and

---

22 "UK green energy investment halves after policy changes", The Guardian, 16 January 2018, www.theguardian.com/business/2018/jan/16/uk-green-energy-investment-plunges-after-policy-changes.

23 The last of the grace periods expired on 31 March 2018.

take advantage of economies of scale. On-site private wire arrangements are also an option, particularly if storage can be economically co-located to allow for smoothing out of the output curve. Storage in itself could also be a game changer if the regulatory and commercial regime for storage can be clarified.

In September 2017, to much governmental trumpeting, the 45-acre Clayhill solar farm in Bedfordshire was officially opened as the United Kingdom's first subsidy-free solar farm. It includes five battery storage units and the government is confident that there will be more such projects being commissioned in the future. Right now, there does not seem to be any reason to think that the government's confidence is misplaced.

United States: California

David Spielberg
Orrick, Herrington & Sutcliffe LLP

1. Introduction

Although there were significant projects earlier, the 'modern' solar era in California began in 2002 with a series of actions by the California Public Utilities Commission that were the beginnings of California's Renewable Portfolio Standard. From those modest beginnings (a goal of 1% of energy usage in 2003 from renewable resources of all types), the California solar industry has expanded at a rapid rate and is the fastest growing part of the fast-growing renewable energy industry in California. In 2016, renewable energy accounted for 29% of the energy consumed in California.[1] Solar energy accounted for 32% of this total, and 9.3% of the total energy from all sources consumed in California in 2016.[2] On peak days, over 16% of the state's total energy use is now supplied just from solar resources.[3] The rapid growth of solar energy and its unique production curve – producing the most power during the middle of the day – have literally changed the shape of the California energy markets and the load profile for California's utilities by shifting peak demand to later in the day and creating a very sharp increase in demand in the early evening.[4]

The first solar projects in California were solar thermal projects,[5] but for the last several years the solar energy market in California has been dominated by photovoltaic solar projects. As of the end of 2016, of the solar energy delivered to the California Independent System Operator (CAISO)-controlled grid grid (which covers approximately 80% of the total load in California), 17,235 GWh came from photovoltaic solar projects, and 2,548 GWh came from solar thermal projects. (Imported solar energy from both photovoltaic solar and solar thermal resources accounted for an additional 3,297 GWh.)[6] California's operating renewable energy capacity, which includes both facilities inside California and projects outside of California that deliver into California, was approximately 27,500 MW as of June

1 See www.energy.ca.gov/renewables/tracking_progress/documents/renewable.pdf.

2 Official estimates of total renewable and solar production vary. Compare www.energy.ca.gov/renewables/tracking_progress/documents/renewable.pdf to www.energy.ca.gov/almanac/electricity_data/total_system_power.html (total renewable energy 25.45%, total solar energy 8.11%). The percentages are significant in either case.

3 See, eg, www.caiso.com/market/Pages/ReportsBulletins/DailyRenewablesWatch.aspx for 13 May 2017 and 14 May 2017.

4 This is the so-called 'duck curve' because a graph of the effect of increasing levels of solar generation resembles a duck. See www.caiso.com/Documents/FlexibleResourcesHelpRenewables_FastFacts.pdf.

5 The Luz Solar Energy Generating Stations (SEGS) were built in multiple phases beginning in 1986. The generating facilities are still operating today, see www.gosolarcalifornia.ca.gov/about/gosolar/california.php.

6 See www.energy.ca.gov/almanac/renewables_data/solar/index.php.

2017, including 6,000 MW of self-generation (distributed generation on the customer's side of the utility energy meter). Of this capacity, 14,700 MW including 5,800 MW of self-generation, was photovoltaic solar. Solar thermal projects represented only 1,300 MW of the total.[7]

Because of its dynamic nature, any discussion of the solar industry in California can only be a snapshot of the industry at a given time. This chapter describes the current structure of the solar energy market in California, the market participants, the legislative and regulatory framework and the typical ways that solar energy is purchased and sold in California. It is reasonable to expect that the basic structure of the industry will remain intact for the foreseeable future. However, the industry, both technology and regulation, can and will change frequently, so the reader is advised to check the referenced sources for updates.

2. Structure of the California market[8]

The solar energy industry in California exists within the framework of the broader energy market. The current market structure is a result of the deregulation of the California electric energy industry in the late 1990s. Under deregulation, California's investor-owned utilities were required to divest at least 50% (some chose to divest 100%) of their generation assets in order to create a competitive electric generation market. The utilities continued to own their transmission and distribution assets, but a new entity, CAISO, was created to manage the transmission grid, ensure open access to transmission, and generally maintain grid reliability.[9] There are about 30 million customers in the area covered by the CAISO-administered grid, which includes about 80% of the total load in California.[10] The rest of the state mostly receives its electricity from publicly owned utilities and public power agencies that manage their own transmission and generation assets.[11]

The CAISO market is a nodal market in which prices at each node (the locational marginal price or LMP) are set through an auction process on a day-ahead and a real-time basis. Market participants can either bid a specific quantity of energy at a stated price (an economic bid) in either or both markets or elect to deliver energy (often the entire forecasted output of the plant) at whatever price clears the market (a self-schedule). Based on the projections of demand and the results of the CAISO's full network model, which produces prices for the production or delivery of energy at particular nodes, the CAISO awards successful participants schedules for the amount of energy they are authorised to deliver to the transmission system in the day-ahead or real-time markets.[12] Because solar energy projects, particularly photovoltaic solar, are intermittent resources, there will inevitably be differences (imbalances) between

---

7 See www.energy.ca.gov/renewables/tracking_progress/documents/renewable.pdf, p3.

8 The following discussion is a greatly simplified description of a portion of the CAISO market that most often affects sellers and purchasers of solar energy. The CAISO tariff that governs this market, with exhibits, runs to almost 3,000 pages.

9 See www.caiso.com/about/Pages/OurBusiness/UnderstandingtheISO/default.aspx.

10 See www.caiso.com/about/Pages/OurBusiness/The-ISO-grid.aspx; and www.caiso.com/about/Pages/OurBusiness/How-power-flows-in-California.aspx.

11 See www.caiso.com/about/Pages/OurBusiness/The-ISO-grid.aspx.

12 See www.caiso.com/market/Pages/MarketProcesses.aspx.

the schedules that are awarded by the CAISO and actual generation from a facility. Charges or payments (depending on whether there is a shortfall or a surplus) for these imbalances are allocated by the CAISO to the party responsible for scheduling the energy from the facility (the scheduling coordinator). The risk of imbalances is often allocated between the parties in power purchase agreements.

The renewable nature of solar energy is tracked through the issuance of renewable energy certificates (RECs). RECs are issued by the Western Renewable Energy Generation Information System (WREGIS) and are, among other things, used by purchasers of solar generation to confirm the renewable character of the energy they are purchasing.[13] A generating facility must be certified by the California Energy Commission as an eligible renewable energy resource in order to be able to register with WREGIS. Both photovoltaic solar and solar thermal generating facilities are eligible for certification. Once a solar generator registers with WREGIS, WREGIS will track the energy generated by its facility and will create a certificate with a unique serial number for each megawatt-hour generated.[14] RECs can be kept bundled with the renewable generation with which they are associated, or they can be unbundled and sold separately from the energy. As a general rule, RECs bundled with eligible renewable energy are more valuable.[15] Load-serving entities and retail electric providers that have obligations to comply with the Renewable Portfolio Standard can do so by acquiring RECs and then retiring them.

3. Market participants

Broadly speaking, there are three categories of market participants in California: sellers of energy and other generator-based products, such as capacity and ancillary services; purchasers of energy and other products; and transmission and distribution entities that own and/or operate the transmission and distribution systems over which energy and other products flow.

Sellers of solar energy include both the owners of solar generation resources and resellers, including utilities and other load-serving entities and energy service providers. Solar generation resources can range in size from residential rooftop systems (that might be owned by investors or the homeowner) of 10 kW or less to generating facilities over 500 MW in size and occupying several square miles.

Purchasers of solar energy include investor-owned utilities (IOUs), publicly owned utilities (POUs) and other load-serving entities that resell the energy to their customers, electric service providers (ESPs) purchasing energy for resale to their customers, and end users of the energy, including retail, commercial and industrial customers. Because of California's Renewable Portfolio Standard (described below), California's three major IOUs – Pacific Gas & Electric, Southern California Edison and San Diego Gas & Electric – which collectively serve about three-quarters of California's electricity demand,[16] have until now been the state's largest purchasers of

---

13 For example, many utility power purchase agreements provide for a lower price (or zero) unless RECs corresponding to the energy purchased are delivered within a specified period of time.

14 See www.wecc.biz/Pages/101.aspx (see 'WREGIS 101').

15 See the discussion in section 4.1 below.

16 See www.energy.ca.gov/pou_reporting/background/difference_pou_iou.html.

solar energy. Procurement by the three IOUs is expected to be more limited in the coming years as they have reached or exceeded their renewable procurement targets and do not anticipate needing additional renewable generation until as late as 2030.[17]

California has more than 40 publicly owned utilities that are operated by local governmental authorities, including cities, irrigation districts, utility districts, public power authorities and community choice aggregators (described below). POUs collectively serve about one-quarter of California's electricity demand.[18] POUs are also subject to California's Renewable Portfolio Standard requirements and as a result are active purchasers of solar energy under long-term contracts. Some of the larger POUs include the Los Angeles Department of Water and Power, the Sacramento Municipal Utility District, the Southern California Public Power Authority, the Northern California Power Authority, the Metropolitan Water District and the Imperial Irrigation District.[19]

Community choice aggregators (CCAs) are a relatively new type of entity that has emerged as a significant purchaser of solar (and other renewable) energy. CCAs are joint powers authorities (an agency formed by two or more other agencies) that are created within the service territory of existing IOUs in order to purchase power for electric customers within the CCA's boundaries. The IOUs continue to own the transmission and distribution infrastructure and remain responsible for billing and collections. Customers are automatically enrolled in a CCA covering their area, but may opt out at any time. CCAs were originally created in the legislation following the energy crisis in 2002 in order to provide an alternative to direct access, which was being frozen.[20] They were not used, however, until Marin County and several cities in Marin County formed a CCA in 2008 in order to provide a greater percentage of renewable energy than was being provided by the incumbent utility. This pattern has been followed by other CCAs, which generally offer a generation mix with a higher percentage of renewable energy than the IOUs are required to provide under state RPS requirements. As of mid-2017, there were eight CCAs in California serving approximately 1,576,000 customers,[21] and another 15 communities representing over 15 million people were considering forming CCAs.[22] With a significant portion of utility load shifting to CCAs, CCAs appear likely to be significant purchasers of utility scale renewable generation, including solar energy, in the coming years.[23]

Electric service providers (ESPs) are non-utility entities that provide electric service to retail customers within an IOU's service territory.[24] ESPs must meet certain qualifications and register with the California Public Utilities Commission (CPUC)[25]

---

17 See https://protect-us.mimecast.com/s/dqkEB3i5qknwSd?domain=newsdata.com (see stories #1, 8, 10, 11, 12).

18 See www.energy.ca.gov/pou_reporting/background/difference_pou_iou.html.

19 See http://cmua.org/about-cmua/members/; http://appanet.files.cms-plus.com/PDFs/Directory%20-%20Statistical%20Report.pdf.

20 Assembly Bill No.117, Chapter 838, 2002 Cal Stat (approved and filed 24 September 2002).

21 See www.newsdata.com/cgi-bin/viewpdf.cgi?iss=cem1447&cid=RH4LG40x2HOq, pp8–9; http://cal-cca.org/wp-content/uploads/2017/01/CalCCA-Quarterly-Update-April-2017.pdf.

22 California Public Utilities Code, §§399.11 *et seq*, http://www.cpuc.ca.gov/uploadedFiles/CPUC_Public_Website/Content/News_Room/News_and_Updates/Retail%20Choice%20White%20Paper%205%208%2017.pdf, p5. See also www.nortonrosefulbright.com/files/chadbourne/publications/pfn_0617_ 0.pdf. p2.

23 See www.newsdata.com/cgi-bin/viewpdf.cgi?iss=cem1447&cid=RH4LG40x2HOq, p3.

24 California Public Utilities Code, §394.

and are subject to the California RPS. However, the current market for ESPs is limited, since the total amount of energy that can be supplied by ESPs is currently limited to specified amounts, because direct retail access has been suspended indefinitely until reinstated by the legislature.[26]

Corporate power purchasers are also becoming important participants in the energy market. Many corporations have adopted sustainability goals that include the purchase of renewable energy. Most corporate power purchases are not direct sales from the generator to the buyer, but are instead 'virtual' power purchase agreements that have a similar economic effect, but involve sales into and purchases from the market.[27] As of late 2016, 451 MW of corporate power purchase agreements had been signed in California in 2016.[28] Exact figures are difficult to confirm, but a substantial amount of this energy is to be supplied from solar energy projects.

4. Legislative and regulatory framework

4.1 Renewable Portfolio Standard

The growth of the renewable energy industry in general, and the solar energy industry in particular, in California has been fuelled by California's Renewable Portfolio Standard (RPS).[29] The RPS was first adopted in 2002,[30] and it has been expanded several times since then to its current form. Under the current RPS, there are escalating requirements for the percentage of each IOU's energy must be supplied from renewable resources as follows:[31]

- 20% of retail sales by 31 December 2013
- 25% of retail sales by 31 December 2016
- 33% of retail sales by 31 December 2020
- 40% of retail sales by 31 December 2024
- 45% of retail sales by 31 December 2027
- 50% of retail sales by 31 December 2030 and each year thereafter.

At least 65% of this renewable energy must be acquired under long-term power purchase agreements of 10 years or more.[32] The RPS also requires that a balanced portfolio of renewable energy be procured, consisting of minimum and maximum amounts of certain 'portfolio content categories' (PCCs).[33] There are three portfolio content categories:

- energy from an eligible renewable resource (ERR) from which the RECs have

---

25 *Id*, §§394, 394.1.

26 *Id*, §365.1.

27 See discussion at section 7.3 below.

28 See www.businessrenewables.org/downloads/brc_nov_2016/State-of-the-market.pdf, p12.

29 This is not to ignore the effect of the federal investment tax credit, which has been instrumental in promoting the growth of the solar industry throughout the United States or, more recently, of corporate or local sustainability initiatives.

30 See www.energy.ca.gov/portfolio/ and Senate Bill 1078, Ch 516, 2002 California State.

31 California Public Utilities Code, §399.15(b)(2)(B).

32 *Id*, §399.13(b).

33 The portfolio content categories are not 'products' in the strict sense of the word. As long as the energy or other product comes from an ERR, the 'product' is "simply 'that which meets the criteria for this category'", California Public Utilities Commission Decision No.11-12-052 (15 December 2011), pp17–18.

not been severed, where the generating resource has its first point of interconnection with a California balancing authority or with a distribution system serving end users within a California balancing authority area, is scheduled into a California balancing authority from an eligible renewable resource without substituting electricity from another source, or is dynamically scheduled into a California balancing authority[34] (PCC-1);
- firmed and shaped energy from an ERR located outside a California balancing authority area that provides incremental electricity and is scheduled into a California balancing authority (PCC-2); and
- energy or other products from an ERR, including unbundled RECs, that do not qualify under either of the other two categories (PCC-3).[35] Beginning in 2017, at least 75% of the renewable energy procured by the IOUs to satisfy their RPS requirements must be PCC-1, and no more than 10% can be PCC-3.[36]

The RPS standard does not just apply to IOUs. All retail energy providers, including POUs, CCAs and ESPs, must also comply with the RPS.[37]

The RPS requires the procurement of renewable energy, not just solar energy, but solar energy represents an ever-increasing portion of the renewable energy used in California. In 2016, solar energy accounted for 32% of the renewable energy generation in California.[38] As of June 2017, almost half of the state's renewable generation (not including self-generation) was either photovoltaic solar or solar thermal.[39] (Although it does not count towards RPS compliance, photovoltaic solar energy accounts for over 95% of renewable self-generation in California.)[40]

4.2 Regulatory agencies

There are three principal state agencies that regulate various aspects of the solar energy industry (in addition to local land use regulation, which is described further below): the California Public Utilities Commission, the California Energy Commission and the California Independent System Operator.[41]

The CPUC regulates the state's investor-owned utilities. It establishes renewable energy procurement targets for retail energy providers (except POUs) and monitors and enforces the IOUs' compliance with the RPS through its oversight of their energy procurement programmes. The CPUC authorises the various solicitations for renewable energy issued by the IOUs from time to time and approves (or sometimes

34 A California balancing authority is a balancing authority that has control over a balancing authority area located primarily in California, which includes the CAISO and several POUs , California Public Utility Commission Decision No.11-12-052 (15 December 2011), pp19–20.

35 California Public Utilities Code, §399.16(b).

36 *Id*, §399.16(c).

37 *Id*, §399.15(b)(1); 20 California Code of Regulations, §§3200–3208 (Enforcement Procedures for the Renewables Portfolio Standard for Local Publicly Owned Electric Utilities).

38 See www.energy.ca.gov/renewables/tracking_progress/documents/renewable.pdf, p2.

39 *Id*, p3.

40 *Id*.

41 The CAISO is actually a not-for-profit corporation but functions like a regulatory agency through its control of most of the transmission system in the state. The CAISO is itself subject to regulation by the Federal Energy Regulatory Commission (FERC).

disapproves) the procurement contracts signed by the IOUs.[42] Although the CPUC does not regulate non-utility generators, its regulatory oversight over the IOUs has a significant impact on the generators seeking to sell to the IOUs. The CPUC also regulates the IOUs' net energy metering programmes.[43]

The CPUC has limited jurisdiction over electric service providers (ESPs), which, although they are not utilities, must still register with the CPUC, provide proof of financial viability and technical ability,[44] and comply with RPS procurement requirements.[45] The CPUC issues general rules and procedures for CCA operations, as well as for interactions between CCAs and the IOUs in whose service territories they operate.[46] The CPUC does not have the authority to regulate publicly owned utilities (POUs). These entities are self-governing and set their own rates and policies.[47]

The California Energy Commission (CEC) is the state's primary energy policy and planning agency. Its stated mission is to reduce energy costs and the environmental impacts of energy use, while ensuring a safe, resilient and reliable supply of energy.[48] Its responsibilities include forecasting energy needs, providing incentives for renewable energy development and enforcing RPS compliance by POUs.[49] Of particular interest to generators of solar energy, the CEC certifies generation resources as RPS-eligible and verifies the eligibility of all energy claimed to satisfy RPS requirements.[50]

The CEC also has jurisdiction over the siting and approval of thermal power plants of 50 MW or more.[51] This includes not only large fossil fuel-fired power plants, but also solar thermal plants of 50 MW or more (which almost all solar thermal plants are). Developers are required to file an application for certification with the CEC, which then triggers an extensive environmental review and public hearing process. (In addition to requirements relating solely to energy projects, the review must comply with the California Environmental Quality Act, which is discussed below.) The certification process is extensive and time consuming, usually taking more than a year, but it does supersede all other state or local permits (including zoning and land use approvals) that might otherwise be required for the project.[52] Changes to a project that is subject to CEC jurisdiction also require CEC approval, which, depending on the nature of the change, can trigger another lengthy review process.

42 Approval of IOU procurement by the CPUC is not legally necessary in order to make the contracts binding on the IOU, but it is necessary for recovery of the procurement costs through rates. As a result, the IOUs require that almost all long-term contracts, especially those of any size, be expressly subject to CPUC approval.

43 California Public Utilities Code, §2827

44 *Id*, §394.

45 Senate Bill 1078, Chapter 516; 2002 California State and Senate Bill X1-2, Chapter 1, 2011 California State.

46 California Public Utilities Commission Decision 12-12-036 Adopting a Code of Conduct and Enforcement Mechanisms Related to Utility Interaction with Community Choice Aggregators, Pursuant to Senate Bill 790 (2012).

47 See www.energy.ca.gov/pou_reporting/background/difference_pou_iou.html.

48 See www.energy.ca.gov/commission/fact_sheets/documents/core/CEC-Core_Responsibilities.pdf.

49 *Id.*

50 See www.energy.ca.gov/commission/fact_sheets/documents/core/REN-Developing_Renewable_Energy.pdf.

51 California Public Resources Code, §25120.

52 See www.energy.ca.gov/commission/fact_sheets/documents/core/SIT-Certifying_Thermal_Power_Plants.pdf.

The CAISO was established as part of the deregulation of the electric industry in the 1990s to manage the transmission system owned by California's IOUs, to administer open access to that transmission system and to ensure grid reliability. The CAISO administers the wholesale energy market, as described in section 2 above.[53]

The CAISO also regulates the process of interconnecting to the CAISO-controlled transmission grid. The CAISO tariff includes extensive provisions regarding the interconnection process which are only briefly summarised here. Interconnection requests are studied in 'clusters'. In order to be included in a cluster, the generation owner must file an application within a specified 'cluster window' (usually no more than a month long) and must post significant deposits. There is a two-stage study process to identify the necessary interconnection facilities, reliability network upgrades and deliverability network upgrades necessary for either full or partial deliverability status.[54]

If the costs and timing of installing the interconnection facilities and the network upgrades are acceptable to the parties, the process will result in the execution of a Generator Interconnection Agreement describing the interconnection facilities and the network upgrades in detail and establishing a construction schedule and the responsibilities for costs (almost always the generator, although some costs may be recoverable).[55] Interconnection to an IOU's distribution system is subject to CPUC jurisdiction and governed by each IOU's tariff issued under Electric Rule 21, which generally describes the interconnection, operating and metering requirements for interconnection to the IOU's electric distribution system.[56] Within a POU's service territory, a generator must follow the interconnection rules set by the POU's governing board.

While the wholesale energy market is deregulated to a large degree, direct retail sales by entities other than utilities, CCAs and ESPs are not allowed, except for direct access customers whose rights were vested prior to the energy crisis or when the Legislature allowed a specific increase in the permitted amount of direct access capacity. As a result, except in certain 'behind the meter' applications that qualify as self-generation, generation owners cannot sell energy directly to end-use customers. CCAs were intended to provide at least a partial substitute for direct access by creating a utility-like entity that could compete with the IOUs, but CCAs apply to all utility users in their territory, not to specific bilateral transactions, such as sales to corporate purchasers.[57]

4.3 Rebates and incentives

There are a wide variety of rebate and incentive programmes in California, including general ones and ones with very specific purposes maintained by POUs and IOUs,

---

53 See www.caiso.com/about/Pages/OurBusiness/The-role-of-the-California-ISO.aspx.

54 See CAISO Fifth Replacement FERC Electric Tariff, Appendix DD (Generator Interconnection and Deliverability Allocation Process as of 8 March 2016).

55 See CAISO Fifth Replacement FERC Electric Tariff, Appendices EE and FF.

56 California Public Utilities Commission Order Instituting Rulemaking 17-07-007 (21 July 2017), p2.

57 Energy is usually sold to corporate customers through 'virtual' or 'synthetic' contracts that do not involve direct sales (as described below).

including net metering.[58] These programmes change over time, and they all have limited amounts of funding.

The California Solar Initiative (CSI) is a state-wide solar rebate programme for California consumers that are customers of the state's three principal IOUs. The goals of CSI were to instal 3,000 MW of solar energy systems, establish a viable solar energy industry by 2016, and place solar energy in 50% of new homes by 2019.[59] CSI is actually multiple programmes and includes the New Solar Homes Partnership (described below). The largest programme, known as the CSI general market programme, provided rebates for the installation of solar energy systems on existing homes, new and existing commercial, agricultural, government and non-profit buildings. It had a declining incentive structure to jump-start growth, with a target of 1750 MW of rooftop solar systems and a budget of $1.95 billion.[60] The programme was successful, providing rebates for over 1,900 MW of solar energy systems,[61] but as of 31 December 2016, funding for the CSI general market programme had been exhausted in all three IOU service territories, and the programme was closed.[62] CSI also includes the Single Family Affordable Solar Homes (SASH) programme, which is open until 31 December 2021 (or until funding runs out), and the Multifamily Affordable Solar Housing (MASH) programme, which is already fully subscribed.[63]

The New Solar Homes Partnership (NSHP) provides incentives for the installation of solar energy systems on new homes. The NSHP's goal is to have 360 MW of new solar energy systems installed by the 2019 CSI target date. The incentive structure has two levels: one for conventional, or market rate, housing and common areas, and a higher incentive for affordable housing and common areas. In order to qualify, both the new home and the new solar energy system must meet NSHP programme requirements, including energy efficiency. Most types of residential buildings are eligible. Solar energy systems must be at least 1 kW and not more than 7.5 kW. (Common areas are not subject to the 7.5 kW limit, but the CEC may ask that larger sizes be justified.) Among other requirements, systems must have at least a 10-year warranty, including a guarantee that degradation will not exceed 15% at the end of the 10-year warranty period. Current funding for the NSHP programme is expected to last until 1 June 2018, with all funds to be committed by that date and disbursed by 31 December 2021.[64] Incentives are currently $0.50/watt for market rate housing and $1.30/watt to $1.50/watt for affordable housing.[65]

A number of POUs also have specific rebate and incentive programmes for their customers. These programmes vary from POU to POU and may include incentive payments based on installed capacity, net metering arrangements and feed-in tariffs for selling excess generation to the POU.[66]

58 See discussion of net metering in Section 7.4 below.
59 California Public Resources Code, §25780(a).
60 See www.cpuc.ca.gov/General.aspx?id=6058.
61 See www.californiasolarstatistics.ca.gov/reports/agency_stats.
62 See www.cpuc.ca.gov/General.aspx?id=6058.
63 See www.gosolarcalifornia.ca.gov/about/csi.php; and www.gosolarcalifornia.ca.gov/affordable/mash.php.
64 See http://docketpublic.energy.ca.gov/PublicDocuments/16-NSHP-01/TN217634_20170518T080845_Final_NSHP_Guidebook_Tenth_Edition_Revised.pdf, pp1–2, 5–10.
65 See http://docketpublic.energy.ca.gov/PublicDocuments/16-NSHP-01/TN217634_20170518T080845_Final_NSHP_Guidebook_Tenth_Edition_Revised.pdf, pp23–26.
66 See http://programs.dsireusa.org/system/program?fromSir=0&state=CA.

4.4 Tax considerations

The principal tax benefits that are available to solar projects in California are the federal tax benefits that are available to solar projects in the United States generally: the federal investment tax credit and the modified accelerated cost recovery system (MACRS). The federal investment tax credit provides a direct credit against federal income taxes equal to 30% of the cost of qualifying property for projects for which construction begins before 1 January 2020. The credit is reduced to 26% for projects that begin construction in 2020 and to 22% for projects that begin construction in 2021. Construction must be completed and the project must be placed in service before 1 January 2024, or the credit is reduced to 10%.[67] The credit is subject to recapture on a *pro rata* basis if there is a change in ownership of the project within five years after the project is placed in service. MACRS allows cost recovery (depreciation) of qualifying property over a five-year period, even though the useful life of the project may be longer than five years.[68] California does not have a state investment tax credit and does not follow MACRS for state tax purposes, but instead uses a different cost recovery method that generally requires that cost recovery occur over a longer period of time.

Real property taxes are limited to 1% of the fair market value of the property on the date it was acquired with only limited increases for general inflation. The assessed value of property for tax purposes is generally increased by the value of new construction and is reassessed when it is sold. However, solar energy projects are exempt from increases in assessment due to new construction. These provisions are discussed in more detail in section 6 below.

As of 1 January 2018 and until 31 December 2030, qualified tangible personal property to be used primarily in the generation, production, storage or distribution of solar energy that is purchased by a person or entity primarily engaged in the business of generating electric power (or purchased by a contractor for the performance of a construction contract with such a person or entity for the construction of solar energy project) is entitled to a partial sales tax exemption from state level sales tax (currently 7.25%). The partial exemption rate is currently 3.9375%, which results in an effective sales tax rate of 3.3125% (7.25% minus 3.9375%), plus applicable local and district taxes, if any.[69] To qualify, the property must be machinery and equipment used in California for these purposes and have a useful life of one year or more. The exemption applies to purchases of up to $200 million of qualifying property in a calendar year.[70]

Sales of solar projects by persons or entities that are not California taxpayers are subject to withholding equal to 3⅓% of the sale proceeds in order to provide assurance that any applicable California tax on the sale will be paid.[71] This is similar to the federal withholding requirement under the Foreign Investment in Real Property Tax Act, and the statute is often called 'Cal-FIRPTA'. Foreign sellers will

---

67 26 USC §48.
68 26 USC §168(e)(3)(B).
69 See www.cdtfa.ca.gov/industry/manufacturing-exemptions.htm#Sellers.
70 California Revenue and Taxation Code, §6377.1.
71 California Revenue and Taxation Code, §18662.

generally be subject to this withholding requirement. The requirement can be waived by delivery of a form certifying that the seller is a California taxpayer.

5. Environmental and land use regulation

Except for solar thermal projects of 50 MW or more, which are subject to CEC jurisdiction, environmental permitting and land use regulation for solar projects are generally governed by local land use and zoning laws in the jurisdiction where the project is located. (Certain federal laws will also apply to solar projects located on federally owned or managed land.)

5.1 California Environmental Quality Act

Review of land use applications for solar projects is subject to the California Environmental Quality Act (CEQA).[72] Under the CEQA, all state and local agencies must consider environmental impacts of any projects (public or private) that require discretionary approval by the public agency as part of their decision-making process and must, absent overriding considerations, "mitigate or avoid the significant effects on the environment of projects that it carries out or approves whenever it is feasible to do so".[73] If more than one agency has discretionary approval over a project, a lead agency will be designated to conduct the environmental review and to prepare the environmental impact report (described below). As a general rule, all projects are subject to CEQA review, although there are certain statutory and categorical exemptions. For example, installation of solar energy systems on rooftops of an existing building or an existing parking lot is exempt from the CEQA review process.[74]

If a project is not exempt, the lead agency conducts an initial study to determine whether the project may have a significant impact on the environment. If the initial study finds no substantial evidence that the project may have a significant impact on the environment, the lead agency can conduct a more limited environmental review and prepare a negative declaration that finds that the project will not have a significant impact on the environment and that no mitigation measures are required. If the initial study shows that there may be significant environmental impacts, but that, with certain project modifications, they can all be mitigated to a less than significant level, the lead agency can adopt a mitigated negative declaration, which means that the agency has concluded that the project, as revised, will not have a significant impact on the environment.[75]

If the initial study does show that a project may have significant environmental impacts, the lead agency is required to prepare an environmental impact report (EIR), which is the heart of the CEQA process. As a general rule, any questions about the significance of an environmental impact should be resolved in favour of preparation

---

72 California Public Resources Code, §§21000 *et seq.*
73 *Id*, §21002.1.
74 *Id*, §21080.35
75 *Id*, §21064.5
76 An EIR is required if there is a 'fair argument' that a project might have a significant environmental impact or that the mitigation measures in a mitigated negative declaration are not sufficient to mitigate all environmental impacts to less than significant: *Friends of B Street v City of Hayward* (1980) 106 Cal App 3d 988; *No Oil, Inc v City of Los Angeles* (1974) 13 Cal 3d 68.

of an EIR.[76] An EIR is supposed to inform the public agencies and the public of a project's significant environmental impacts and to propose mitigations and alternatives which may reduce or avoid them.[77] A draft EIR is first prepared and circulated for public comment. Based on comments on the draft EIR, the lead agency will prepare a final EIR, which must be certified by the lead agency before it can take action to approve or disapprove a project. If certain environmental impacts are unavoidable (such as the removal of agricultural land from production in order to build a solar energy facility), the agency must make a finding of overriding considerations in order to approve the project. Otherwise, most mitigation measures recommended in an EIR will be included in the final conditions of the land use approval, since agencies are required to mitigate environmental impacts "whenever it is feasible to do so".[78]

5.2 Zoning

Although quite extensive in its requirements, the CEQA primarily deals with the process of reviewing permit applications, not with the substance of the land use being proposed. Except for solar thermal projects of 50 MW or more, which are subject to CEC jurisdiction, local governments regulate projects on a substantive basis through their regulation of land use.

Under the California Planning and Zoning Law,[79] local governments (counties and cities) must adopt a general plan, which is a "comprehensive, long-term general plan for the physical development of the county or city"[80] and serves as a guide for future development, but does not directly regulate land use.[81] Local governments also adopt zoning laws, which define allowable land uses and development standards within specific zones or districts. Zoning laws must be consistent with the general plan adopted by a local government.[82] A use may be permitted within a certain zone if it meets the requirements specified in the zoning ordinance or it may only be permitted upon approval by the local agency of project-specific conditions (a 'conditional use'). If a property's zoning does not allow construction of a solar project, the project sponsor can seek a variance from the zoning or a re-zoning of the property. Variances are granted in special cases where zoning regulations would deprive the property of uses enjoyed by other nearby properties in the same zone, and provide restrictions on use to ensure that the variance does not grant any special privilege. However, a variance does not permit use that is otherwise not allowed in the applicable zone.[83] Similarly, if a proposed re-zoning conflicts with the local agency's general plan, the request must be denied.[84]

Each local government has its own permitting processes, but they typically

77 14 California Code of Regulations, §15121.
78 See California Public Resources Code, §21002.1(b).
79 California Government Code, §§65000 *et seq.*
80 *Id*, §65300.
81 *Id*, 65302 provides the required components of a general plan.
82 *Id*, §65800.
83 OPR, California Planning Guide: An Introduction to Planning in California (2005). See www.opr.ca.gov/docs/California_Planning_Guide_2005.pdf.
84 OPR, California Planning Guide: An Introduction to Planning in California (2005). See www.opr.ca.gov/docs/California_Planning_Guide_2005.pdf.

require that a completed project application be submitted to the applicable public agencies and reviewed within a specified time frame, including the CEQA review. The final decision on a project may be appealed by administrative or judicial avenues. The requirements and time frames for appeals will vary between local governments.

5.3 Agricultural land preserves

The California Land Conservation Act of 1965, commonly known as the 'Williamson Act', provides property tax breaks to owners of farmland in exchange for a 10-year contract with a local government that the subject land will not be developed or otherwise converted to another use.[85] The purpose of the Williamson Act is to promote voluntary land conservation.[86]

If a solar energy project is proposed on property covered by a Williamson Act contract, the property owner may: seek a finding that the solar project is compatible with the agricultural use of the property; institute a non-renewal process; seek a cancellation of the Williamson Act contract; or agree to rescind the Williamson Act contract and replace it with a solar use easement.[87]

If the local government determines that development of a solar project on contracted land is a 'compatible use' which does not significantly compromise, displace or impair the current or future agricultural use and capability of the site or result in significant removal of adjacent contracted land from agricultural or open-space use, it may approve the project with the Williamson Act contract remaining in place.[88] However, given the broad land coverage of solar projects (as opposed to wind projects), it is unlikely that the local government will be able to make this finding.[89]

An owner may unilaterally initiate non-renewal of the Williamson Act contract. The process for non-renewal is straightforward, but lengthy. The owner must file a notice of non-renewal at least 90 days prior to the contract renewal date, and the contract will wind down over the remaining term, which is usually nine years.[90] During this time, the contract restrictions will continue to apply and property taxes will gradually return to the full rate by the end of the non-renewal period.

An owner may also seek cancellation of a Williamson Act contract prior to the nine-year term required for non-renewal. However, cancellation is subject to discretionary approval by the local government with jurisdiction over the property and is permitted only if it is either consistent with the Williamson Act or in the public interest.[91] Certain local governments may require both findings. Cancellation can also be quite costly. There is a statutorily required cancellation fee of 12.5% of the cancellation value of the property.[92] Local governments may also require payment of additional fees, such as application fees and fees to recover the cost of

---

85 California Government Code, §§51240, 51243 and 51244.
86 *Id*, §51200.
87 *Id*, §§51190 *et seq*; and *id*, §51255.1.
88 *Id*, §51238.1(a).
89 *Id*, §51238.1 makes reference to 'electric facilities' as a compatible use, but it is likely that this provision only intended to cover facilities, such as poles and wires, providing electricity for agricultural operations.
90 California Government Code, §§51245 and 51244.5.
91 Government Code §51282(a).
92 *Id*, §51283.

the agency's services with respect to cancellation.[93] Some of these fees may be waived in certain limited circumstances.[94]

Until 1 January 2020, the parties to a Williamson Act contract may also agree to rescind the contract and simultaneously enter into a solar use easement, with a term of at least 20 years and extinguishable only by non-renewal, which restricts land use for certain solar facilities or other alternative renewable energy facilities.[95] In order for this to occur, the Department of Conservation must determine that the land is eligible for rescission if the land is "marginally productive or physically impaired" and is not located on lands designated as prime farmland, unique farmland, or farmland of statewide importance by the Farmland Mapping and Monitoring Programme.[96] A rescission fee of 10% of the fair market value of the land is also required.[97]

6. Real estate rights

Real estate issues in California are for the most part similar to those encountered in other US jurisdictions. A project sponsor needs to have exclusive control of the entire surface area on which the project will be built. This is usually accomplished through direct ownership or long-term lease of the project site. On smaller projects, such as rooftop installations, the site occupancy rights may be built into the energy services agreement or similar agreement with the site owner (who is often also the purchaser of the solar energy). On larger projects, a sponsor may obtain title insurance to protect itself against defects in title, although this may not be cost-effective for smaller projects.

If mineral rights have been separated from the surface ownership, appropriate arrangements must be made for limiting the surface use by mineral rights holders. Unlike some other states, California does not have a 'reasonable accommodation' requirement with respect to severed mineral rights. If the instrument conveying the mineral rights does not expressly limit the mineral holder's surface entry rights, the holder will have strong implied rights over the entire surface of the property to access and develop its mineral estate.[98] Investors and lenders will require that severed mineral rights be resolved prior to their investment or financial closing.

Mechanic's liens are secured real property interests for the benefit of the contractors involved in the construction of a project to ensure that contractors, subcontractors, labourers and suppliers are compensated for their goods and services. As a matter of public policy, mechanic's liens cannot be waived in advance. Mechanic's liens may be waived by a contractor and its subcontractors upon delivery of a lien waiver in exchange for payment.[99] Lien waivers may be conditional

---

93 *Id*, §§51281.1 and 51287.

94 *Id*, §51283(c).

95 *Id*, §§51255.1, 51190(c) and 51191.2.

96 Section 1 of Statutes 2011, Chapter 596 (SB 618). Note that the specific criteria for eligibility is provided in Government Code §51191. Also note that this determination is not subject to CEQA, but other aspects of developing a project may be subject to CEQA. Government Code §51191(d).

97 Government Code §51255.1(c)(2).

98 Callahan v Martin, 3 Cal 2d 110, 127 (1935).

99 Civil Code, §§8120 *et seq* provides the requirements for a waiver or release of a mechanic's lien.

(conditioned on the payment being made) or unconditional (usually only given if the payment has been made).[100]

California provides strong protections to contractors, subcontractors, labourers and suppliers, by providing for the priority of mechanic's liens over other encumbrances. If proper procedures are followed, a mechanic's lien will have a priority as of the date the services or materials are first provided by any contractor or subcontractor, rather than when the lien is recorded.[101] If work of any kind (including grubbing and clearing or setting grading stakes) is begun before a deed of trust or mortgage securing financing is recorded, all mechanic's liens that are filed will have priority over the deed of trust or mortgage.[102]

Property taxes in California are generally limited to 1% of the fair market value of the property on the date it was acquired with only limited increases for general inflation. Property taxes are only subject to reassessment if there is a new construction or a change in ownership. However, certain 'active solar energy systems' are excluded from new construction property tax reassessment once construction is completed.[103] A solar project is considered an 'active solar energy system' if "upon completion of the construction of a system as part of a new property or the addition of a system to an existing property, uses solar devices, which are thermally isolated from living space or any other area where the energy is used, to provide for the collection, storage, or distribution of solar energy".[104] This exclusion remains in effect only until there is a subsequent change in ownership.[105] Changes in ownership can occur in less than obvious ways. Creation and termination of leases on taxable real property with terms of 35 years or more (including renewal options), or transfers of a lessee's or lessor's interest with a remaining term of 35 years or more (including renewal options), are deemed to be changes of ownership and subject to reassessment.[106] Acquisition or transfer of more than 50% of the ownership interests in an entity can also be a change of ownership.[107]

7. Power sales arrangements

The types and structure of power sales arrangements vary with the type of purchaser. In the commercial context, most agreements for the purchase and sale of solar energy are entered into with one of four general categories of purchaser: utilities, including IOUs, POUs and CCAs; power marketers and resellers, including ESPs; corporate purchasers; and end users of on-site generation (behind the meter). Residential users are a type of on-site generation.

100 *Id*, §§8132 and 8134.
101 *Id*, §8458(a).
102 *Id*, §8458(a).
103 Revenue & Taxation Code, §73.
104 *Id*, §73(b)(1).
105 *Id*, §73(f).
106 *Id*, §61(c) and 64(c).
107 *Id*, §64(c). There are other rules relating to transfers of ownership interests in legal entities not discussed in this section, which can be found in Section 401 of the assessor's handbook.

7.1 Utility purchasers

Power purchase agreements between a solar generator and a utility typically involve the purchase and sale of the entire output of a project, which generally gives the utility the right not only to known products, such as energy, capacity and ancillary services, but also to any future products (such as a new form of environmental credit) that may be developed during the term of the agreement.[108] Power purchase agreements with the IOUs will also contain certain 'non-modifiable' provisions adopted by the CPUC relating to matters such as CPUC approval of the agreement, WREGIS and the tracking of RECs, RPS compliance, environmental attributes and governing law.[109]

The product sold will include the 'green attributes' of the project, which includes all RECs and other environmental attributes of the project, but excludes tax credits and financial incentives such at the federal investment tax credit and state rebate programmes. (The definition of 'green attributes' is a non-modifiable provision.) Sellers are also required to be certified as an eligible renewable energy resource (ERR) by the CEC and are required to represent in the power purchase agreement that the project is an ERR and that its output qualifies under the requirements of the RPS. However, if there is a change in law that would cause these representations to be untrue, the seller will not be in default as long as it uses commercially reasonable efforts to comply with the change in law.

Some utility contracts further refine the concept of commercially reasonable efforts by creating a 'compliance expenditure cap' which limits the amount a seller is obligated to spend on an annual and/or cumulative basis in order to comply with a change in law. The compliance expenditure cap is generally limited to certain specified areas, such as RPS compliance. Remedies vary if the cost of complying with a change in law exceeds the compliance expenditure cap, but generally the purchaser has the option of contributing the extra funds necessary for compliance, waiving the obligation or, in some cases, terminating the agreement on a 'no fault' basis. These agreements also contain provisions relating to the registration of RECs with WREGIS, transferring the RECs from seller to purchaser, and compensating for shortfalls in REC deliveries.

Even though solar resources are intermittent, most utility contracts include a minimum energy production requirement on an ongoing basis. Sellers are required to deliver a certain minimum percentage (generally between 70% and 85%) of the expected annual production of the facility, measured either annually or over a two-consecutive-year period. Remedies vary from paying damages for the shortfall (usually based on the market price of replacement energy) to termination of the agreement in some cases.

Projects that are under development or construction will often be required to meet certain interim milestones, as well as a final milestone of the guaranteed commercial operation date. The interim milestones are intended to ensure that the

108 This is not always the case, and some contracts specifically delineate what products are and are not being purchased.

109 See, eg CPUC Decisions 07-11-025, 08-04-009, 08-08-028, 10-03-021, 11-01-025 and 13-11-024.

project will be completed on time. Milestones may include items such as issuance of permits, start of construction, financial closing and, of course, commercial operation. Delay damages are generally payable if a project has not achieved commercial operation by its guaranteed commercial operation date, but they may also be payable if other milestones, such as the start of construction, are not achieved on schedule. Damages for failure to achieve an interim milestone can sometimes be recovered if a project is completed on schedule. Ultimately, if a facility does not achieve commercial operation by an outside date (usually, 6–12 months after the guaranteed commercial operation date), the power purchase agreement is subject to termination by the utility.

The intermittency of solar generation can create imbalances between energy forecasted or scheduled to be delivered from a project and actual energy deliveries. As noted above, it is the responsibility of the scheduling coordinator for the facility to manage and ultimately bear the cost of these imbalances. Because a utility purchaser usually has a better ability to deal with imbalances than a single project, most contracts provide that the purchaser (the utility) will be the scheduling coordinator, although occasionally the seller will act as scheduling coordinator. If the buyer is the scheduling coordinator, the seller's exposure to imbalance charges will generally be limited to those caused by its own acts or omissions. Some contracts also give the purchaser a right to curtail the output of the project for economic reasons (eg, to avoid a negative LMP at the delivery point), but the utility has to pay for the lost output at the contract rate. (In some contracts, the utility is entitled to some amount of curtailment at no charge.) Utilities also limit their exposure to over-generation by not accepting energy deliveries in excess of a project's nameplate capacity and/or by reducing the price once annual generation exceeds a certain threshold (eg, 115% to 120% of expected annual production).

The credit and security provisions of utility contracts are often one-sided – the seller must post substantial security (eg, equal to a year's expected revenues for a 20-year contract) and maintain it through the life of the contract, even though the utility's exposure to a seller default decreases the later one gets into the term of the agreement. By contrast, IOUs and POUs in California almost never post collateral as security for their obligations. Because they are new and do not yet have substantial assets, CCAs present a different credit risk profile. The standard for security to be posted by CCAs is evolving to include financial covenants and similar matters, including in some cases posting liquid security such as a letter of credit.

7.2 Power marketers and resellers

With the exception of ESPs, who do sell to retail customers on a limited basis, power marketers and other resellers generally operate in the wholesale market, like utilities. Owners of generation will sell their output (including energy, capacity and ancillary services) to power marketers, who will then try to resell it in the wholesale markets. The power purchase agreements are often (but not always) simpler than a utility contract and frequently do not include provisions relating to the construction of the project. The contracts are usually for a shorter term than a utility contract. They may also include guaranteed energy production requirements similar to the IOUs. In some cases, a power marketer will be required to post security (cash, letter of credit or parent guarantee) for its obligations.

7.3 Corporate purchasers

Ordinary business corporations were not traditional participants in the energy markets, but, in recent years, there has been a sharp increase in the number of power purchase agreements signed by corporate buyers, such as Amazon, Apple and Google. This change has been driven by both corporate environmental and sustainability goals and a desire to control the long-term cost of energy.

Corporate power purchase agreements generally include the purchase and sale of bundled electricity and RECs. In jurisdictions where direct retail access is permitted, corporate power purchase agreements can be 'physical' agreements that provide for the physical delivery of electricity to the purchaser. However, with limited exceptions, direct retail access is not available in California, so the agreements that are signed are 'virtual' or 'synthetic' power purchase agreements, which are based on the concept of financially settling differences between floating prices in the local electricity market and contracted prices under the power purchase agreement. Both forms of agreement support the principle of 'additionality', ie, they support the construction of additional renewable energy projects, which results in an overall increase in the amount of renewable energy that is being generated and put on to the grid, even if it is not used directly by the corporate purchaser.

Virtual power purchase agreements often take the form of a 'contract for differences'.[110] Under a contract for differences, the buyer agrees to purchase renewable energy and RECs from a project for a fixed price. The seller sells the project's electricity (without the associated RECs) into the market (keeping the proceeds from that sale) and transfers the RECs to the buyer. The buyer purchases electricity from its local utility for its own use, and the buyer and seller settle the difference between the contract (fixed) price and the applicable real-time market (floating) price on a periodic basis based on the amount of power sold. Financial settlements typically require the buyer to pay the seller the difference between the two prices if the market price is less than the fixed price and the seller to pay the buyer the difference between the two prices if the market price is greater than the fixed price.

Corporate power purchase agreements include many of the same issues as utility power purchase agreements, but they also include some unique elements as a result of the nature of the transaction. The virtual nature of the agreement can create 'basis risk' between the energy price at the project's point of interconnection and the energy price at the point at which the buyer is pricing the energy (often a liquid trading hub). The seller usually bears the basis risk in corporate transactions. Many corporate power purchase agreements permit economic curtailment by the buyer to some degree, with the buyer paying for the curtailed energy (except sometimes if the LMP for the energy at the delivery point is negative). Unlike utility power purchase

110 Virtual agreements can also take the form of hedging agreements, such as a fixed-for-floating swaps or collars that provide both a floor and a ceiling on contract prices.

agreements, corporate purchasers are usually willing to provide security for their performance, most often in the form of a letter of credit or a parent guarantee (if the corporate parent is sufficiently creditworthy). Sellers are almost always required to provide performance security, as in a utility power purchase agreement.[111]

7.4 On-site generation

Because of the physical constraints of locating generating facilities at commercial or industrial site or at a residence, these projects tend to be smaller (less than 1 MW for commercial and industrial systems and 10 kW or less for most residential systems). Residential (typically rooftop) systems are essentially consumer products and are subject to consumer protection laws as well as the rules governing the energy markets. Both on-site commercial and industrial contracts and residential power purchase agreements generally require the end user to pay for all energy generated by the system, but allow the customer to offset these costs by selling any excess energy to the local utility through net energy metering.

California leads the United States in distributed or on-site generation,[112] due in no small part to net energy metering (NEM). NEM is a special billing arrangement for retail customers (both residential and commercial) with behind-the-meter generation that offsets a customer's energy bill for any excess electricity sold back to the grid. In order to qualify for NEM in California, a customer must generate at least some of their electricity through solar (or wind) energy on premises, and the generating system's peak capacity output must be 1,000 kW or less.[113] Approval is subject to an application process with the customer's utility. Savings from net metering can be kept by the customer or shared among the parties.

7.5 Particular issues in power sales arrangements

Power sales arrangements in California are subject to the same general concerns and considerations as power sales arrangements in other jurisdictions, such as assurances of project revenue, curtailment rights, creditworthiness of the power purchaser, excuses for failure to deliver or receive energy, guaranteed minimum energy production requirements and limitations on assignability. The following items should also be considered in connection with sales of solar energy in California:

- Change in law – most solar power purchase agreements include a requirement that the project be an 'eligible renewable energy resource' and otherwise comply with California laws governing RPS qualification. Given the long tenor of most power sales agreements, a seller may have significant exposure to the cost of complying with changes in law over the life of the contract. This risk is often managed through a compliance expenditure cap that applies to some or all of these requirements and caps required compliance costs on an annual and/or cumulative basis.

---

111 Corporate PPAs: Market Trends and Opportunities (17 May 2016), http://s3.amazonaws.com/cdn.orrick.com/files/Corporate-PPAs-Market-Trends-and-Opportunities.pdf.

112 California DG Statistics, http://californiadgstats.ca.gov/.

113 Net Energy Metering in California (CA.gov), www.gosolarcalifornia.ca.gov/solar_basics/net_metering.php.

- Commencement of delivery term – a power sales agreement for a utility-scale solar project will generally have a number of requirements relating to the commencement of the delivery term in addition to completing all interconnection requirements and achieving commercial operation of the project, such as execution of *pro forma* agreements with the CAISO related to metering and similar matters. If the project is also required to be qualified to deliver resource adequacy benefits (see next bullet point), an additional 60 to 90 days should be allowed in the project schedule to complete the testing and certification required to be able to provide resource adequacy benefits.
- Resource adequacy requirements – although solar energy facilities generate intermittently, they do qualify for a limited amount of resource adequacy benefits (a type of installed capacity product) under the CAISO tariff.[114] However, in order to qualify, a project usually must have 'full capacity deliverability status' under its interconnection agreement, which means that power from the project is deliverable under most conditions.[115]
- Definition of environmental attributes – the definition of environmental attributes in most power sales agreements has become standard and follows the definition approved by the CPUC. Delivery of environmental attributes is usually tracked by the delivery of certificates from WREGIS as described in section 2 above. Additional requirements for tracking or proving the delivery of environmental attributes should be avoided because currently there is no viable system other than WREGIS for tracking environmental attributes.
- Renewable energy category – California divides renewable energy into three categories as described in section 4.1 above. Some power sales agreements require that only certain categories of renewable energy be delivered. This will usually not be a problem for a project located in California that is interconnected to the CAISO grid or another balancing area authority in California.

8. The future of solar energy in California

California is committed to increasing the proportion of energy that it gets from renewable resources, and solar energy will be a big part of that future. As noted above, solar energy has become an increasingly larger part of the renewable energy mix in California, and this trend will probably continue into the future as the cost of solar energy continues to decline.

The solar energy industry may be given further impetus for growth if there is an increase in the required amount of renewable energy under the RPS. The RPS currently requires procurement of 50% of energy consumption from renewable resources (including solar) by 2030. Legislation was introduced in 2017 that would have increased RPS requirements to 100% by 2045, but it was not adopted by the end of the legislative term and was returned to committee.[116] If this legislation or

114 California Independent System Operator Corporation Fifth Replacement FERC Electric Tariff, §40.8.1.6.

115 *Id*, §§40.4.6 *et seq.*

116 See California Senate Bill 100 (De Leon) (2017), re-referred to committee on 11 September 2017.

something similar is enacted, California will be the second state in the United States (after Hawaii) to adopt a 100% RPS requirement. Adoption of this standard will require a great expansion of renewable resources to meet it, and solar energy will be one of the principal ways of doing so.

The advent of other energy technologies, particularly energy storage, could have a major effect on the solar energy industry as well. The combination of solar energy and energy storage will allow solar generators to change their production profiles to provide a product that more closely resembles the output of conventional generation by being dispatchable, and/or to absorb excess solar generation during the middle of the day for later use, thereby enabling net output to more closely match peak demand, such as the steep evening ramp reflected in the 'duck curve'.[117] If certain requirements are met, the energy storage portion of a solar energy project can also qualify for the federal investment tax credit.

The future may also bring new entrants into the market. As noted above, corporate buyers and CCAs are relatively new market participants, but they are already changing the solar energy market. Community solar programmes (where multiple individuals or businesses participate in commonly owned solar facilities) are just beginning to appear in California, but may become more important as they grow. The number of corporate buyers could also increase significantly if direct access for retail sales is restored.

In short, the signs are positive for solar energy in California. Solar energy has a bright future in the Golden State.

The author would like to thank Christie Lim, Peter Richmond, Gil Shefer and Helen Aki for their invaluable assistance in preparing this chapter.

---

117 See discussion above at note 5.

United States: Texas

Giji M John
Orrick, Herrington & Sutcliffe LLP

I must say as to what I have seen of Texas, it is the garden spot of the world. The best land & best prospects for health I ever saw is here, and I do believe it is a fortune to any man to come here. There is a world of country to settle.
(Davy Crockett, letter to his children, 9 January 1836).

1. Introduction

Over 180 years after its writing, Davy Crockett's letter to his children and ode to his adopted Texas[1] continues to serve as a siren song to frontiersmen. While cowboys, picturesque sunsets, and Friday Night Lights may define Texas legend, Texas' reality is defined by its modern frontiersmen – those engaged in its industries of agriculture, finance, healthcare, technology, transportation and energy. And of these industries, energy is Texas' calling card to the world.

In Texas, energy is synonymous with oil and gas exploration, production and transportation. Houston's downtown skyscrapers paint this reality, bearing the name of oil and gas giants, both old and new – Shell, ExxonMobil, Chevron, Cheniere, etc. Nevertheless, Texas' energy landscape also has a different vista – renewable energy, which rose in the late 1990s with the adoption of renewable energy mandates by the Public Utility Commission of Texas.

While wind energy capacity in Texas has reached market-leading heights,[2] solar energy development in Texas has trailed. Texas ranks only seventh among US states in solar energy capacity with approximately 1,847 MW of installed capacity.[3] Nevertheless, solar energy in Texas is making marked and accelerating gains, with 228 MW of solar energy capacity having been installed in the third quarter of 2017. Since 2015 and the third quarter of 2017, Texas has steadily moved up in installed solar energy capacity relative to other US states. Pushing ahead with these projects are US, European and Asian companies actively engaged in solar development all over the world.

---

1 Davy Crockett was a native of Tennessee, having served in a number of public positions in that state.
2 Texas has installed wind energy capacity of approximately 21,450 MW. American Wind Energy Association, 3Q 2017 Market Report (public version available at www.awea.files.cms-plus.com/3Q%202017%20AWEA%20Market%20Report%20Public%20Version.pdf).
3 See Solar Energy Industries Association (SEIA) Data Sheet (www.seia.org/solar-industry-data).
4 SEIA predicts additional capacity of 1,907 MW over the next five years. See "Solar Spotlight: Texas", SEIA (14 December 2017) (www.seia.org/sites/default/files/2017-12/Federal_2017Q3_Texas.pdf). Additionally, ERCOT's 2016 Long-Term System Assessment predicts increasing penetration of solar energy in each of its seven load and generation forecasts (including scenarios of decreasing load). Available at: www.ercot.com/content/wcm/lists/89476/2016_Long_Term_System_Assessment_for_the_ERCOT_Region.pdf.

Indeed, the potential for solar energy development is high in Texas.[4] Texas has, among its many advantages to solar energy developers:

- tremendous insolation;[5]
- access to robust electricity transmission;
- well-established real estate structures based on royalties and shared benefits;
- a streamlined permitting regime; and
- robust and mature options for power offtake.

This chapter provides a broad overview of the many facets of developing a utility-scale solar energy project[6] in the Texas market – including real estate, electrical interconnection, permitting and related concerns, power offtake, construction and financing – and attempts to provide solar energy developers with a high level primer in navigating them.[7]

2. Overview of Texas market

Electricity in Texas is a dynamic, sub-hourly ecosystem of entities generating, transmitting, distributing and consuming electricity. Overseeing this ecosystem is the Public Utility Commission of Texas (PUCT) – Texas' primary body for regulating electricity activities[8] and for providing competitive market oversight and rule enforcement for the electric industry.[9] Of particular importance in this ecosystem for solar energy projects is the bulk power market, in which wholesale electricity transactions between solar energy projects and retail electric providers (REPs) occurs.[10] This market is administered by three different grid operators in Texas – Electric Reliability Council of Texas, Inc (ERCOT), Southwest Power Pool (SPP) and Midwest Independent System Operator, Inc (MISO).

ERCOT is wholly contained within Texas, manages electric power flow to 24 million customers, and ensures reliability for 46,500 miles of electric transmission lines and in excess of 570 electric generating units.[11] In addition to the Northwest Panhandle and the Northeastern areas of Texas, SPP manages all or a portion of the electricity grid in the states of Arkansas, Iowa, Kansas, Louisiana, Minnesota, Missouri, Montana, Nebraska, New Mexico, North Dakota, Oklahoma, South Dakota and Wyoming,[12] and is responsible for ensuring the reliability of 65,755 miles of electric transmission lines and 790 electric generating units.[13] In addition to the Eastern area of Texas, MISO manages all or a portion of the electricity grid in the states of Arkansas, Illinois, Indiana, Iowa, Kentucky, Louisiana, Minnesota,

---

5 See generally NREL's Solar Maps (www.nrel.gov/gis/solar.html).

6 While a tremendously diverse and interesting topic in its own right, this chapter does not address development of distributed generation solar energy projects.

7 Potential solar energy developers should take care to consult their technical, economic, accounting and legal advisers as to how to specifically address issues raised in this chapter.

8 See https://stateimpact.npr.org/texas/tag/public-utility-commission-of-texas/.

9 See www.puc.texas.gov/agency/about/mission.aspx.

10 Note that while Texas is a deregulated market in allowing end-users to choose between different providers of electricity, electricity projects are not allowed to sell power directly to end-users. Instead, projects must sell electricity at wholesale to REPs, which then in turn sell to end-users.

11 See www.ercot.com.

12 See www.ferc.gov/market-oversight/mkt-electric/spp.asp .

13 See www.spp.org/about-us/fast-facts/.

Mississippi, Missouri, Montana, Nebraska, North Dakota, South Dakota, Wisconsin, and the Canadian province of Manitoba, and is responsible for ensuring the reliability of 65,800 miles of electric transmission lines and 6,610 electric generating units.[14]

A solar energy project's location determines which of these three grid operators will govern not only the project's interconnection, but in many cases (as discussed further in 'power offtake' below) its power offtake arrangements.

3. Real estate

Acquiring real estate is generally the first step in developing a solar energy project, and will dictate a number of development concerns, including:

- which grid operator and interconnection regime the project will have to navigate;
- which power offtake options are available to the project;
- which permits will apply to the project's construction and operation; and
- which mitigation measures may be necessary in light of protected species and/or historical and cultural disturbances.

Fundamentally, solar energy projects need real estate rights to site and utilise solar panels, inverters and related generation equipment; and to site and utilise electric transmission lines (both electric collection systems and lines to interconnect with the local utility).

Solar energy projects generally acquire the bulk of required land rights to site and utilise solar panels, inverters and related generation equipment; and to site and utilise electric transmission lines through lease agreements with local landowners. Additional real estate rights may be procured through a variety of agreements with third parties (eg, easements, crossing agreements and surface use waivers). Each are briefly described below.

3.1 Land lease

A typical Texas solar energy lease agreement divides the lease tenor into a development period of between two and four years and a commercialisation period of approximately 30 years. The former period provides the developer with the time necessary to constitute a developable project[15] – in this period, the developer may attempt to secure additional leases and other land rights, pursue the interconnection process, begin initial permitting and seek initial financing arrangements. During the development period, the developer pays the landowner a fixed annual fee. If at the end of the development period, the developer has not opted to start the commercialisation period, the lease terminates.

If the developer begins construction activities on the project, the commercialisation period under the lease commences. During the commercialisation

14 See www.misoenergy.org/AboutUs/Pages/FactSheet.aspx.

15 We also note that, rather than constituting a separate development term, the landowner and developer may agree to a more formalised lease option with the later lease to be entered into upon the actual commercialisation period.

period, the landowner is paid both a fixed annual fee, and after the project is placed in service, a variable fee based on a percentage share of the net revenue of the project's product sales with a minimum floor payment. Landowners receive additional payments in the event of certain construction-related activities: construction of roads, damage to certain property (eg, bushes, fences, etc) and payments meant to compensate landowners for lost hunting revenues.

3.2 Easements and crossing agreements

In addition to easements granted in land leases, developers may also obtain standalone easements with other landowners to enable placement and use of electric transmission lines over such landowners' property. Generally, payments under these easements are for a fixed fee and are secured by options that are exercised by the developer when the project proves to be viable.

Relatedly, portions of a solar energy project may need the rights to cross easements held by third parties, as opposed to the project needing easements in its own rights. In those cases, developers may seek crossing agreements with those third parties in order to prevent unauthorised and potentially unlawful use of such landowners' easements.

3.3 Oil, gas and minerals

Texas produces over one-third of US crude oil and approximately one-quarter of US natural gas,[16] thereby putting the management of conflicting oil, gas and other mineral interests with solar energy projects squarely within developers' 'must do' list. In Texas, oil, gas and mineral interest holders may make use of the surface in order to develop the minerals. In order to prevent unwanted interference with solar energy projects, developers are advised to obtain surface use waivers from such holders.

4. Electrical interconnection

Interconnecting a solar energy project to the electric utility grid in Texas is an iterative process, and begins with the filing of a Generator Interconnection or Change Request (GINR) application with the connecting utility and ERCOT.[17,18] The GINR application includes the size, location and technology type of the project as well as the point of interconnection into the connecting utility's system. ERCOT will review the GINR application, and notify the project within seven business days if the GINR application is incomplete. Once it determines that the GINR application is materially complete, ERCOT will assign an interconnection queue position. It is this queue position that, other factors being equal, gives that project priority rights over later projects seeking to interconnect at a particular point.

After the initial interconnection request, the interconnection process is

16 See www.eia.gov/state/data.php?sid=TX.

17 In the interest of brevity and because ERCOT interconnection procedures govern the largest portion of Texas, only ERCOT's procedures with respect to interconnection of 10 MW capacity energy projects are surveyed. A summary of MISO's interconnection procedures can be found at: www.misoenergy.org/Planning/GeneratorInterconnection/Pages/ProceduresRequirements.aspx. A summary of SPP's interconnection procedures can be found at: http://sppoasis.spp.org/documents/swpp/transmission/studies/Attachment_V_GIP.pdf.

18 See ERCOT Planning Guide, Section 5.2.1 (Generation Resource Interconnection or Change Request).

dominated by a series of studies, including any of the following: a steady-state security screening study, a full interconnection study (FIS), transient stability studies, a facility study, or an economic study.[19] Each of these studies is designed to estimate the electric impact of interconnection at the interconnection point as well as to estimate the costs to both the developer and the connecting utility of such interconnection. Within 180 days of the completion of the FIS, the project and the connecting utility sign a Standard Generation Interconnection Agreement (SGIA) in order to obtain transmission service for the project,[20] and that signed SGIA is subsequently filed with the PUCT.

5. Permitting, reliability standards and reports

Texas has long distinguished itself in the energy project developer community by its relatively light permitting regime. While a solar energy project in Texas is nominally subject to federal, state and local requirements, those requirements are generally less onerous than in other US states. Moreover, those projects located within ERCOT are exempted from a number of FERC permitting requirements since ERCOT-based transactions take place wholly within the state of Texas, and are not inter-state commerce transactions, the latter being subject to FERC jurisdiction.[21] In addition to traditional permitting requirements, Texas solar energy projects are also required to comply with technical reliability standards and may be required to deliver certain traditional reports necessary in connection with obtaining certain classes of financing.

5.1 Permits

A typical, but non-exhaustive list of permits for a Texas solar energy project might include the following:

- local county permits (sometimes in the forms of agreements) with respect to road use, access road driveways and conditional use, and construction of certain on-site facilities;
- local municipal district permits with respect to water;
- Texas Historical Commission determinations that the project should proceed with no impact to historic properties;
- a US Army Corps of Engineers Section 404 permit;
- Federal Aviation Administration Determinations of No Hazard (as well as related notices);
- Texas Commission on Environmental Quality permits with respect to pollutant discharges associated with construction activities, air quality for concrete batch plants, public water systems, waste management, characterisation, handling and disposal, and above ground/underground storage tanks;
- Texas Department of Transportation permits for highway work; and
- a FERC self-certification notice of exempt wholesale generator status.

---

19 See ERCOT Planning Guide, Section 5.4 (Study Processes and Procedures).

20 See ERCOT Planning Guide, Section 5.5.1 (Standard Generation Interconnection Agreement). A form of the SGIA can be found at: www.ercot.com/content/services/rq/re/SGIA.DOC.

21 See www.ferc.gov/industries/electric/indus-act/rto/ercot.asp.

5.2 Reliability requirements

The North American Electric Reliability Corporation (NERC) monitors and regulates the bulk reliability of electric transmission systems and interconnected electric generating units across the United States. NERC has delegated its authority in this regard among eight regional entities. In Texas, the Texas Reliability Entit, Inc (TRE) serves as NERC's regional reliability entity for the portion of the grid managed by ERCOT, the Western Electricity Coordinating Council (WECC) serves this function for the portion of the grid in the extreme Western portion of Texas, SPP serves this function for the portion of the grid managed by SPP, and the Midwest Reliability Organisation (MRO) serves this function for the portion of the grid managed by MISO. Solar energy projects are required to register as a generator owner/operator with NERC and its applicable regional entity prior to commercial operation. The projects will be required to comply with standards promulgated by NERC during operations.

5.3 Reports and studies

While not fitting neatly into permits or other required filings, there are a number of reports and studies that may be required in order to obtain financing to develop a Texas solar generation project. These may include: Phase I Environmental Site Assessments (and if additional study is required due to more intensive environmental issues, a Phase II Environmental Site Assessment), protected species and habitat reports, wetlands reports, transmission congestion reports, cultural resources surveys, geotechnical surveys, and microwave surveys.

6. Power offtake

Traditionally, renewable energy power purchase agreements (PPAs) have been the tool through which local utilities have fulfilled their requirements under renewable energy portfolio standards (RPS). Various state RPSs and related utility integrated resource planning were the primary instruments propelling the adoption of renewable energy (primarily wind energy) in the mid- to late-2000s and into the early 2010s. Having satisfied their RPS requirements in 1999 and again in 2005 (after an increase in requirements), Texas utilities are arguably not the primary long-term power offtake vehicle in the Texas market.[22] Instead, and as a result of the highly liquid nature of the ERCOT, MISO and SPP bulk power markets, power offtake options are expanding beyond traditional PPAs and into financial and other structured products with an increasingly diverse group of offtaker counterparties.

In 2009, ERCOT transitioned from a zonal market to a nodal market.[23] The new nodal market more accurately reflected the costs of transmission and congestion for any given unit of electricity. The shift into the nodal market created more trading and pricing points for electricity, which greatly increased price transparency and liquidity of trades. This transparency and liquidity has enabled the creation of financial products – corporate and industrial PPAs, renewable energy hedges and proxy revenue swaps, which as a general rule did not exist prior to 2009. While each of these

---

22 See "The Future of Renewable Energy is in Texas", CBS Money Watch (27 July 2017).
23 See www.ercot.com/about/profile/history.

products were generally launched to facilitate more wind energy transactions, they are increasingly being deployed for solar energy generation projects.

6.1 Power purchase agreements: general provisions

Common to all financeable power offtake agreements (whether in the form of the traditional utility PPA or the aforementioned financial products) are the following provisions:

- a fixed price for energy in favour of the project;
- a tenor that is in excess of the anticipated third party financing;
- easily calculable credit support posting requirements of the project, which do not exceed a finite amount; and
- robust rights of the project and the financing party to cure defaults.

Power offtake agreements may also contain provisions requiring:

- the achievement of certain development- and construction-related milestones with liquidated damages payable for failure to achieve those milestones;
- provisions requiring minimum mechanical availability and/or minimum energy output with liquidated damages payable for failure to achieve such required minimums;
- required credit support posting by offtakers; and
- detailed provisions regarding dispatch and curtailment of the project – although note that inclusion of these provisions varies widely as to which type of power offtake agreement is at issue.

6.2 Corporate and industrial power purchase agreements

Corporate and industrial (C&I) PPAs can either be 'physical' PPAs requiring the transmission of electricity directly from the energy project to the C&I; or they can be 'financial' PPAs whereby electricity is transmitted from the project to the grid and the project and the C&I customer swap revenue streams – the project exchanges its variable revenue stream received from sale of electricity to the grid for a fixed revenue stream from the C&I customer. In addition, the project conveys renewable energy credits (RECs) associated with its actual energy generation to the C&I customer, allowing the C&I customer to claim the 'green' benefits associated with the renewable energy.

6.3 Renewable energy hedges

Renewable energy hedges are another financial product[24] available in the Texas renewable energy PPA market. Like financial C&I PPAs, electricity is transmitted from the energy project to the grid and the project and the bank exchange revenue

---

24 Technically speaking, the renewable energy hedge does require a transfer of physical power of a contractually fixed volume of energy at a liquid trading hub. Nevertheless, we group the renewable energy hedge with other financial products since the hedge does not share the more traditional elements of a utility PPA of settling energy at the point of electrical interconnection and based upon the actual generation of a project.

streams. Unlike C&I PPAs, RECs are not generally conveyed to the bank. As a further differentiator from C&I PPAs, renewable energy hedges require a fixed volume of renewable energy, generally sized to a P95 or P99 certainty. As such, a project can potentially be short of the required fixed volume and would be required to purchase replacement electricity in the bulk power market to deliver to the bank.

6.4 Proxy revenue swaps

Among the various new financial products available in the Texas renewable energy PPA market, the proxy revenue swap (PRS) is the latest, having only been deployed in 2016. A PRS has many similarities to both C&I PPAs and renewable energy hedges. And like the renewable energy hedge, the PRS is sized to a fixed quantity. The PRS's differentiating feature is that it is sized based upon meteorological data corresponding to the particular project site and expected operational efficiency factor for the project rather than actual generation. Like the renewable energy hedge, the energy project is liable to make the swap counterparty whole to the extent its actual generation is less than the predicted electricity generated.

7. Construction

Construction of an energy project in Texas, while challenging and laborious in its own right, does not present especially unique challenges as compared to other US states. Utility-scale solar energy projects are generally constructed by third party contractors; and the terms of that construction are agreed through engineering, procurement and construction (EPC) contracts. EPC contracts contain terms with respect to payment, payment milestones, liquidated damages for delay, liquidated damages for failure of projects to perform as mechanically warranted and customary provisions with respect to additional representations, warranties, covenants and indemnities.

Nevertheless, there are a few Texas-specific provisions of note. First, under 34 Texas Administrative Code §3.291, the EPC contract must be structured as a separated contract in order to avoid sales tax being imposed on the entire contract. To that end, the purchase price must be broken out into:

- non-taxable labour (equipment, fees, services, insurance);
- non-taxable incorporated materials;
- taxable labour; taxable incorporated materials; and
- sales tax charged on taxable labour and taxable incorporated materials.

Second, Texas has a nuanced approach to mechanics' and materialmen's liens and related retainage. Under many US jurisdictions, contractors performing construction on real estate projects have vested liens and security interests – so-called 'mechanics' and materialmen's liens' – upon completing such work. These mechanics' and materialmen's liens generally have superior priority rights as against most other debtors' rights, other than rights in favour of taxing authorities. Texas' state constitution provides a mechanics' and materialmen's lien in favour of prime contractors (although not for subcontractors) that is automatic and self-executing.[25]

25 See Texas Constitution, Article XVI §37.

Texas also provides for statutory mechanics' and materialmen's liens in favour of both contractors and subcontractors provided that they file lien affidavits within the prescribed period.[26]

Notably, however, the risks relating to mechanics' and materialmen's liens are double-sided. While mechanics' and materialmen's liens in the case where owners fail to pay owed amounts allow contractors and subcontractors to recoup their proper payments, their imposition in the case where owners make full payments is problematical. In order to protect owners from mechanic's and materialmen's liens from subcontractors in the case where owners have made full payment to prime contractors (ie, the prime contractors have not made payment to the subcontractors), construction contracts in most US jurisdictions allow the project owner to retain anywhere from 5% to 20% of the construction contract price as 'retainage' in order to ensure that all subcontractors are properly paid. By way of further nuance, in Texas, Section 53.101 of the Texas Property Code requires that the project owner retains 10% of the construction contract price or the value of the work. Failure by project owners to retain such amounts entitles subcontractors and suppliers to mechanics' and materialmen's liens for at least the amount of the required retainage.[27]

Third, a Texas statute[28] prohibits indemnity provisions in construction contracts that require indemnitors to indemnify indemnitees for the indemnitees' negligence. This provision applies to construction contracts for which an indemnitor is provided or procures a consolidated insurance programme or property and casualty insurance. Claims for death or bodily injury of an indemnitor's employee are specifically excepted by statute. There are 11 additional statutory exclusions: consolidated insurance programmes; independent breach of contract or warranty actions; loan and financing documents (other than construction contracts to which lenders are a party); general agreements of indemnity required by sureties; workers' compensation benefits and protections; agreements subject to Chapter 127 of the Civil Practice and Remedies Code; licence or access agreements with railroad companies; indemnity provisions applying to copyright infringement claims; construction contracts pertaining to single-family homes, townhouses and duplexes; public works projects of municipalities; and joint defence agreements entered into after a claim is made.

Finally, a Texas statute[29] makes choice of law clauses voidable by the contractor or subcontractor if the real property that is the subject of the contract is in Texas and the choice of law clause makes the contract subject to another state's law, litigation in the courts of another state, or arbitration in another state.

8. Financing

Financing of the development, construction, operation and maintenance of solar

---

26 See Texas Property Code Chapter 53. The filing period for prime contractors is the 15th day of the fourth calendar month after the month in which the work was completed. The filing period for subcontractors is the last day of the last month in which work was performed.

27 See Texas Property Code §53.105.

28 See Texas Property Code §151.102.

29 See Texas Business and Commerce Code §272.001.

energy projects may take place through a combination of investment of sponsor equity, monetisation of tax benefits and non- or limited-recourse debt financing. Such financing parties generally scrutinise a solar project for a number of factors, including (among others): overall project economics; cohesive provisions among project contracts for allocating liquidated damages for delay and for deficiency; the general terms and conditions of the various project contracts; and overall execution ability of the particular developer. Texas solar energy projects may also present negative pricing risk and node to hub basis risk.

Negative pricing risk is the risk that a project will be required to pay to transmit its energy to the grid rather than receiving payment. Negative pricing occurs during periods when energy production exceeds available demand. Wind energy projects have been particularly affected by negative pricing risk in Texas since, as a general rule and with the notable exception of wind projects along the Texas Gulf Coast, their energy production peaks late at night and in the very early hours of the morning when electricity demand is lower. By contrast, solar energy production in Texas is generally coincident with peak loads (ie, higher insolation results in not only higher solar energy production but also higher demand for air conditioners). While negative pricing risk for Texas solar projects would not currently seem to be a large concern for financing parties, the risk likely increases with continuing buildout. Notably, California's solar energy production is significantly higher than that of Texas; and solar projects in that state carry significant negative pricing risk.

Second, as power offtake agreements (and in particular financial products) continue to move to financial settlements based on liquid trading hubs rather than projects' electrical interconnection points (nodes) to the grid (which are more common in utility PPAs), financing parties continue to scrutinise the risk in basis between such hubs and nodes. That basis risk is linked to transmission congestion, and the increase in projects built will generally increase basis risk in the absence of additional transmission build out. In recent years, wind projects in certain parts of West Texas have come under greater financial scrutiny as a result of outsized basis risk. Notably, hedge providers have begun to offer certain short-term basis risk hedging projects in the market and those products may provide comfort to financing parties in some degree.

8.1 Sponsor equity

Sponsor equity is an omnibus term for equity investments made by entities that are neither lenders nor traditional tax equity investors looking to monetise tax benefits. Sponsor equity includes the equity invested by the original developer of the project, but can also include equity investments made by third parties that did not develop the project (so-called 'cash equity investors'). Cash equity investors can be any combination of traditional utilities, private equity investment vehicles and sovereign wealth funds. Cash equity investments are generally structured as joint ventures between the developer and the cash equity investor, with various provisions around governance, timing of investment and exit being so negotiated.

8.2 Monetisation of tax benefits

Under 26 US §48, solar energy projects are eligible to earn the federal investment tax credit (ITC), which allows owners to deduct 30% of the basis that is invested in eligible property for projects placed in service prior to the end of 2019. For projects placed in service in 2020, the ITC drops to 26%; for projects placed in service in 2021, the ITC drops to 22%; and for all projects placed in service in 2022, the ITC reduces to a permanent 10%.[30]

In addition to the ITC, solar energy projects qualify for two additional tax benefits: a five-year accelerated depreciation through the Modified Accelerated Cost Recovery System (MACRS);[31] and bonus depreciation for projects placed in service in 2018 of 40% and for projects placed in service in 2019 of 30%.

While the tax benefits of the ITC, MACRs and bonus depreciation are available to owners of solar energy projects, most projects do not carry significant tax liabilities so as to be able to efficiently utilise such tax advantages. As such, projects generally try to monetise these tax benefits in favour of third party investors, who in turn provide needed capital to develop, construct and operate the project. Monetisation of these tax benefits can be via a number of structures (ie, the partnership flip, inverted leases or sale-leaseback transactions). In each of these transactions, care must be taken to ensure that third party investment is made at the appropriate stage of the development and construction cycle so as not to be too late to be eligible for the tax benefits. These transactions are also structured to include sufficiently tight project covenants requiring the developer to continue to own, operate and maintain the project so as to maximise the third party investor's return of and on investment. Finally, these transactions are structured to prohibit any transfer of ownership within the first five years after the date that the project is placed in service so as to avoid the risk that the tax benefits will be 'recaptured' by the IRS.

8.3 Debt financing

In addition to sponsor equity and the monetisation of tax benefits, solar energy project developers may also choose to incur debt financing in order to complete the capital stack to enable the development, construction and operation of a project. Debt financing of utility-scale solar energy projects will generally take place in the traditional project finance, rather than enterprise finance markets. The former is based on principles of non-recourse or limited recourse lending, and therefore carry a number of requirements designed to protect the creditors' interest in the project, as the sole source of collateral for their loans.

In addition, debt financing must take place so as to allow full monetisation of the tax benefits. While traditional project financing has entailed loans taken by the project directly with an all assets lien in favour of the lenders during the construction and/or operational period, the current tax equity market generally disfavours the incurrence of debt financing at the project level during the period in which tax equity investors are invested in the project. So while debt financing and an all assets lien may be put into place during the construction period – prior to the point in

---

30 See http://programs.dsireusa.org/system/program/detail/658.
31 See www.seia.org/initiatives/depreciation-solar-energy-property-macrs.

which the project is placed in service and the tax equity investment is made – operational debt financing must instead be placed with an entity upstream of the tax equity investor's investment (so-called backleveraged financing).

Backleveraged financing does not place liens directly onto the solar energy project, but instead approximates the first lien position by taking a pledge of the ownership interests in the entity that holds the sponsor's equity. Backleveraged financing will contain robust provisions as to the operation of the project, including the requirements to operate the project in accordance with applicable laws and prudent industry practices, and will also contain robust debt service and cash maintenance covenants, etc.

9. Conclusion

ERCOT's 2016 Long-Term System Assessment predicts increasing penetration of solar energy in each of its seven load and generation forecasts (including scenarios of decreasing load),[32] and an increasing number of financial products are coming to market to enable solar energy development. These two factors suggest a bright long-term outlook for the development of Texas solar energy projects. In the short-term, however, two developments have the potential to create significant headwinds for development of solar projects in Texas.

First, the Tax Cuts and Jobs Act of 2017 was passed into the law at the end of 2017. The Act reduced the federal corporate income tax rate from 35% to 21%. That reduction inherently makes the benefits of the ITC less valuable to tax equity investors now carrying less overall tax liability, and thereby increases the cost of funding to solar energy projects.

Second, on 22 January 2018, President Trump adopted recommendations of the International Trade Commission to impose a 30% tariff on certain imported solar panels from Asia.[33] The tariffs reduce over a four-year period and the first 2.5 GW are excluded in each of those four years.[34]

Solar project developers are assessing the impacts of both the corporate income tax rate reductions and the tariffs on their returns and pricing economics. It is therefore too soon to say how this combination of actions will impact Texas solar development other than to say that development will likely be less lucrative than it was in prior years. And if early (and mostly anecdotal) market indications hold, the negative impact may be little more than a speed bump in a longer-term wave of solar energy development in Texas.

---

32 See www.ercot.com/content/wcm/lists/89476/2016_Long_Term_System_Assessment_for_the_ERCOT_Region.pdf.

33 See www.greentechmedia.com/articles/read/breaking-trump-admin-issues-a-30-solar-tariff#gs.

34 *Id.*

The future for solar power

Dan Wells
Foresight Group

1. Introduction

As shown elsewhere in this book, solar power has become one of the most important forms of new power generation being added to grids worldwide. We believe this is a trend which will continue and that in future solar will be a fundamental component of energy systems, and in some places dominate the generating mix. The year 2016 represented a significant inflection point, as for the first time renewable energy generation represented a majority of global capacity additions onto the grid worldwide.[1]

In order to understand the role solar will play in future energy systems it is necessary to appreciate that there are three key attributes of solar power which drive its relationship with the wider grid. First, solar energy – in addition to other renewable sources such as wind – is cheap, quick to build and clean (ie, zero emissions). These characteristics mean it will continue to outcompete other energy sources. Second, solar is variable in its output, creating the requirement for a more 'flexible' grid infrastructure such as energy storage and peaking power plants to offset renewables' intermittency. Finally, solar is more widely distributed in its location around the grid than traditional fossil fuel plants, which creates the need for investment in transmission and distribution infrastructure.

The result of the rise to pre-eminence of renewable energy (and solar in particular) with these characteristics layered in, is that for the first time in over a century energy systems worldwide are being fundamentally redesigned as we move away from the centralised, fossil fuel intensive system towards a future characterised by a system where power is generated at a large number of smaller facilities, more widely spread around the network. Instead of principally flowing in a linear manner from centralised generation assets to a widespread network of consumers, the new style of energy distribution means that electricity is generated at a number of different points optimal for renewable production and moves in increasingly complex patterns around the grid.

While the march of solar appears unstoppable when viewed from a high level, this overall dynamic will mask some nuances that are important to understand. First, the roll out of solar will play out in different ways around the world. There is a stark comparison between the developed world where solar power is being added incrementally to an existing though ageing set of established infrastructure, and

1 Source: BNEF Global Renewable Energy Outlook.

developing countries where in many cases solar is being installed in areas lacking an incumbent system. Second, there are potential limiting factors specific to the roll out of solar, and to some degree the value of adding further solar to the grid lessens the more solar is installed, principally because of limits to production times. However, the emergence of other technologies such as energy storage will help to address the limitations of solar power ensuring that its value to the system is increased.

Solar is in fact only one instrument in the overall 'symphony orchestra' of the modern power grid. Its interaction with other generating technologies, flexible resources such as energy storage and with modern, efficient transmission and distribution infrastructure is complex. As we reach greater levels of penetration of solar into the energy mix these interactions will evolve in ways we cannot currently foresee. However, for now, what we can be sure of is that significantly more solar power will be installed for many years to come.

2. The three characteristics of solar power which will shape the grid's future

First, it is necessary to set out the key characteristics of solar power that shape its relationship with the wider energy system.

2.1 Solar power is cheap, quick to build and clean

Solar power is now exceptionally cheap to build. The price of the crystalline silicone modules required for PV solar decreased from US$82/W to US$0.41/W by the end of 2016. This trend is set to continue; it has been forecast that the levelised cost of electricity (LCOE[2]) from solar PV will drop by 66% from now until 2040.[3] We will explore the reasons for these cost reductions in more detail below, but at the construction stage an equally important factor is the speed with which solar power plants can be built. While a conventional fossil fuel or hydro power plant may take four to five years (or in the case of the nuclear plants, even decades) for planning, development and construction, solar parks can be fully built relatively quickly, often in six to 12 months.[4] Once built, solar power is also extremely cheap to operate and as a result will increasingly out-compete other forms of generation. This is partly because solar power has zero feedstock costs – sunshine is free – and solar parks have few moving parts. The reduction of operating costs due to the industry's many years of experience of running solar power plants is also a significant factor.

Solar, of course, also has a near zero-emissions profile which makes it a key pillar of international decarbonisation efforts. Taken together with the cost and speed of installing solar, this explains why it has risen to prominence on the international energy agenda.

2.2 Solar power's output is not constant

By its nature solar power is intermittent and variable. This occurs in several ways and

---

2 BNEF definition: "Long-term offtake price to achieve the required equity hurdle rate for a developer, considering its total capital operating and finance costs over the life of the project".

3 "Solar modules to get even cheaper and more efficient", BNEF, 30 May 2017.

4 "Utility-Scale Solar Photovoltaic Power Plants", International Finance Corporation, 2015.

over different timescales; most obviously solar power only produces energy during the daytime, but its output also varies considerably on a day-by-day basis in relation to cloud cover at any given time. Solar output also varies on a seasonal basis; in Germany, for instance, solar power produces roughly six times as much energy during the summer as compared to the winter.[5] As a consequence, solar power creates the need to build 'flexibility' into energy systems in order to to absorb excess power, to produce replacement power at short notice or to reduce demand at a given time. This flexibility can take the form of energy storage, fast response peaking power, demand response, or interconnections with other energy systems.

2.3 Solar power is relatively highly distributed on the grid

Renewable energy generation sources such as solar tend to be more spread out around the margins of power systems as compared with conventional, centralised fossil fuel generation. This is partly a function of the modular nature of solar – solar power can work effectively at any scale. This can be highly advantageous as smaller plants – sub 50 MW for instance – can be developed much more easily and quickly from a permitting and planning point of view. The consequence of this is that solar sites are often located in the outer 'Distribution' elements of the grid, rather than the central 'Transmission' system. Consumers of power, both residential and commercial, may also use solar to generate power for their own use, selling any excess back to the grid. As a result of this, electrons are beginning to flow in multiple directions on the system, rather than just in a linear direction from generator to consumer. On a larger scale, solar power is of course better suited to sunnier locations, which may not be in close proximity to centres of load (power demand). These factors result in a requirement for substantial investment into the transmission and distribution infrastructure, and the emergence of a more decentralised system.

The importance of these characteristics will become clear as we set out a vision of how the energy system may develop in future, and the role for solar in this regard.

3. Extrapolating the current trajectory of solar

As a result of the above factors the amount of solar power, as well as other forms of renewable energy such as wind, are high on agendas of energy market participants around the world. US$10.2 trillion is earmarked to be invested into new power generation capacity worldwide before 2040, of which 86% will be from zero-emission sources.[6]

The original catalyst for the emergence of solar was policy driven and borne out of an increasing imperative for the reduction of carbon emissions. The international decarbonisation agenda was manifested in a series of international agreements on greenhouse gas emission reduction. Dating back to the establishment of the UN Framework Convention on Climate Change (UNFCCC) in 1994, the international decarbonisation agenda is most evidenced by the 2016 Paris Agreement (the United

5 "Wind and Solar – better together?", BNEF, 2 May 2016.
6 "New Energy Outlook 2017: Global Overview", BNEF, 15 June 2017.

States has signalled an intent to withdraw from the Paris Agreement, making it one of only two countries in the world not to be party to the agreement, the other being Syria). The green agenda has percolated from these international agreements down to national and supra-national law setting across the world. The European Union, for instance, has a target to cut greenhouse gas emissions to 80% of 1990 levels by 2050, and, in order to achieve this, a significant commitment to renewable energy generation, and the efficient management of that energy, is needed.

Solar power's environmental credentials have been an important factor in its rise; conversely for fossil fuel – and coal in particular – concern over both carbon dioxide emissions and the production of other localised pollutants have compounded the decline of the sector in terms of share of generation. In the United Kingdom alone, five coal plants with an aggregated capacity of more than 5 GW retired between December 2015 and October 2016. A further 7.4 GW is set to retire by 2020, and the remaining 5.5 GW to come offline by 2025. The contribution to the United Kingdom's reduction of carbon emissions that these removals will have is material, however, it does leave a potentially substantial gap in the supply market which government must ensure is successfully and seamlessly bridged. Despite the stated intent of removing itself from the Paris Agreement, the United States is following a similar pattern,with 18 GW of coal retired in 2015, 13 GW in 2016 and a further 8 GW expected to retire in 2017. Coal-fired generators produced 30% of the total generation in the United States in 2016, down from over 50% in 2003.

While the background to the rise of solar was this emergent set of policy frameworks, what caused the explosion in deployment was the subsequent reduction in costs of equipment, as referred to above. This decrease was made possible by a combination of three elements – economies of scale, technological innovations, and manufacturing experience – all of which are likely to aid in further reductions in the coming years. According to a 2017 industry survey conducted by Bloomberg New Energy Finance (BNEF) around the PV manufacturing chain, cost improvements to solar modules are set to continue to improve.[7] Technological updates are also due to keep coming; polysilicon uses half as much energy today as it did five years ago, and advances suggest that efficiencies in energy use will continue to improve.

The consequence of these factors is that the short, medium and even long-term prospects of solar look very optimistic. Of the $10.2 trillion requirement for investment in energy infrastructure referred to above, solar power will likely make up US$2.8 trillion and investment growth in the sector is forecasted globally at 2.3% on average per year. By 2040, solar could account for 32% of the total installed capacity of energy generation worldwide, compared with just 5% today.[8] Indicators of what the future may hold can be seen in recent statistics – while the current installed global share of capacity from solar may seem small at circa 5%, growth in installed capacity has been exponential in the last five years, and in 2016 more GW of solar was added globally (75 GW) than of any other technology.[9]

---

7 "Solar Modules to Get Even Cheaper and More Efficient", BNEF, 30 May 2017.
8 "New Energy Outlook 2017: Global Overview", BNEF, 15 June 2017.
9 "Global Trends in Renewable Energy Investment 2017", BNEF/UNEP, 2017.

4. Contrasting the development of solar in different regions globally

4.1 Developed versus developing world

As an energy source, and an alternative to traditional methods of energy production, solar plays an important but very different role in all corners of the world.

While in developing countries solar technology is solving rudimentary issues such as providing power in schools and hospitals, in the developed world it is playing a much larger part in working towards carbon emission reduction targets that are becoming more ambitious as technologies continue to be developed. Although just one part of a much larger picture, solar is a prime example of how fast technology can evolve to make these goals achievable, and alongside wind turbine technology it has pioneered the transition to a renewables-focused grid and a decentralised network.

In the developed world, energy experts are navigating established infrastructure to install solar retrospectively into existing networks, encountering hurdles unique to the new, intermittent style of solar energy generation such as production downtimes, storage, grid outages, challenges resulting from the inability to store electricity, transmission losses/leakage and loss of power in transportation.

Solar technology has reached a level of mainstream in developed countries such that it is now being included as a matter of course into general construction. Many new-build home developments embed residential solar as an energy source, and commercial buildings are doing the same. Efficiencies such as connectivity points and optimal positioning can be taken into consideration at early stages of construction ensuring that solar generation can be used to its full potential. When fitting modules to buildings retrospectively, compromises must often be made on the value of changing the infrastructure or sacrificing optimal panel efficiencies. We are also seeing an increased number of solar modules integrated into building materials, for example Tesla's roof tiles with PV technology, or canvases with PV fibres threaded through.

Due to resource and financial constraint, a number of countries in the developing world lack the basic infrastructure to be able to supply enough power to their whole population. In June 2016 over 18,000 villages in India had no access to electricity; at the same time President Modi committed to provide electricity across the whole of India within 703 days through off-grid electrification schemes. Although an optimistic target, this feat would have been nearly impossible by installing conventional grid infrastructure. However, the rapid reduction in the costs of solar panels, wind turbines and battery storage systems, have opened up new opportunities to supply electricity directly to unelectrified villages by establishing microgrids.

In the developing world, the accessibility of renewable energy has the power to significantly change the lives of local people. The scarce, and in some cases total absence of, reliable access to electricity through power lines or traditional systems (primarily due to constraints in resources and financial support) prohibits much of the developing world from being able to offer support services to their communities such as adequate hospitals, power in schools, and access services such as banking and telecommunication networks.

Elon Musk, CEO of Tesla and Solar City, believes that declining costs in such technologies will allow the world's poorest citizens to leapfrog connectivity to a centralised network, and enter straight into a decentralised network, comparing it to the introduction of mobile phones to areas without telephone lines. This allows places such as hospitals to have small installations on their sites that cost little to install and maintain, but that can provide the consistent power required to keep the lights on, critical machines running, and even increase the possibility of offering more sophisticated procedures to the communities they serve.

The cost of decentralised energy infrastructure is more feasible for developing nations than the traditional central network. In reality, the introduction of solar power to developing nations catapults them through several decades, straight into the 21st century, not only with positive implications on power supplies to infrastructure such as schools and hospitals, but also because it often brings people into the banking system for the first time (solar systems are often paid for by customers over time via mobile phone-based payment systems; hence for the first time such people have access to credit).

5. Obstacles in the path of further solar build out

Because the entire solar fleet in any given location produces power at the same time – and such a time may not correspond with peak demand – generally the value of each incremental unit of solar added to the grid reduces with the increase of installed modules. California is a 'sweet spot' for solar generation with its sunny weather and high irradiance levels, yet the peak power demand for the state is in the evening, after the sun has gone down. If solar were to generate power at all hours, each component would bring with it extra value, however, as modules are only meaningful during times when demand is not at its peak, much of the energy produced is wasted, bringing the value of the panels down. Additionally, in most power markets generators are stacked for dispatch by their running costs, and so solar and wind tend to undercut existing coal, nuclear and gas prices. Therefore, marginal return of solar or wind further decreases as capacity is increased because at times of peak resource (ie, high irradiation or wind speed), the market demand is spread across a greater supply.

A further challenge to the rise of solar is the complexities it presents to system operators. Historically, supply and demand balancing within systems was reasonably easy to predict and manage, and was very much one-sided. Grid operators had only to be concerned about managing peaks in demand, as supply could be easily controlled as fossil fuels can be stored in natural form and burnt at a rate that matches expected demand. However, it is no longer that simple for three reasons: the increase in intermittent generation, the diminishment of existing flexible sources, and greater peaking demand in developed economies.

A combination of increasingly uncertain energy generation sources and less predictable demand profiles has added a new level of complexity to demand/supply balancing, which must be continuously managed across a variety of time frames. For simplicity, there are three main time horizons, over which network operators must look to smooth fluctuations in demand and supply in order to minimise imbalances.

All of the factors mentioned above mean that total energy demand is becoming harder to forecast and is generally more volatile. Peaks in demand have become much harder to match, and it is also becoming evident that times of high resource in renewables will rarely coincide with times of peak demand.

Energy storage is often identified as a potential solution to the intermittency of renewables. What is not often articulated is the degree to which this intermittency can vary. More precisely, energy storage is required to address challenges presented at three very different timescales – all of which solar has an impact on – and the ability of storage to address each of these differs markedly.

First and foremost, there is a requirement for second-by-second stabilising. In order to operate efficiently, a grid's frequency is effectively the balance of extremely short-term (sub-second) fluctuations in demand and supply. During peak production periods, balancing this fluctuation is not a problem for renewable energy plants as energy is readily available, however, during periods when production is not possible, appropriate storage solutions need to be deployed. Traditionally, this 'frequency regulation' role was provided by large, spinning fossil fuel turbines. As coal plants in particular are retired, system inertia decreases and the need for other tools to regulate frequency – such as battery storage – is increased.

The second type of reserve power that needs to be considered is diurnal supply which is balancing fluctuations across hours or days. It is difficult to forecast a peak in wind generation, but solar has a regular peak in the middle of the day at which irradiation levels are optimum for producing power, should it be needed at that time. As has been mentioned earlier in the chapter, the hurdle in this situation is the ability to store any excess energy that is produced at peak times to have in reserve for times when consumption demand is higher.

Finally, we need to consider the seasonal profile of energy demand. Throughout seasons of a year there are periods of greater and lesser consumption so the reliance on baseload power varies. For example, in the winter electricity consumption is much higher due to an increased reliance on heating, a greater use of appliances such as ovens and kettles, and a decrease in daylight hours. The problem in this case is that while wind output may be greater it is hard to forecast, and by nature there is a significant decrease in daylight hours over winter, so grid controllers rely much more heavily on consistent and predictable methods of energy production such as thermal power.

While technology exists to address the imbalance in the smaller time periods of second-by-second and diurnal supply, what is currently lacking is a solution for seasonal disparity in supply and demand. This is a question that comes into play in the discussion of long-term 'deep carbonisation' and could be the main underlying reason that we may never make the leap from 80% to 100% of energy generation being renewable. Traditional fossil fuel generating plants have the capabilities to regulate these disparities which have not yet been translated into intermittent generation methods.

6. The energy symphony orchestra; solar is just one instrument

As the above illustrates, demand is increasing for infrastructure to manage energy efficiently in addition to generating it. While solar is an integral part of the energy landscape of the future, it is important to note that it is one instrument in a symphony orchestra, working alongside other technologies and processes such as wind, storage, flexible generation and distribution to name just a few. These technologies and processes are needed to balance the generation times and capabilities of solar, to aid in how the energy produced on solar farms is stored, managed and distributed, and to assist the transition to, and adoption of, a decentralised grid.

Similar patterns are expected in the wind sector with the levelised cost of energy (LCOE) of onshore wind set to drop by 2040 due to the improved efficiency of turbines, and the streamlining of procedures such as operations and maintenance (O&M).

Another critical instrument in this energy symphony is the use of wind turbine technology to generate power. Although not as fast as solar, the cost of wind technology has also fallen. Since 2007, the cost of wind turbines has fallen from €1.32/W to €0.86/W in 2017, which is roughly 35% and equivalent to a learning rate of circa 9%.

The real advances in the generation of energy through wind are better reflected through turbine size and efficiency, as opposed to unit price as they are with solar. The size and efficiency of these turbines is critical, as there are many fewer individual machines attributed with creating a larger volume of power than there are on a solar farm, where there can be thousands of modules producing small amounts of electricity.

These reductions in equipment costs, alongside improving efficiency of turbines, and declining financing and development costs are expected to reduce the global benchmark LCOE for onshore wind by nearly 47% by 2040.

As previously mentioned, the development of viable storage solutions is crucial to the future balance of the energy grid, storing energy for use in peak demand times, or when production is not possible. At present the most attention is being focused on lithium-ion batteries but there is a wide range of other technologies available, some of which, such as pumped hydro, are well established, and others which are at various stages of development, including compressed air, flow batteries and use of hydrogen gas.

Lithium-ion batteries have been marked as the next 'hero' in energy storage, and to-date have been the prevailing technology of choice for utility-scale battery storage projects, helping to manage supply-side variability at grid level. Nearly 645 MW of utility-scale energy storage projects were commissioned globally in 2016, up from just 25 MW commissioned in 2012. The lithium-ion battery energy storage market is expected to become a $20 billion per year market in 2040, which even at current battery prices would imply around 33 GW deployed each year.[10]

Furthermore, 530 million new electric vehicles are expected on the road by 2040, the vast majority of which will likely be powered by lithium-ion batteries. Global EV lithium-ion battery demand is expected to grow from 47 GWh in 2017 to 1,294 GWh

---

10 "Dataset: Utility-scale lithium-ion battery system costs for greenfield developments", BNEF.

by 2030. If this growing fleet of grid connected lithium-ion EV batteries could be feasibly harnessed by grid operators, it could unlock a huge potential of storage capacity worldwide. This growing fleet of electric vehicles connected to grid infrastructure networks could potentially offer storage capabilities to network operators worldwide.

As well as their capabilities in commercial use, it is becoming increasingly popular for lithium-ion batteries to be found in a domestic situation. This ownership of energy storage by consumers has positive connotations for the grid as a whole as it helps to level peak demand volatility if people are using their own reserves in these peak times.

This widespread use of lithium-ion batteries has become possible due to the significant decrease in price since 2010. The average global price of one of these battery packs in 2016 was $273/kWh, down from $1,000/kWh in 2010. BNEF has forecast that the cost of lithium-ion battery packs will more than halve again to $73/kWh by 2030 based on an observed learning rate of 19%.

Pumped-hydro storage is the process of pumping water uphill from one reservoir into another reservoir at times of low demand and high supply of electricity. When the system starts to become strained, the water is released downhill through pipes to drive turbines and generate electricity. Pumped-hydro is a crucial instrument in the larger orchestra of the future energy grid as it is used largely for managing infrequencies until large power stations can come online. This means that the imbalances it is managing are typically hourly, or minute-by-minute loads.

Producing hydrogen using electrolysis and then storing the hydrogen either in fuel cells or by integrating it into existing natural gas networks, offers a solution for storing energy over longer time periods. Some natural gas networks can tolerate a dilution of up to 9% of the fuel make up comprising hydrogen, with the remainder being natural gas. Japan is one of the pioneering countries in the technology, channelling investment into research and development of hydrogen storage systems, such as the system in Fukushima. Developed by Toshiba Corporation and a group of others, the 10 MW system will convert renewable energy into a large volume of hydrogen for long-term storage. However, despite the early signs of progress in the sector the economics are still some way from stacking up. A recent BNEF study found that the cost of storing 1 MWh in Germany for three months through hydrogen technologies is still €244/MWh, which is far higher than the cost of generation for baseload fossil fuel generators.

In addition to hydrogen, flow batteries and compressed air systems are also capable of storing energy over long periods. Flow batteries are liquid batteries that hold their charge almost indefinitely. Compressed air energy storage stores large amounts of air at high pressure in underground caverns, which can be heated to drive power generators. Like hydrogen, these two technologies are currently still very costly, again making the economics difficult to attract investment.

These different technologies have a role to play in the three timescales set out above of second-by-second, diurnal and seasonal balancing. Given its fast response, lithium-ion is very effective in addressing the first of these but it is still too costly to provide an effective multi-hour back-up at scale, especially given its degradation profile. This diurnal requirement is currently met primarily from pumped hydro and 'flexible generation', being small-scale power plants embedded into the distribution network that are able to offer a range of power and balancing services to the grid. These projects can come online within minutes making them a great source of reserve or 'back-up' power for grid systems. The most common technology currently being used are gas-fired reciprocating engines. These peaking power plants help to smooth fluctuations in demand and supply, either through the provision of balancing services to grid operators, or simply by exploiting the growing volatility in hourly wholesale power prices. Flow batteries, compressed air and hydrogen have roles to play in this application too.

Another instrument in the energy symphony is the development in long-distance transmission infrastructure, which are the lines that transport energy from high production areas to high usage areas. One feature of the decentralised grid is that electricity generation sites are often close in proximity to the end power user due to historic inefficiencies in transporting the electricity across long distances. This has changed with developments in long-distance transmission infrastructure. Modern cables are significantly more efficient than they once were and use high voltage direct current (HVDC) rather than the alternating current that is often used to transport electricity across shorter distances. HVDC becomes the cheaper alternative to AC after approximately 500km and can be more easily connected to weaker grids, providing a bridge between networks operating at different electricity frequencies. HVDC systems can also accommodate devices that allow the flow to be controlled more precisely than can be allowed for with AC, enabling greater grid flexibility and offering possible solutions for supply/demand infrequencies.[11]

The technology is in a place now where theoretically it should be possible to transport large amounts of energy from places of high generation (solar in North Africa) across significant distances to places of high demand (Europe) meaning that intermittent generation projects could be located where conditions are optimum, rather than where the demand is high. A project has recently been commissioned in the United States which will transport electricity using HVDC over 1,100 km from Oklahoma's wind farms for use in Arkansas and Tennessee.

It is also worth noting that while we have focused on PV solar, other forms of solar power could have a significant role to play in future. One such technology is Concentrated Solar Power (CSP). CSP has not experienced rapid growth for several reasons – it has not benefited from the rapid cost declines associated with solar PV, CSP plants typically take longer to construct and require a prolonged period of optimisation in order to maximise power output, and CSP is less modular in nature meaning that plants usually need to have substantial scale in order to operate efficiently. However, there are signs that CSP is gaining traction. A contract has been signed for Australia's first CSP plant, which will also have the ability to store energy generated during the day for use at peak times, so the plant is not expected to be fully operational until 2020.[12]

11 Case Study: High Voltage Transmission for Renewables, BNEF.
12 "Australia CSP contract uses daytime load to access peak prices", New Energy Update: CSP, 2017.

Efforts are underway to facilitate CSP at smaller scales – Sandria National Laboratories, for instance, have designed small-scale receivers that can be printed for 1 MW generating sites in India which will be used to serve small, rural communities.[13]

7. Conclusion: the long-term future of solar

As mentioned earlier in the chapter, the decentralised nature of the modern grid means that the whole system has to work together in a much more cohesive way than it did previously. As energy consumers, we will only be able to get the most out of one aspect if it is synchronised correctly with the next, the prime example being solar production and storage for use in low supply and high demand periods. This simple fact makes it difficult to predict the future of one element in isolation, and encourages us to look at the orchestra in its entirety.

What will happen with renewables in the very long term? The truth is we are not entirely sure.

There is a debate as to how far the renewables market can develop, but while previously the debate has been whether it can achieve a market share of over 50% renewables with technologies such as wind turbines, solar panels and hydro generators, the debate is now focused around whether we can ever achieve a 100% renewable grid, or whether we will struggle to get past 80%, relying on nuclear or fossil fuel for the remaining 20% with carbon capture and storage.

The question mark hanging over that 20% stems from uncertainty as to whether the grid can cope with 100% variable baseload, or whether the predictable and reliable nature of fossil fuels is required in order to retain balance in a decentralised network.

The point is that we simply do not know. The advancement of technologies in the last 20 years is testament to the fact that a lot can change in a very short space of time, and technologies such as nuclear fusion remain a possibility which, if successful, provide what is essentially a guarantee of unlimited clean energy, removing any issues of volatility, intermittent generation and storage.

For now, the most effective approach is to keep building as much solar as possible, recognising that it is one of the cheapest, cleanest forms of new generation, while continuing to focus efforts on developing technologies and systems to accommodate solar power such as energy storage and high efficiency transmission and distribution infrastructure. This will enable solar power to continue its development as a fundamental technology in modern life; while today we are seeing the integration of PV solar into building materials, in future we may see it as a pervasive characteristic throughout society as it is built into the outer layers of automobiles, ships and even roads, for instance. Solar power will therefore become increasingly woven into the fabric of our lives.

---

13 "3D printing enables new high-efficiency CSP receivers", New Energy Update: CSP, 2017.

About the authors

Elena Aldave
Associate, Pérez-Llorca
ealdave@perezllorca.com

Elena Aldave Yániz is the current associate at Pérez-Llorca's London office. Elena joined Pérez-Llorca in 2015 and specialises in corporate and commercial law. Elena advises national and international clients on all types of corporate transactions, especially on corporate restructurings, mergers and acquisitions.

Apart from Spanish which is her native tongue she is fluent in English and French.

Dario Bertagna
Vice president, Capital Dynamics
DBertagna@capdyn.com

Dario Bertagna is a renewables and solar energy expert having spent most of his career developing and advising funds investing in renewables, clean infrastructure and energy assets. He is currently a vice president in the clean energy infrastructure team at Capital Dynamics in London.

Prior to joining Capital Dynamics, Dario held investment management positions at Bluefield Partners LLP and at Platina Partners where he worked across multiple renewable energy transactions and project financings in the United Kingdom and Europe. His previous experience also includes roles at the international energy developer MedEnergy and in the Energy Corporate Finance team at Jefferies International.

Dario holds a BA in Economics and International Trades from the University of Turin and a Masters in European Business from ESCP Europe.

Elisabeth Blunsdon
Of counsel, Orrick, Herrington & Sutcliffe LLP
eblunsdon@orrick.com

Lis Blunsdon is an English qualified solicitor with more than 20 years' experience in the electricity sector. She has acted for developers, utilities, lenders and governments in negotiating power offtake agreements for facilities deploying a diverse range of power technologies. In recent years she has been heavily involved in the UK renewables sector with a particular focus on solar.

Foued Bourabiat
Founder, Bourabiat Asssociés
f.bourabiat@bourabiatassocies.com

A graduate of King's College London (LLM) and the University Paris 1 Panthéon-Sorbonne (post-graduate degree in English and American business law), Foued Bourabiat worked for top-tier law firms in Paris (Hogan Lovells LLP and Herbert Smith Freehills LLP) for almost 10 years before founding Bourabiat Associés in 2012. He has succeeded in establishing his firm as the go-to firm for complex transactions in Algeria. He is recognised by all the major international legal guides as a leading lawyer in Algeria.

His experience includes assisting and advising major investors and credit institutions in respect of investments and financings in the MENA region, including in Algeria. He has been involved in complex cross-border financing transactions and cutting-edge project finance deals, notably in the energy sector. He is a member of the committee of experts in charge of assisting and advising the Ministry of Environment and Renewable Energy on the legal reforms relating to the renewable energy sector.

Keith Bullen
Partner, DLA Piper
Keith.Bullen@DLAPiper.com

Keith Bullen is an English-qualified solicitor and a partner in DLA Piper's projects and finance group. Keith has been active in the Middle East project finance sector for nearly a decade and has worked on many of the region's largest utilities projects. He has a strong track record in advising procurers and offtakers. In Saudi Arabia alone he has advised on more than 10 GW of power projects.

Keith has advised multiple procurers on the structuring of their first solar PV, CSP and wind projects. Converting conventional IPP models to suit the requirements of renewable energy and unbundled utility markets has been a significant portion of Keith's workload in recent years. These government mandates give Keith a detailed insight into energy policy and regulatory change.

Keith has also advised on projects outside of public procurement frameworks, including captive power projects and solar PV systems under net metering schemes.

Alberto Cuellar
Senior renewable energy consultant – solar, Mott Macdonald
alberto.cuellar@mottmac.com

Alberto Cuellar is an MSc qualified Industrial Engineer with a MSc in Renewable Energy and international education in Spain, the United States and Germany. Alberto has in-depth technical expertise with over eight years' international experience as project engineer and project manager in the solar industry both in Concentrating Solar Power (CSP) and Photovoltaic (PV) projects advising developers, investors and banks. He has worked in a wide range of roles across different project development stages from techno-economical feasibility to detailed engineering and construction and operation monitoring.

John Deacon
Partner, Orrick, Herrington & Sutcliffe LLP
john.deacon@orrick.com

John is a partner in the Energy and Infrasturcture team at Orrick. With over 30 years' experience in the preparation and negotiation of the contracts for design, procurement, construction and operation of energy projects, John has been recognised for many years by *Chambers Directory* as a "Leading Individual" for renewable energy. John has worked on renewable energy projects across Europe, Africa, Asia and the Americas. John holds a Bachelor of Laws degree and a Master of Science degree from the department of engineering at King's College London.

William Evans
Partner, Elgar Middleton Infrastructure & Energy Finance LLP
williamevans@elgarmiddleton.com

William Evans is a partner in Elgar Middleton's London office. He specialises in infrastructure and renewable energy with 17 years' experience from projects across the United Kingdom and Ireland.

William leads each and every transaction he works on. This includes sell-side, buy-side and debt financing mandates acting for investors, major corporates and developers. In renewable energy, William has worked on solar, wind, anaerobic digestion and biomass projects. While in infrastructure, he has worked on PPP/PFI, transport and student accommodation.

Elgar Middleton was founded in 2012 with a specific goal to act as an independent financial advisory boutique for its clients. The partnership has offices in London and Sydney and has advised on over 50 discrete renewable energy projects in the past four years enabling its clients to raise over $US3 billion of capital in that period.

Niklas Ganssauge
Partner, CMS Germany
niklas.ganssauge@cms-hs.com

Niklas Ganssauge joined CMS in 2005 and became a partner in 2012. He is a member of the energy law group of CMS Germany and has special expertise in the field of renewable energy.

Niklas advises project developers, plant manufacturers, investors and banks on the planning, construction, acquisition and financing of wind and solar farms in Germany and abroad. He has special expertise in drafting and negotiating EPC, supply, maintenance and service agreements as well as construction contracts. His practice also covers all real estate law aspects of energy projects as well as advising on M&A transactions in the energy industry. Niklas is a lecturer at the Frankfurt School of Finance and Management (renewable energy finance) and at the Bucerius Law School (international construction agreements).

Nasif Hamed
Associate, Pérez-Llorca
nhamed@perezllorca.com

Nasif Hamed joined Pérez-Llorca in 2014, after having practised law at Vector Cuatro for two and a half years. Prior to his experience there, he was a trainee lawyer at Clifford Chance (Spain).

Nasif specialises in corporate and commercial law. His main practice covers mergers and acquisitions, as well as corporate restructuring, including share capital increase and decrease. In addition, Nasif has advised on national and cross-border transactions related to investment and divestment in infrastructure projects and renewable energy projects.

Nasif lectures for the courses on Corporate Law at Universidad Carlos III (Madrid) and also at Universidad Pontificia Comillas ICAI-ICADE.

Mona Hammadi
Associate, DLA Piper
Mona.Hammadi@DLAPiper.com

Mona Hammadi is an English-qualified solicitor. She is a Bahraini national holding both undergraduate and post-graduate degrees from UK universities. She is a member of DLA Piper's projects and finance group. Mona has practised in the region for more than eight years. She has worked on IPP, IWP and IWPP procurements and financings throughout the region and has particular expertise in Saudi Arabia. As the regional focus has shifted from conventional power to renewables, she has been actively involved in advising procurers on a number of 'first' renewables public procurements and initiatives in the region.

While Mona's practice has primarily focused on public procurement and project finance in the utilities sector, she also advises on infrastructure, PPP projects and a broad range of banking and finance transactions. Clients in those transactions include international and regional financial institutions, corporates, sponsors and developers.

Munir Hassan
Head of clean energy, CMS Cameron McKenna Nabarro Olswang LLP
munir.hassan@cms-cmno.com

Munir Hassan is head of clean energy at CMS in London, helping to determine the firm's strategy on renewables and clean generation. Munir has over 20 years of experience advising the power sector on commercial arrangements, M&A transactions, electricity sector restructurings and reforms, price-regulated energy networks, regional trading arrangements, establishment of regulatory frameworks and wholesale/retail supply arrangements. He has advised on technologies across the power space, including on offshore and onshore wind, solar, tidal, biomass, energy from waste, tidal and tidal lagoon, wave power, CCGT and CHP, coal-fired projects, electricity transmission networks and electricity distribution networks. He has advised extensively on both the sector in the United Kingdom and power projects and market reforms across numerous jurisdictions around the world.

Brian Hegarty
Partner, Ashfords LLP
b.hegarty@ashfords.co.uk

Brian Hegarty, a partner at Ashfords LLP since 1997, has 32 years' experience as a commercial property lawyer and has specialised in the energy sector for the last 15 years. His team advises UK PLCs and private companies on solar projects from site acquisition, planning financing and development through to operation refinancing and sale.

Brian and his team have a reputation for delivering insightful and commercial solutions tailored to their clients in the energy sector and particularly solar projects. He has a clear understanding of his clients' markets. Brian's clients say that he understands their challenges and their sectors – and his focus is on finding solutions and getting the results they want.

Maya Ito
Partner, Nishimura & Asahi
m_ito@jurists.co.jp

Maya Ito is a partner at Nishimura & Asahi and works mainly in the area of cross-border project finance, as well as energy projects, transactions and regulations. She joined the firm in 2002 after graduating from Keio University (LLB) in 1999 and attending the Legal Training and Research Institute of the Supreme Court of Japan. She graduated from Columbia University School of Law (LLM) in 2007. Since then she has gained extensive experience in energy projects, project finance, structured finance, international finance and general corporate law, advising a range of clients, including multinational corporations, commercial banks, multilateral financing institutions, project developers and sponsors, and other parties on a range of international and cross-border matters.

Giji M John
Partner, Orrick, Herrington & Sutcliffe LLP
gjohn@orrick.com

Giji M John is a partner in the Houston, Texas office of Orrick, Herrington & Sutcliffe LLP. Mr John focuses on the development and financing of wind energy and solar energy projects in addition to other energy assets. He has been involved in virtually every aspect of the project development process, including initial real estate acquisition, long-term power offtake negotiations, engineering, procurement and construction negotiations to and through debt and equity financing and operations. He has particular expertise in the negotiation of long-term power offtake agreements, including physical power purchase agreements and financially settled transactions under the ISDA, EEI, WSPP and NAESB trading forms and other bespoke arrangements. Mr John has been ranked in *Chambers Global, Chambers USA, Legal 500 US* and *Law 360.*

James Larmuth
Renewable energy consultant – solar, Mott Macdonald
james.larmuth@mottmac.com

James Larmuth has diverse project experience in technical design and advisory services for concentrated solar power (CSP) and photovoltaic (PV) projects across Sub-Saharan Africa. While his current focus in construction and operations monitoring of CSP projects across South Africa, his earlier experience includes research, design and construction of a South African developed CSP technology, with special focus on heliostat development.

Lynia Lau
Partner, Ernst & Young
lynia.lau@eylawgc.com

Lynia Lau has more than 20 years' experience in advising large-scale state-owned enterprises in the development, acquisitions, government regulations, sales and purchases, finances and operations of large-scale infrastructure projects. She specialises in LNG, oil & gas, power, refinery, highway, tunnel, bridge, terminal and other public utilities.

She is the chief lawyer leading a consortium of four law firms advising the governments of Hong Kong, Macau and Guangdong Province on the Hong Kong–Zhuhai–Macau Bridge project.

Lynia is the visiting professor of HK Financial Services Institute and HK WEN WEI Management College and a contributing editor for the monthly journal IELR Regional Developments published by Sweet & Maxwell.

Lynia is fluent in English, Mandarin, Cantonese, French and Japanese.

Sadayuki Matsudaira
Counsel, Nishimura & Asahi
s_matsudaira@jurists.co.jp

Sadayuki Matsudaira is a counsel at Nishimura & Asahi and works mainly in the area of energy projects, transactions and regulations, as well as M&As and joint ventures. He joined the firm in 2002 after graduating from the University of Tokyo (LLB) in 2001 and attending the Legal Training and Research Institute of the Supreme Court of Japan. He graduated from the University of Michigan (LLM) in 2011. He is admitted to practise in Japan (since 2002) as well as in the state of New York (since 2012). Mr Matsudaira obtained the Energy and Resources Deal of the Year in ALB Japan Law Awards 2016. Since 2016, he is the Regional Vice Chair, Asia Pacific of the Energy Group of Lex Mundi.

Simon Middleton
Partner, Elgar Middleton Infrastructure & Energy Finance LLP
simonmiddleton@elgarmiddleton.com

Simon Middleton is a partner in Elgar Middleton's London office. He specialises in renewable energy and power with 25 years' experience from projects across Europe, Africa, the Americas, Asia and Australasia. Simon has worked on the equity buy and sell side, debt advisory and lending during his career.

In renewable energy, Simon has worked on solar, wind (onshore and offshore) biomass, anaerobic digestion and battery storage projects. He also has extensive experience of coal, gas and peaker power plants.

Elgar Middleton was founded in 2012 with a specific goal to act as an independent financial advisory boutique for its clients. The partnership has offices in London and Sydney and has advised on over 50 discrete renewable energy projects in the past four years enabling its clients to raise over $US3 billion of capital.

Ghalia Mokhtari
Avocat à la Cour, Mokhtari Avocats
gmokhtari@mokhtari-avocats.com

Ghalia Mokhtari is a specialist in projects and banking law with expertise in the energy, infrastructure and real estate sectors. She regularly advises borrowers, financial institutions, funds, energy companies, corporations and governments on a range of transactions, including projects, public-private partnerships (PPPs), acquisitions, construction issues and general Moroccan regulation matters.

Before founding Mokhtari Avocats, Ghalia worked with the real estate and project finance teams of Linklaters in Paris before heading the projects and financing team of an independent North African law firm in Casablanca. She has also worked for major energy developers, investors and financial institutions in Europe and Africa.

Ghalia regularly publishes articles on the energy sector for legal journals and teaches project finance and business law at a leading business school in Casablanca.

Carlo Montella
Partner, Orrick, Herrington & Sutcliffe LLP
cmontella@orrick.com

Carlo Montella is a partner at Orrick and Deputy Business Unit Leader of the Energy and Infrastructure practice.

Carlo is a transactional lawyer with extensive experience in project development, asset acquisitions and financings in the energy and infrastructure sectors. He has strong expertise and consolidated background in assisting international sponsors, multinational energy companies, large private equity funds, project developers, borrowers and debt issuers, investment banks and other capital providers and investors who value his contribution thanks to his vast experience, practical know how and willingness to go the extra mile for his client. Carlo regularly advises on energy sectors mergers, acquisitions, joint ventures, development, structuring and financing of large-scale energy and infrastructure projects, both in Italy and internationally (including UK and Eastern European countries).

Carlo has a significant experience in the electric power sector, wind, solar, biomass, geothermal and other renewable energy projects and has also represented clients in transactions involving electric power and natural gas transmission. He has advised clients on the most large-scale, complex and high profile energy transactions in the Italian market and his clients are among the most active energy developers and investors in the market.

Carlo is Leading Individual in "Energy, Italy" for Chambers Europe 2018, won the "Energy Law Firm of the Year" award at the Legalcommunity Energy Awards 2018 and the "Lawyer of the Year Renewables" award at the Legalcommunity Energy Awards 2017. The Italian Energy practice is ranked in band 2 in the *Legal 500 EMEA 2017* guide.

Jose Muñoz
Senior renewable energy consultant – solar, Mott Macdonald
jose.munoz@mottmac.com

Jose Muñoz is a MSc qualified engineer with over ten years experience in the solar photovoltaic (PV) industry, having worked in a wide range of roles in the solar sector, from research and development up to consultancy but also including construction monitoring of PV plants, Operation & Maintenance activities and training programmes. While in the past his specialisation lay in the characterisation and analysis of PV systems and components (including different PV modules technologies), currently he performs advisory services acting as Project Manager for PV projects developed across the globe, including due diligence and construction monitoring works.

Urias Martiniano Garcia Neto
Partner, Tomanik Pompeu (in association with LRNG Law Firm)
urias@tomanikpompeu.adv.br

Urias Martiniano Garcia Neto is a partner of Tomanik Pompeu. He holds a degree in law and postgraduate studies in civil procedure law, infrastructure contracts and energy law. His fields of practice focus on energy regulation and infrastructure projects, with an emphasis on advisory, strategic litigation in the judicial and arbitration spheres in the Brazilian electric sector.

Urias started his career in the energy sector in the Brazilian Chamber of Electric Energy Commercialisation (CCEE), where he worked for several years, before dedicating himself to private practice at Tomanik Pompeu, working in association with LRNG – Landi, Rodrigues, Nakano e Giovannetti Advogados in Sao Paulo.

Urias is also a member of the Energy Law Commission of the Brazilian Bar Association and author of several articles on the electric energy sector.

Tilman Lorenz Petersen
Associate, CMS Germany
tilman.petersen@cms-hs.com

Tilman Lorenz Petersen joined CMS in 2016. Tilman is a member of the energy law group of CMS Germany. Tilman is primarily involved in advising clients on the development, construction, financing and operation of renewable energy projects.

Tilman advises project developers, plant manufacturers, investors and banks focusing on renewable energy projects in and beyond Germany. He has special expertise in drafting all relevant contracts in this respect and in handling claims under FIDIC-based contracts.

Ricardo Pineiro
Partner, Foresight Group
rpineiro@foresightgroup.eu

Ricardo is a partner at Foresight Group and has led the firm's solar investments in the UK since joining in 2011, having overseen the acquisition of over 60 solar power plants totalling more than 800MW.

Ricardo acts as fund manager for the Foresight Solar Fund Limited, a close-end investment company listed in the London Stock Exchange since 2013 and investing in a diversified portfolio of large-scale solar assets in the UK and in Australia.

Before joining Foresight Group, he spent four years in the project finance division at Espirito Santo Investment, with a special focus on transport, energy and oil and gas.

Fernando Quicios
Partner, Pérez-Llorca
fquicios@perezllorca.com

Fernando Quicios is the current resident partner at Pérez-Llorca's London office, where he regularly advises funds, financial entities and companies on their investments and divestments in Spain. Fernando has advised sponsors, suppliers of equipment, civil works suppliers and providers of finance in various transactions in the renewables fields (mainly solar, thermosolar and wind) both in Spain and in other Southern European and Latin American jurisdictions.

Fernando has also advised in the restructuring of the indebtedness of energy projects, both in Spain and abroad.

Fernando is an officer of the Insolvency Committee at the International Bar Association, and is recommended by a number of directories (*Chambers and Partners*, *Legal 500*, etc).

Rajen Ranchhoojee
Special legal consultant, Orrick, Herrington & Sutcliffe LLP
rranchhoojee@orrick.com

Rajen Ranchhoojee is admitted to practice in South Africa and registered as a special legal consultant with the Washington DC Bar. He specialises in cross-border projects and transactions, with particular expertise in energy and infrastructure projects in Africa.

Rajen advises governments, Fortune 500 companies, state utilities, private equity firms, donor funding institutions as well as leading local and international companies with interests. Rajen has advised on over 50 renewable energy projects in Africa, predominantly in South Africa, acting as project counsel, sponsor counsel, contractor's counsel and lenders' legal adviser.

His Africa-specific experience spans over 14 countries and includes the Democratic Republic of Congo, Ghana, Kenya, Malawi, Mauritius, Mozambique, Namibia, Nigeria, Reunion, South

Africa, Tanzania, Uganda, Zambia and Zimbabwe. He has been recognised by *Chambers Global* as a leading South African lawyer for Projects & Energy.

João Ribeiro da Costa
Partner, LRNG Law Firm (in association with Tomanik Pompeu)
joao.ribeiro@lrng.com.br

João Ribeiro da Costa is a partner at LRNG – Landi, Rodrigues, Nakano e Giovannetti Advogados in São Paulo. He holds a degree in law and postgraduate studies in capital markets. João worked in leading law firms in Portugal and Brazil, focusing his practices mainly in corporate issues, M&A transactions, capital markets and regulatory issues, especially multinational deals.

In particular, he has advised foreign investors on a number of transactions in Brazil in a variety of sectors, including energy, engineering and industrial activities, construction, financial services and IT. In addition, he has been involved in a number of transactions for Brazilian companies accessing international capital markets.

He also has experience as managing director of a multinational group with a business in Brazil.

João is a member of the Portuguese and Brazilian Bars and the Portuguese and Spanish Chambers of Commerce.

George Rigo
Partner, Orrick, Herrington & Sutcliffe LLP
grigo@orrick.com

George Rigo is an M&A lawyer, based in Orrick's Paris office. Specialising in corporate law, he is devoted to advising European and Asian industrial groups and holding companies on their mergers, acquisitions and joint ventures, notably in the energy, food and car manufacturing sectors and services.

George has very strong expertise in the renewable energy sector, working alongside major players in the market and advising on key projects in France. He mainly advises on the financing and acquisition of windfarm and solar projects and lately he has counselled 123 Ventures, Impax, Nord LB, Platina Partners, EOS Holding and ERG on different transactions.

George's extensive experience on cross-border transactions includes negotiating and drafting international contracts or commercial partnerships with an emphasis on litigation issues succeeding the implementation of such agreements.

Anthony Riley
Partner, Orrick, Herrington & Sutcliffe LLP
ariley@orrick.com

Anthony Riley is a partner at Orrick, Herrington & Sutcliffe (UK) LLP. He advises on M&A transactions across the renewable energy sector with a particular focus on solar projects. In the UK market he has advised on the acquisition of consented and operational projects for a range of private and listed funds. Over the course of the last five years he has advised more than 20 buyers and sellers of solar projects located throughout the United Kingdom with a combined installed capacity of well in excess of 1 GW. He is also currently advising in connection with the acquisition of solar projects across Europe, Africa, the United States, Latin America and Australia.

Edward Rivera
Foreign attorney, Nishimura & Asahi
e_rivera@jurists.co.jp

Edward Rivera is a foreign attorney at Nishimura & Asahi and works mainly in the area of cross-border project finance. His practice is currently focused on solar projects and construction projects. He joined the firm in 2017 having graduated from Columbia Law School (JD) and Johns Hopkins SAIS (MA). He is a registered member of the New York Bar.

Wagner Silva Rodrigues
Partner, LRNG Law Firm (in association with Tomanik Pompeu)
wagner@lrng.com.br

Wagner Silva Rodrigues is one of the partners in charge of the tax department of LRNG – Landi, Rodrigues, Nakano e Giovannetti Advogados. He holds a degree in law and postgraduate studies in tax law, applied economics, controlling and finance. His fields of practice include tax advising and tax litigation, with a focus on energy and infrastructure projects.

Wagner started his career in one of Brazil's leading firms, where he worked for several years, before dedicating himself to LRNG Law Firm, of which he is one of the founding partners. Wagner led the firm's legal team in many leading tax disputes in Brazil, mainly involving renewable energy.

Wagner is also a member of the Tax Law Brazilian Institute (Instituto Brasileiro de Direito Tributário – IBDT).

Andrew Sawbridge
Trainee solicitor, CMS Cameron McKenna Nabarro Olswang LLP
andrew.sawbridge@cms-cmno.com

Andrew Sawbridge is a trainee solicitor at CMS in London, practising in the firm's Energy, Projects and Construction department. Andrew has acted on a range of renewable power projects in Europe, including advising on offshore wind, solar and battery storage projects. Andrew also has experience in contentious matters having acted on a number of judicial reviews, an appeal to the Competition and Markets Authority and other commercial litigation with a focus on energy matters.

Abhishek Saxena
Partner, Phoenix Legal
abhishek.saxena@phoenixlegal.in

Abhishek Saxena is a co-founding partner based at the New Delhi office of Phoenix Legal, a full-service law firm with offices in New Delhi, Mumbai and Chennai.

Abhishek focuses on cross-border deals and has advised on a number of complex and prestigious projects in diverse sectors such as financial services, IT, industrial, oil and gas, power, automotive, infrastructure, manufacturing, retail, hospitality and aviation. His practice includes corporate transactional work including joint ventures and M&A, projects (including solar power), structuring and tax advisory, foreign investment and regulation, bankruptcy. He represents large and reputed Indian companies, foreign investors and multinational corporations (including Fortune 100 companies) in various sectors.

Abhishek's services have been widely recognised and he has been regularly recommended as a leading lawyer for business by various international legal publications including *Chambers & Partners (Global and Asia Pacific), Legal 500, IFLR 1000, Who's Who Legal* and *Asialaw*. Abhishek was listed by Asian Legal Business as one of "ALB's Hot 100" listing the best legal minds in Asia. He has also been included in "The A-List – India's leading 100 lawyers 2016" published by *Indian Business Law Journal*.

Nathan Schmidt
Counsel, Nishimura & Asahi
n_schmidt@jurists.co.jp

Nathan Schmidt is a counsel at Nishimura & Asahi and has extensive experience representing developers, lenders and EPC companies in connection with the development of Japanese solar PV projects as well as other renewable energy projects. Mr Schmidt also represents lenders and borrowers in a broad range of banking and finance matters, including real estate financings, leveraged financings, restructurings, project financings and structured finance transactions. He graduated from Duke University School of Law (JD/LLM) in 2003 and practised in New York City with a focus on structured finance transactions until he joined Nishimura & Asahi in 2008.

David Spielberg
Partner, Orrick, Herrington & Sutcliffe LLP
dspielberg@orrick.com

David Spielberg is a partner in the energy and infrastructure group in Orrick's San Francisco office. He has over 35 years of experience in energy transactions of all types, including all stages of project acquisition and disposition, development, construction and financing, and has been involved in the solar energy industry since the 1990s. He has represented clients in connection with the development, construction and financing of thousands of megawatts of solar energy projects all over the United States, including some of the earliest and largest solar energy projects in California. His practice covers all types of renewable energy, including wind, geothermal and biomass projects, in addition to solar energy projects. He is a graduate of the Massachusetts Institute of Technology (electrical engineering) and Stanford Law School.

Ting Su
Senior legal manager, Ernst & Young
ting.su@eylawgc.com

Ting Su is in the energy and infrastructure projects team. She has advised on projects, M&A and project finance matters, in particular the LNG projects including gas sales contracts, LNG sales and purchase agreements, terminal use agreements, and EPC for LNG import terminals, gas pipeline, power station, tunnel and highway-related projects and joint ventures.

Ting has also provided general advice in the large-scale energy and infrastructure projects, assisting the PRC state-owned enterprises in outbound investments, financial structuring, project operation and maintaining arrangements.

She has also advised several PRC state-owned enterprises and the PRC banks in their overseas litigation and disputes, in particular the debt collections and enforcement of foreign judgments.

Ting has got double degrees in law and English and is a contributor to the *International Energy Law Review* published by Sweet & Maxwell.

Oscar Velasco
Senior renewable energy consultant – solar, Mott Macdonald
oscar.velasco@mottmac.com

Oscar Velasco has 11 years' experience related to photovoltaic grid-connected, off-grid and hybrid projects. His expertise includes wide knowledge of the components such as PV modules, PV inverters, Battery inverters, Energy Management system (EMS), Batteries, Diesel Generators, as well as the project design, construction, operation & maintenance and due diligence advisory. He has international experience through having been involved in projects worldwide in countries including Spain, Kenya, Ghana, Mozambique, Nigeria, Brazil and Nicaragua.

Dan Wells
Partner, Foresight Group
dwells@foresightgroup.eu

Dan Wells is a partner at Foresight Group and has responsibility for the firm's European and US new energy infrastructure activities. Dan has 17 years of experience and has worked in sustainability-driven investment management strategies in Europe, North America and Asia. He has written on the topic in publications including the *Financial Times*.

Prior to Foresight, Dan worked at Sindicatum, a dedicated clean energy investing firm, from 2007 to 2012. At Sindicatum he initially focused on raising c.$300 million of primarily US institutional capital before overseeing the fund deployment in South East Asia, China and India.

Between 2000 and 2007 Dan worked for EY advising clients, primarily in energy-related themes. Dan is a Chartered Accountant; he holds an MSc in Sustainable Leadership from CSEM (distinction) and a BA in Modern History from the University of Bristol.

Belén Wert
Associate, Pérez-Llorca
bwert@perezllorca.com

Belén Wert joined Pérez-Llorca in September 2017 after having worked for a year and a half in Administrative Law. She advised Spanish and foreign clients on all matters related to energy law. Belén specialises in administrative law. She mainly advises on energy and regulated sectors.

In 2010 Belén finished her degree in law at Universidad CEU San Pablo and she prepared for the competitive state exams for State Advocate until 2016. She works in Spanish, English and Italian.

Matthew Williams
Partner and co-head of Power Sector Practice, Orrick, Herrington & Sutcliffe LLP
matthew.williams@orrick.com

Matthew Williams is co-head of Orrick's Power Sector Practice and a member of the firm's Global Advisory Board. His practice focuses on transactional and regulatory work in the energy sector with extensive experience on project finance, M&A, commodity trading and general commercial matters. Matthew has established a market leading power sector M&A practice and over the last 20 years has represented parties involved in multiple acquisitions and disposals of conventional and renewable power assets in the UK, Europe and Africa. He has a particular interest in the intersection between new technology and the energy markets. He acts for a broad range of clients including independent power project developers, utilities, infrastructure funds, investment banks and private trading houses. *Chambers UK* and *Legal 500* recognise Matthew as a leading individual in his field. Matthew holds an MA in Jurisprudence from Oxford University and was formerly an in-house lawyer with the international power project development team at National Power plc.

Yolanda Yong
Senior associate, Orrick, Herrington & Sutcliffe LLP
yyong@orrick.com

Yolanda Yong is a senior associate at Orrick, Herrington & Sutcliffe (UK) LLP. She has worked on renewable energy projects including offshore wind projects, energy from waste, biomass and solar energy projects for lenders, contractors, developers and investors. Yolanda advises on M&A transactions and project documents across the renewable energy sector with a particular focus on solar. She has acted for a number of the leading solar yieldcos and investment funds in the United Kingdom on the acquisition and sale of numerous development and operational projects. Yolanda holds a BA in Law from the University of Cambridge.

About Globe Law and Business

Globe Law and Business was established in 2005 as an imprint of Globe Business Media Group. We set out to create law books which are sufficiently high level to be of real use to the experienced professional, yet still accessible and easy to navigate. Most of our authors are drawn from Magic Circle and other top commercial firms, both in the UK and internationally.

In Autumn 2015, we had an exciting turn in our history when Globe Law and Business was bought by a management team which includes my colleague Jim Smith and me, Sian O'Neill.

We love books, and our hardback titles are carefully produced, with the utmost attention paid to the editorial, design and production processes. We hope this results in high-quality titles which are easy to read, and a pleasure to own.

Our customers and authors are of paramount importance – we are committed to devoting time and attention to each book as well as everyone involved in the publishing process.

We'd very much like to hear from you with your thoughts on our titles or ideas for improving what we offer. Please do feel free to email me on sian@globelawandbusiness.com with your views.

Sian O'Neill
Managing director
Globe Law and Business

www.globelawandbusiness.com